AF356381

Radar Signal Detection, Recognition and Identification

Radar Signal Detection, Recognition and Identification

Guest Editors

Janusz Dudczyk
Piotr Samczyński

Basel • Beijing • Wuhan • Barcelona • Belgrade • Novi Sad • Cluj • Manchester

Guest Editors

Janusz Dudczyk
Institute of Communications
Systems, Faculty of
Electronics
Military University of
Technology
Warsaw
Poland

Piotr Samczyński
Faculty of Electronics and
Information Technology,
Institute of Electronic Systems
Warsaw University of
Technology
Warszawa
Poland

Editorial Office
MDPI AG
Grosspeteranlage 5
4052 Basel, Switzerland

This is a reprint of the Special Issue, published open access by the journal *Sensors* (ISSN 1424-8220), freely accessible at: https://www.mdpi.com/journal/sensors/special_issues/radar_signal_detection.

For citation purposes, cite each article independently as indicated on the article page online and as indicated below:

Lastname, A.A.; Lastname, B.B. Article Title. *Journal Name* **Year**, *Volume Number*, Page Range.

ISBN 978-3-7258-3323-8 (Hbk)
ISBN 978-3-7258-3324-5 (PDF)
https://doi.org/10.3390/books978-3-7258-3324-5

Cover image courtesy of Janusz Dudczyk

Contents

About the Editors

Janusz Dudczyk

Janusz Dudczyk was awarded an M.Sc. in Electronics Engineering at the Military University of Technology of Warsaw, Poland, with distinction, studying Electronics and Cybernetics. In 2004, he was awarded a Ph.D. from the Military University of Technology (Electromagnetic Compatibility), and in 2016, he became a Habilitated Doctor and Associate Professor. His primary research interests include the sophisticated process of specific emitter identification (SEI), radar signal processing, and the application of artificial intelligence to radar signal recognition. Also, his research interest is in the areas of EMC, the utilization of relational modelling to the emitter database design of ELINT systems, and applying the radiated emission, heterogeneous UAV payloads, and SEI methods in electronic warfare. He is the author and co-author of more than 160 scientific articles and has authored works which have appeared in national and international journals on radar signal processing techniques, unintentional radiated emission, fractal techniques and theory, database modelling for ELINT systems, SEI methods, hypergeometric divide classifier methods and techniques, and data fusion in decision-making processes based on ANNs. He has experience as an expert of the National Centre for Research and Development in Poland and as a certified Project Manager for State Security and Defence at the Polish Ministry of Science and Higher Education, European Defence Agency, and Ministry of the Interior and Administration Republic of Poland. Further, he serves as the Vice President of the Polish Chapter "Hussars Roost" of the Association of Old Crows (AOC). He has been frequently honored with distinctions, awards, and medals for civilian and military merits by the President of the Republic of Poland, the Minister of National Education, and the Minister of National Defence.

Piotr Samczyński

Prof. Piotr Samczyński received his B.Sc. and M.Sc. degrees in electronics and PhD and D.Sc. in Telecommunications from the Warsaw University of Technology (WUT), Warsaw, Poland, in 2004, 2005, 2010, and 2013, respectively. In 2023, he was awarded the title of Full Professor by the President of the Republic of Poland. Since 2024, he has been a Full Professor at the WUT; between 2018 and 2023, he has served as an Associate Professor at the WUT, and since 2014, a member of the WUT's Faculty of Electronics and Information Technology Council. Before this, he was an Assistant Professor at WUT (2018–2010), a research assistant at the Przemyslowy Instytut Telekomunikacji S.A. (PIT S.A.) (2010–2005), and the head of PIT's Radar Signal Processing Department (2010–2009). He is the co-founder of XY-Sensing Ltd., where, since 2018, he has held the position of CEO.

Prof. Samczyński's research interests are in the radar signal processing, passive radar, synthetic aperture radar, and digital signal processing. He is the author of over 280 scientific papers.

Piotr Samczyński was involved in several projects for the European Research Agency (EDA), Polish National Centre for Research and Development (NCBiR), and Polish Ministry of Science and Higher Education (MiNSW), including projects on SAR, ISAR, and passive radars. For his work, he was honored in 2020 with the Bronze Cross of Merit awarded by the President of the Republic of Poland. He was also honored in 2021 with a Bronze Medal "for merits for the defense of the country", awarded by the Minister of National Defence. The award was for special merits in the development of modern military thought, military technology and professional work, which significantly contributed to the development and strengthening of the defense system of Poland.

Preface

The main objective of radar signal recognition and identification is to identify distinctive signal patterns and establish methods for distinguishing them. In the literature, a concept widely used to describe pattern recognition and classification is source emission pattern-separating surfaces in the measurable feature space. Because radar emitter recognition and classification are based on defining the location of an emission source from said separating surfaces, it is essential to highlight a very significant fact: radar metrics need to be defined in the measurable feature space of a signal, and the specific features of a radar signal must be extracted in order to establish a distinctive radar signal pattern.

Recently, there have been rapid developments in electronic warfare systems. Different methods of electromagnetic environment observation have been used to analyze radar signatures; these methods increase the quality of algorithms that can recognize objects and targets automatically.

The first paper in this Special Issue is titled "Intra-Pulse Modulation Recognition of Radar Signals Based on Efficient Cross-Scale Aware Network"; radar signal intra-pulse modulation recognition can be achieved through convolutional neural networks (CNNs) and time–frequency images. The author proposes a lightweight CNN known as a cross-scale aware network (CSANet) to recognize intra-pulse modulation based on three types of time-frequency images. The cross-scale aware (CSA) module, designed as a residual and parallel architecture, comprises a depthwise dilated convolution group (DDConv Group), a cross-channel interaction (CCI) mechanism, and spatial information focus (SIF). The DDConv Group produces multiple-scale features with a dynamic receptive field, while CCI fuses the features and mitigates noise in multiple channels, and SIF has awareness of the cross-scale details of TFI structures. Furthermore, by employing image preprocessing techniques, i.e., adaptive binarization, morphological processing, and feature fusion, a novel time–frequency fusion (TFF) feature based on three types of TFI is developed.

In the field of electronic warfare, noise radars are examples of LPI and LPE systems that have gained substantial prominence in the past decade despite exhibiting a common drawback of limited Doppler tolerance. The prevalence of Low Probability of Interception (LPI) and Low Probability of Exploitation (LPE) radars presents an ongoing challenge for defense mechanisms, necessitating constant progress in establishing protective strategies. The sophisticated Electronic Intelligence (ELINT) systems that employ Time Frequency Analysis (TFA) and image processing methods may intercept APCN and estimate important parameters of APCN waveforms, such as bandwidth, operating frequency, time duration, and pulse repetition interval, as shown in the article entitled "On a Closer Look of a Doppler Tolerant Noise Radar Waveform in Surveillance Applications ". The contributions to this work offer enhancements to noise radar capabilities while facilitating improvements in ESM systems.

In recent years, radar emitter signal recognition has enjoyed a wide range of applications in electronic support measure systems and communication security. An increasing number of deep learning algorithms have been used to improve the recognition accuracy of radar emitter signals. However, complex deep learning algorithms and data preprocessing operations have a huge demand for computing power, which cannot meet the requirements of low power consumption and high real-time processing scenarios. To solve this problem, the authors "Efficient FPGA Implementation of Convolutional Neural Networks and Long Short-Term Memory for Radar Emitter Signal Recognition" propose a resource reuse computing acceleration platform based on field programmable gate arrays (FPGAs) and implement a one-dimensional convolutional neural network

(CNN) and long short-term memory neural network model for radar emitter signal recognition. As the research conducted shows, the energy efficiency ratio and real-time performance of the radar emitter recognition system is showing significant improvement.

Another featured paper, "An Application of Analytic Wavelet Transform and Convolutional Neural Network for Radar Intrapulse Modulation Recognition", analyzes the possibility of using Analytic Wavelet Transform and Convolutional Neural Networks for the purpose of recognizing the intra-pulse modulation of radar signals. This article considers selected types of intra-pulse modulation, namely linear frequency modulation (LFM) and the following phase-coded waveform (PCW) types: Frank, Barker, P1, P2, and Px. The authors of this article also consider the possibility of using other time–frequency transformations, such as Short-Time Fourier Transform or Wigner–Ville Distribution. Finally, the results of the simulation tests carried out in the Matlab environment are presented.

Developments in radio detection and radar technology have made hand gesture recognition possible. In heat map-based gesture recognition, feature images are large and require complex neural networks to extract information. Machine learning methods typically require large amounts of data; hand gestures recognition achieved with radar is time- and energy-consuming. Thus, the paper "Low Complexity Radar Gesture Recognition Using Synthetic Training Data" proposes a low computational complexity algorithm for hand gesture recognition based on a frequency-modulated continuous-wave radar and a synthetic hand gesture feature generator. The experimental results demonstrate that the average recognition accuracy of the model on the test set can reach a high level when the synthetic data are used as the training set and the real data are used as the test set. This indicates that the generation of synthetic data can make a meaningful contribution to the pre-training phase.

In the paper entitled "Recognition of Targets in SAR Images Based on a WVV Feature Using a Subset of Scattering Centers", a robust method for feature-based matching with potential for application in synthetic aperture radar (SAR) automatic target recognition (ATR) is proposed. The scarcity of measured SAR data available for training classification algorithms leads to the replacement of such data with synthetic data. As attributed scattering centers (ASCs) extracted from the SAR image reflect the electromagnetic phenomenon of the SAR target, this is effective for classifying targets. Experiments on synthetic and measured paired labeled experiment datasets, which are publicly available, are conducted to verify the effectiveness and robustness of the proposed method. The proposed method can be used in practical SAR ATR systems trained using simulated images.

In the next article of this Special Issue, titled "Estimation and Classification of NLFM Signals Based on the Time–Chirp Representation", a new approach to the estimation and classification of nonlinear frequency modulated (NLFM) signals is presented. These problems are crucial in electronic reconnaissance systems, whose role is to indicate what signals are being received and recognized by the intercepting receiver. The methods presented in the paper belong to the time–chirp domain, which is relatively rarely cited in the literature. The author proposes using polynomial approximations of nonlinear frequency and phase functions describing signals. This allows the cubic phase function to be applied as an estimator of phase polynomial coefficients. The proposed classification method was examined for three standard NLFM signals and one LFM signal. The results of the simulation research revealed good estimation performance and error-free classification for the SNR range encountered in practical applications.

Further, the article "Radar Detection-Inspired Signal Retrieval from the Short-Time Fourier Transform" presents a novel adaptive algorithm for multicomponent signal decomposition using the Short-Time Fourier Transform. The proposed algorithm enhances signal retrieval by adaptively identifying and clustering high-magnitude regions in the time-frequency (TF) domain. Through these clusters, the algorithm creates TF masks that focus on relevant signal modes, effectively reducing noise and interference across a range of waveform types. Unlike traditional techniques, which may suffer from interference or lose components in the TF domain, this method offers an improved reconstruction quality, even for complex, non-stationary signals.

One contribution, "Multi-Sensory Data Fusion in Terms of UAV Detection in 3D Space", focuses on the problem of detecting unmanned aerial vehicles that violate restricted airspace. The main purpose of the research is to develop an algorithm that enables the detection, identification, and recognition in 3D space of a UAV violating restricted airspace, which is of great importance today both in civilian and military contexts. The proposed method consists of multi-sensor data fusion and is based on conditional complementary filtration and multi-stage clustering. Based on the review of the available UAV detection technologies, three sensory systems classified into the groups of passive and active methods are selected. The UAV detection algorithm is developed on the basis of data collected during field tests under real conditions from three sensors: a radio system, an ADS-B transponder, and a radar equipped with four antenna arrays. The efficiency of the proposed solution was tested on the basis of rapid prototyping in the MATLAB simulation environment with the use of data from the real sensory system obtained during controlled UAV flights. The results of UAV detections confirmed the effectiveness of the proposed method and the theoretical expectations. The proposed solution has huge potential and is suitable for use in defense-drone systems.

In the article "Pedestrian and Animal Recognition Using Doppler Radar Signature and Deep Learning", the authors used radar technology to identify pedestrians. When distinct parts of an object move in front of a radar, micro-Doppler signals are produced that may be utilized to identify the object. As the authors claim, using a deep-learning network and time–frequency analysis, a method for classifying pedestrians and animals based on their micro-Doppler radar signature features is possible. The proposed approach was evaluated on the MAFAT Radar Challenge dataset. Encouraging results were obtained, with a high Area Under the Curve value achieved for the public and the private test sets. The proposed DNN architecture, in contrast with more common shallow CNN architectures, is one of the first attempts to use such an approach in the domain of radar data.

As is well-known, technological advancement in battlefield and surveillance applications switch radar investigators to increase efficacy. There are also many common theories and models known to improve the process of target detection in Doppler-tolerant radar. However, more effort is still needed to minimize the noise below the radar threshold limit to accurately detect the target. In the paper entitled "Multi-Criteria Decision Making to Detect Multiple Moving Targets in Radar Using Digital Codes", a digital coding technique is discussed to mitigate noise and to create clear windows for desired target detection. Moreover, digital code combination criteria are developed using discrete mathematics, and all designed codes have been tested to investigate various target detection properties such as the autocorrelation, cross-correlation properties, and ambiguity function using mat-lab to optimize and enhance static and moving targets in the presence of the Doppler in a multi-target environment.

The authors of "An X-Band CMOS Digital Phased Array Radar from Hardware to Software" propose an X-band element-level digital phased array radar utilizing fully integrated complementary metal-oxide-semiconductor transceivers. This approach is proposed to achieve a low-cost and compact-size digital beamforming system. A radar demonstrator with scalable subarray modules simultaneously realizes range sensing and azimuth recognition for pulsed radar configurations. Phased array technology features rapid and directional scanning and has become a promising approach for remote sensing and wireless communication. In addition, element-level digitization has increased the feasibility of complicated signal processing and simultaneous multi-beamforming processes.

Janusz Dudczyk and Piotr Samczyński

Guest Editors

Article

Intra-Pulse Modulation Recognition of Radar Signals Based on Efficient Cross-Scale Aware Network

Jingyue Liang [1], Zhongtao Luo [2,*] and Renlong Liao [2]

1   Hunan Nanoradar Science and Technology Co., Ltd., Changsha 410205, China; jyliang@nanoradar.cn
2   School of Communication and Information Engineering, Chongqing University of Posts and
     Telecommunications, Chongqing 400065, China; roxlong@foxmail.com
*   Correspondence: luozt@cqupt.edu.cn

Abstract: Radar signal intra-pulse modulation recognition can be addressed with convolutional neural networks (CNNs) and time–frequency images (TFIs). However, current CNNs have high computational complexity and do not perform well in low-signal-to-noise ratio (SNR) scenarios. In this paper, we propose a lightweight CNN known as the cross-scale aware network (CSANet) to recognize intra-pulse modulation based on three types of TFIs. The cross-scale aware (CSA) module, designed as a residual and parallel architecture, comprises a depthwise dilated convolution group (DDConv Group), a cross-channel interaction (CCI) mechanism, and spatial information focus (SIF). DDConv Group produces multiple-scale features with a dynamic receptive field, CCI fuses the features and mitigates noise in multiple channels, and SIF is aware of the cross-scale details of TFI structures. Furthermore, we develop a novel time–frequency fusion (TFF) feature based on three types of TFIs by employing image preprocessing techniques, i.e., adaptive binarization, morphological processing, and feature fusion. Experiments demonstrate that CSANet achieves higher accuracy with our TFF compared to other TFIs. Meanwhile, CSANet outperforms cutting-edge networks across twelve radar signal datasets, providing an efficient solution for high-precision recognition in low-SNR scenarios.

Keywords: intra-pulse modulation recognition; convolutional neural network (CNN); time–frequency images (TFIs); cross-scale aware (CSA)

Citation: Liang, J.; Luo, Z.; Liao, R. Intra-Pulse Modulation Recognition of Radar Signals Based on Efficient Cross-Scale Aware Network. *Sensors* **2024**, *24*, 5344. https://doi.org/10.3390/s24165344

Academic Editor: Ram M. Narayanan

Received: 5 June 2024
Revised: 5 August 2024
Accepted: 16 August 2024
Published: 18 August 2024

1. Introduction

Radar signal modulation recognition can be classified into two categories: inter-pulse modulation recognition and intra-pulse modulation recognition [1]. Early studies focused on inter-pulse characteristics for signal recognition [2–5]. With the advancement of radar technology and the increasing complexity of radar systems, traditional inter-pulse feature analysis becomes insufficient [6]. In recent years, intra-pulse feature analysis has attracted growing attention and has become a valuable tool in the field of radar signal recognition, offering significant advantages in terms of recognition efficiency and accuracy [7].

The traditional way of intra-pulse modulation recognition is based on manual feature design and pattern matching. This approach has some shortcomings, e.g., the complexity of feature extraction, the limitation to single signal recognition, the inadequacy in terms of efficiency, and the poor performance under low-SNR conditions [8]. With the advancement of deep learning (DL), signal recognition based on neural networks has become a promising solution for intra-pulse modulation recognition, as it offers the potential for intelligently recognizing complex, multi-class radar signals [9–13].

Generally, intra-pulse modulation recognition based on DL uses either the signal sequences or the signal time–frequency images (TFIs) as its input. Employing the signal sequence as input means extracting the intra-pulse characteristics using the designed network. For instance, ref. [14] proposes a modified convolutional neural network that uses the signal sequence as input. Ref. [15] designs its algorithm by combining a convolutional neural network (CNN) with a long short-term memory (LSTM) network, which also adopts signal sequences as input. In [16], an omni-dimensional dynamic-convolution-layer-based network (OD-CNN) with a focal loss function is designed and applied to classify radar intra-pulse modulations based on signal sequences.

In contrast, using the TFIs as input implies the signal is preprocessed before the network. In [17], the Choi–Williams distribution (CWD) is employed as an input to an improved deep residual network (ResNet). Ref. [18] transforms the radar signal's bicubic interpolation Wigner–Ville distribution (WVD) matrix into a square matrix. This matrix is then utilized to train a CNN for signal recognition.

However, there are two primary challenges to improving the performance of intra-pulse modulation recognition using TFIs and CNNs. The first challenge is the severe contamination of TFIs at low SNRs. Reference [19] employs an improved convolutional denoising autoencoder (CDAE) to de-noise TFIs and then utilizes a CNN to identify 10 radar signals at a -6 dB SNR, achieving a recognition rate exceeding 88%. However, this approach requires an additional noise estimation network and does not extend its application to lower-SNR scenarios.

The second challenge lies in the limitations of CNNs to effectively extract features [20], particularly due to the local characteristics of convolutional layers, which makes it difficult for them to capture global information [21]. To overcome this, some researchers have proposed increasing the number of layers within CNNs, allowing for the extraction of more complex features [22]. However, adding more layers introduces redundancy into the network and, consequently, increases computational time and complexity.

This paper proposes to design a lightweight CNN that makes use of multiple TFIs jointly. The key challenge is how to extract signal features comprehensively under low SNRs, since the contextual information carried by the TFIs differs in amount, range, etc., for various modulation types of radar signals. A transformer is able to address contextual information, but it generally requires a large number of network parameters [23]. In this paper, to solve this problem, we use a combination of three different types of TFIs. They have more nonchiasmatic information, which complements the defects between them. A series of meticulous preprocessing techniques are used for noise reduction and image sizing. Then, channel fusion technology is utilized to fuse TFIs to form a time–frequency fusion feature (TFF) as the object of deep neural network learning.

In this paper, we propose a lightweight cross-scale aware network (CSANet) to recognize the modes of intra-pulse modulation based on TFIs; it consists of cross-scale aware (CSA) modules, convolution layers, and fully connected layers. The CSA module employs spatial and channel attention mechanisms on feature blocks across different scales and has a residual structure to avoid the problems of exploding and vanishing gradients. Furthermore, to ensure the network remains lightweight and computationally efficient, we consider multiple aspects to design the CSA module, including depthwise convolution, a parallel branch architecture, and channel size adjustment. Experiments demonstrate that the CSA module can direct the attention of the CNN towards global features while also recognizing the time–frequency structure of radar signals.

2. Signal Model and System Overview

2.1. Signal Model

Intra-pulse modulation modes of radar signals mainly include frequency modulation, phase modulation, combined modulation, etc. [24]. Table 1 shows the 12 typical modulation modes of radar signals that are used in this paper, where A, T, and f_c denote the amplitude, pulse width, and carrier frequency, respectively, B is the bandwidth, k is the slope of the frequency modulation, φ is the primary phase, Δf denotes the frequency interval, c is a random code that controls the frequency modulation, N is the number of codes, T_s is the width of a code, and M is the count of sub-pulses within one group. Further, $g_T(t) = 1/\sqrt{T}\mathrm{rec}(t/T)$, where the rectangular function $\mathrm{rec}(t') = 1$, and $t' \in [0, 1]$.

Considering a noisy environment, we model the received signal as

$$y(t) = x(t) + n(t) \tag{1}$$

where $y(t)$ is the received signal, $x(t)$ is the radar signal, and $n(t)$ is the additive noise, which is usually considered as white Gaussian noise.

Table 1. The formulas of typical radar signals.

Modulation	Formula
CW (Continuous Wave)	$x(t) = A\mathrm{rec}(t/T)e^{j2\pi f_c t}$
LFM (Linear Frequency Modulation)	$x(t) = A\mathrm{rec}(t/T)e^{j[2\pi(f_c t + \pi k t^2 + \varphi)]}$
NLFM (Nonlinear Frequency Modulation)	$x(t) = A\mathrm{rec}(t/T)e^{j2\pi \int_{0-}^{t} T^{-1}(f)dt}$ $T(f) = T\int_{-\infty}^{f} W(u)du / \int_{-B/2}^{B/2} W(v)dv,$ $W(f) = 0.63 + 0.46\cos(2\pi f/B), f \in [-\frac{B}{2}, \frac{B}{2}]$
BPSK (Binary Phase Shift Keying)	$x(t) = A\sum_{i=1}^{N} e^{j(2\pi f_c t + \varphi)} g_{T_s}(t - iT_s)$ $\varphi = 0, \pi$
QPSK (Quadrature Phase Shift Keying)	$x(t) = A\sum_{i=1}^{N} e^{j(2\pi f_c t + \varphi)} g_{T_s}(t - iT_s)$ $\varphi = 0, \frac{1\pi}{2}, \pi, \frac{3\pi}{2}$
FSK (Frequency Shift Keying)	$x(t) = A\sum_{i=1}^{N} e^{j2\pi(f_c + c\Delta f)t} g_{T_s}(t - iT_s)$ $c = 0, 1$
4FSK (Four-Frequency Shift Keying)	$x(t) = A\sum_{i=1}^{N} e^{j2\pi(f_c + c\Delta f)t} g_{T_s}(t - iT_s)$ $c = 0, 1, 2, 3$
FRANK	$x(t) = A\sum_{i=1}^{N} e^{j(2\pi f_c t + \varphi_{i,j})} g_{T_s}(t - iT_s)$ $\varphi_{i,j} = \frac{2\pi}{M}(i-1)(j-1), i, j = 1, 2, \ldots, M$
P1	$x(t) = A\sum_{i=1}^{N} e^{j(2\pi f_c t + \varphi_{i,j})} g_{T_s}(t - iT_s)$ $\varphi_{i,j} = \frac{-\pi}{M}[M - (2i-1)][(j-1)M + (j-1)]$ $i, j = 1, 2, \ldots, M$
P2	$x(t) = A\sum_{i=1}^{N} e^{j(2\pi f_c t + \varphi_{i,j})} g_{T_s}(t - iT_s)$ $\varphi_{i,j} = \frac{\pi}{2M}[M + 1 - 2i)][M + 1 - 2j)]$ $i, j = 1, 2, \ldots, M$
P3	$x(t) = A\sum_{i=1}^{N} e^{j(2\pi f_c t + \varphi_i)} g_{T_s}(t - iT_s)$ $\varphi_i = \frac{\pi}{M}(i-1)^2, i = 1, 2, \ldots, M$
P4	$x(t) = A\sum_{i=1}^{N} e^{j(2\pi f_c t + \varphi_i)} g_{T_s}(t - iT_s)$ $\varphi_i = \frac{\pi}{M}(i-1)^2 - \pi(i-1), i = 1, 2, \ldots, M$

2.2. System Overview

In this paper, we design an intra-pulse modulation recognition system for radar signals that consists of feature extraction and a CSANet classifier, as illustrated in Figure 1. Our system contains three primary steps:

(1) Time–frequency analysis: Initially, we apply time–frequency analysis techniques to the radar signals to obtain the TFIs. Specifically, this paper utilizes three distinct types of time–frequency features: FSST (Fourier synchrosqueezed transform) [25], *SPWVD* (smoothed pseudo Wigner–Ville distribution) [26], and HHT (Hilbert–Huang transform) [27].

(2) Image preprocessing: Subsequently, image preprocessing approaches, including binarization and cubic interpolation clipping, are conducted on the TFIs. Then, the obtained time–frequency features are fused into the TFF feature.

(3) Feature fusion and model training: Finally, we construct TFF feature datasets from various signals and scenarios, which are divided into training and test sets for the CSANet. The CSANet is applied to recognize the 12 types of radar signals.

In the following, Section 3 presents time–frequency analysis and the feature extraction process, and Section 4 details the architecture of our CSANet.

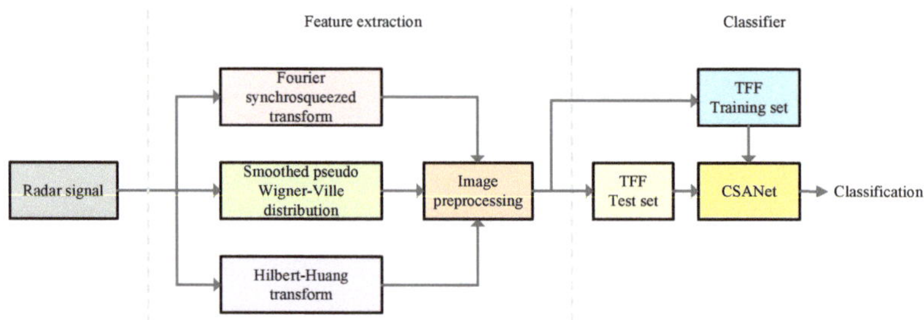

Figure 1. Structure diagram of the recognition system.

3. Time–Frequency Analysis and Feature Extraction

This section discusses time–frequency analysis techniques and the feature extraction process. Sections 3.1–3.3 introduce three time–frequency analysis techniques, respectively. Section 3.4 presents the methods of TFI preprocessing and TFI fusion.

3.1. Cohen Class Time–Frequency Distribution

As a typical case of a quadratic time–frequency distribution, a Cohen class time–frequency distribution typically employs a kernel function to smooth the quadratic function of signals [28]. This process requires a balance between time–frequency resolution and a cross-term. The Cohen class time–frequency distribution can be formulated as

$$C(t,f) = \frac{1}{4\pi^2} \int_{-\infty}^{+\infty} \int_{-\infty}^{+\infty} \int_{-\infty}^{+\infty} x\left(u + \frac{\tau}{2}\right) x^*\left(u - \frac{\tau}{2}\right) \phi(\tau,v) e^{-j2\pi(tv-uv+f\tau)} \, dud\tau dv \quad (2)$$

where t, f, τ, v, and u denote the time, frequency, time delay, frequency shift, and center of the correlation function, respectively, and $\phi(\tau,v)$ represents the kernel function. Different kernel functions lead to different kinds of Cohen class time–frequency distributions.

The Cohen class time–frequency distribution is equivalent to the Wigner–Ville distribution (WVD) when $\phi(\tau,v) = 1$. The WVD is characterized by its high time–frequency resolution. However, when the signal comprises multiple components, the WVD can produce cross-terms. Interference among the signal's components can result in the mixing of their characteristics, potentially obscuring the distinct features of the individual components [29].

A smoothed pseudo-Wigner–Ville distribution (*SPWVD*) suppresses the cross-term interference by smoothing the WVD with two window functions. One window function operates in the time domain, while the other is applied in the frequency domain. A *SPWVD* can be formulated as

$$SPWVD(t,f) = \int_{-\infty}^{+\infty} h(\tau) \int_{-\infty}^{+\infty} g(u-\tau)x(u+\frac{\tau}{2})x^*(u-\frac{\tau}{2})e^{-j2\pi t}dud\tau \quad (3)$$

where $h(\tau)$ and $g(u)$ are the smoothing window functions.

The *SPWVD* effectively seeks a balance between the high time–frequency resolution and the suppression of cross-term interference. Figure 2 shows the *SPWVD* of radar signals with various modulation modes when SNR = 10 dB. As can be seen, each modulation mode exhibits distinctive behavior.

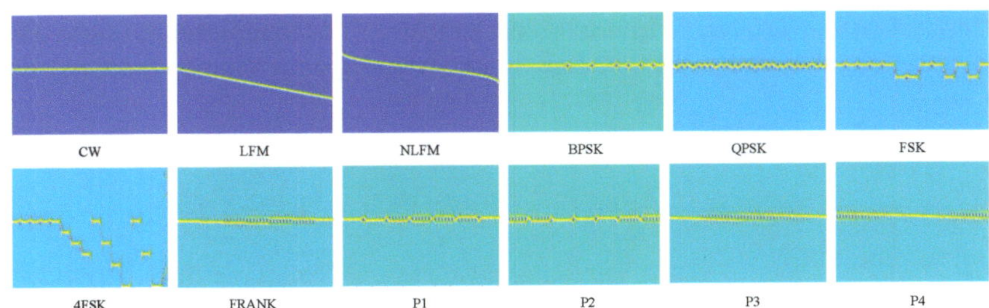

Figure 2. *SPWVD* of various radar signals for SNR = 10 dB.

3.2. Fourier Synchrosqueezed Transform (FSST)

The short-time Fourier transform (*STFT*) is formulated as

$$STFT(t, f) = \int_{-\infty}^{+\infty} x(\tau)\eta^*(\tau - t)e^{-j2\pi f\tau}d\tau \tag{4}$$

where $(\cdot)^*$ denotes the conjugate of a complex number, and η denotes the window function. Based on the *STFT*, the FSST is formed by a synchronous compression transform, defined as

$$T_f(t, \omega) = \frac{1}{\eta^*(0)} \int_{-\infty}^{+\infty} STFT(t, f)\delta[\omega - \hat{\omega}_f(t, f)]df \tag{5}$$

where ω represents the frequency of the correction function, and δ is the impulse response. The term $\hat{\omega}_f(t, f)$ is the local instantaneous frequency, given by

$$\hat{\omega}_f(t, f) = \text{Re}\left(\frac{1}{j2\pi} \frac{\partial_t STFT(t, f)}{STFT(t, f)} \right) \tag{6}$$

where ∂_t denotes the partial derivative, and $\text{Re}(\cdot)$ denotes the real part. FSST compresses the time–frequency curve along the frequency dimension, thereby concentrating the signal energy in the time–frequency domain. This concentration effectively minimizes the noise's impact. One example is presented in Figure 3.

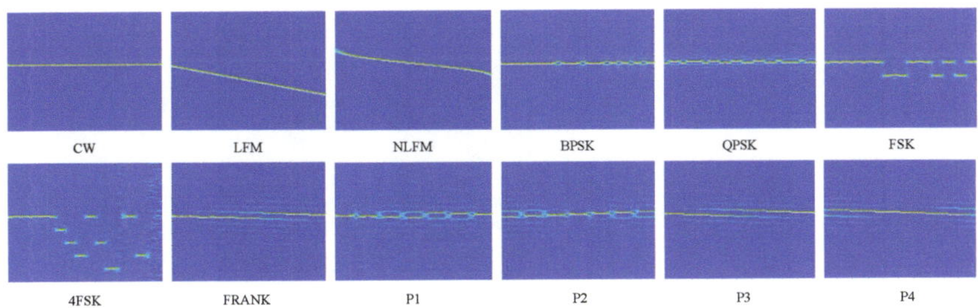

Figure 3. FSSTs of various radar signals for SNR = 10 dB.

3.3. Hilbert–Huang Transform (HHT)

HHT combines the Hilbert transform and adaptive signal decomposition to form a time–frequency feature. For instance, the empirical mode decomposition (EMD) or the variational mode decomposition (VMD) is employed to decompose radar signals into a collection of sub-signals, i.e., intrinsic mode functions (IMFs) [30]. Subsequently, the Hilbert method is utilized to derive time–frequency characteristics. Unlike EMD, VMD is a non-iterative signal processing technique. By iteratively searching for the optimal solution of the variational modes, VMD refines the optimal central frequencies and bandwidths of the IMFs adaptively. Therefore, VMD is much more robust to sampling and noise than EMD [31].

VMD processing includes two steps. Firstly, based on the input signal $x(t)$, the set of IMFs u_k is calculated by the decomposition algorithms [32]. Then, for each IMF, its Hilbert transform is calculated as

$$d_i(t) = \frac{1}{\pi} \int_{-\infty}^{+\infty} \frac{u_i(\tau)}{t - \tau} d\tau. \tag{7}$$

Hence, the instantaneous frequency is

$$\omega_i(t) = d\left(\frac{d_i(t)}{u_i(t)}\right)/dt. \tag{8}$$

Figure 4 shows one example of an HHT.

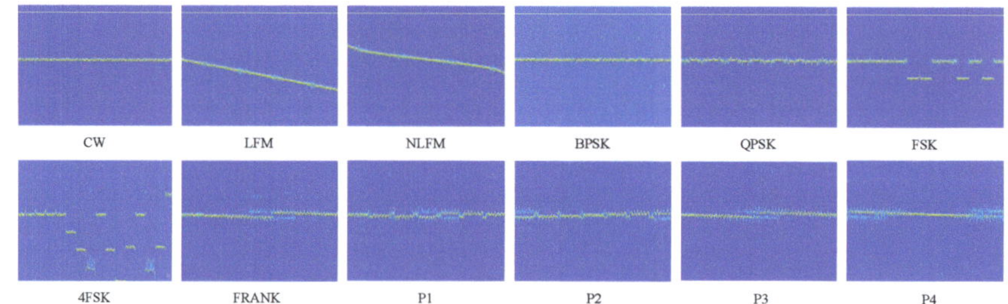

Figure 4. HHTs of various radar signals for SNR = 10 dB.

3.4. Time–Frequency Feature Preprocessing and Fusion

Using the FSST, *SPWVD*, and HHT methods, we obtain three types of time–frequency images. It is necessary to preprocess these images for more accurate identification. At present, there are two types of methods: image reconstruction based on neural networks and denoising based on traditional signal processing technology. However, some methods may not perform well under low-SNR conditions. For example, we employ CDAE to process *SPWVD*, and we use different noise variances to train the CDAE to reconstruct images [19]. The results are shown in Figure 5.

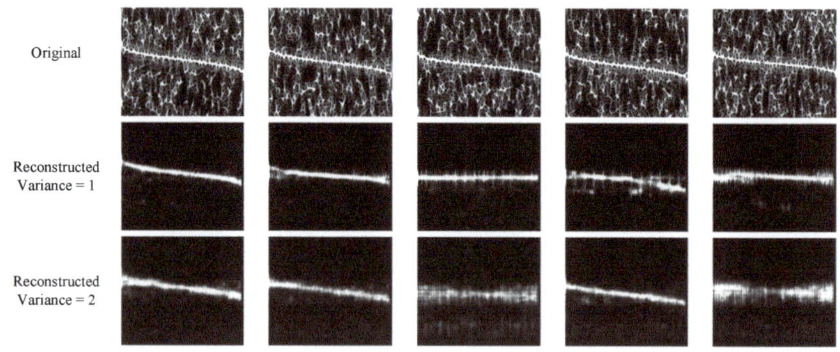

Figure 5. *SPWVD* of NLFM signal reconstructed based on CDAE; SNR = −8 dB.

We introduce an image preprocessing step designed to reduce the impact of noise and the computational complexity for deep neural networks. As depicted in Figure 6, the preprocessing process encompasses the following steps: (a) converting the original TFIs to grayscale, (b) applying adaptive threshold binarization [33], (c) employing morphological operations to fill in missing points and remove noise-induced outliers, and

(d) employing bicubic interpolation to resize the images to 256×256 pixels to make them suitable for input into the CSANet mode. This series of preprocessing filters out a lot of noise and mainly preserves the outline of the time–frequency ridge, which is the key to radar signal recognition [34].

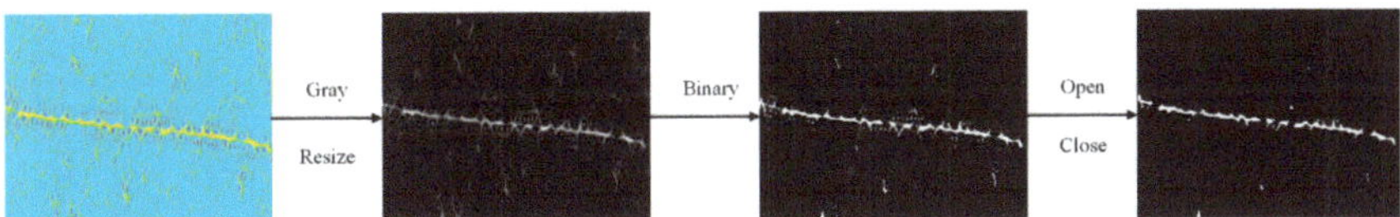

Figure 6. *SPWVD* image preprocessing of the NLFM signal for SNR = −8 dB.

Given that *SPWVD*, FSST, and HHT are based on different principles and show different time–frequency distributions, the use of feature fusion to enhance feature extraction performance, especially at low SNRs, is a promising approach. *SPWVD* belongs to the Cohen class of time–frequency distributions and is a quadratic, nonlinear time–frequency distribution with high time–frequency resolution, but it is susceptible to the influence of cross-terms. FSST is a linear time–frequency distribution based on the time window, which enhances the time–frequency concentration of *STFT* through the synchronous compression operator. However, due to the fixed window and basis function in the analysis, it performs poorly in matching multi-component and time-varying signals. HHT consists of two steps: variational mode decomposition (VMD) and Hilbert amplitude spectrum analysis (HAS). It is fully adaptive and is capable of processing nonlinear and non-stationary data, but it suffers from the issue of mode aliasing. These three types of time–frequency distributions each have their advantages and belong to different categories, possessing non-cross information. Therefore, combining these three types of time–frequency analyses can enhance the robustness of the integrated features.

After the preprocessing of three classes of time–frequency images, we construct the TFF as a multi-channel two-dimensional image by concatenating the images along the channel dimension. Although image fusion increases the computational load for both the time–frequency analysis and the initial layer of the neural network, the TFF has been demonstrated to effectively enhance recognition performance.

4. Cross-Scale Aware Network (CSANet)

After feature extraction, we need to design a network to recognize the type of radar signal modulation. CNN, as a classical type of neural network proposed by Yann Le-Cun [35], has been widely used for radar signal modulation recognition. To improve the accuracy of signal modulation recognition under low SNRs, complex deep CNNs have been employed in recent years. For instance, ResNet, which uses the residual structure and easily constructs networks with dozens or even hundreds of layers [36], performs well in recognizing radar signals. However, there is a need for further improvement in recognition accuracy. Additionally, complex networks often face difficulties when applied to lightweight platforms due to their computational demands.

In this paper, we design CSANet, which offers high recognition accuracy and low computational complexity. The CSANet architecture is presented in Figure 7a. CSANet extracts the TFF image features using four convolution (Conv) layers, two maximum pooling (MP) layers, and three CSA modules. Then, the extracted features are flattened and connected to a linear layer, and classification results are obtained. The CSA module is an essential component of the proposed CSANet. The following details the operation process of the CSA module and its constituent components.

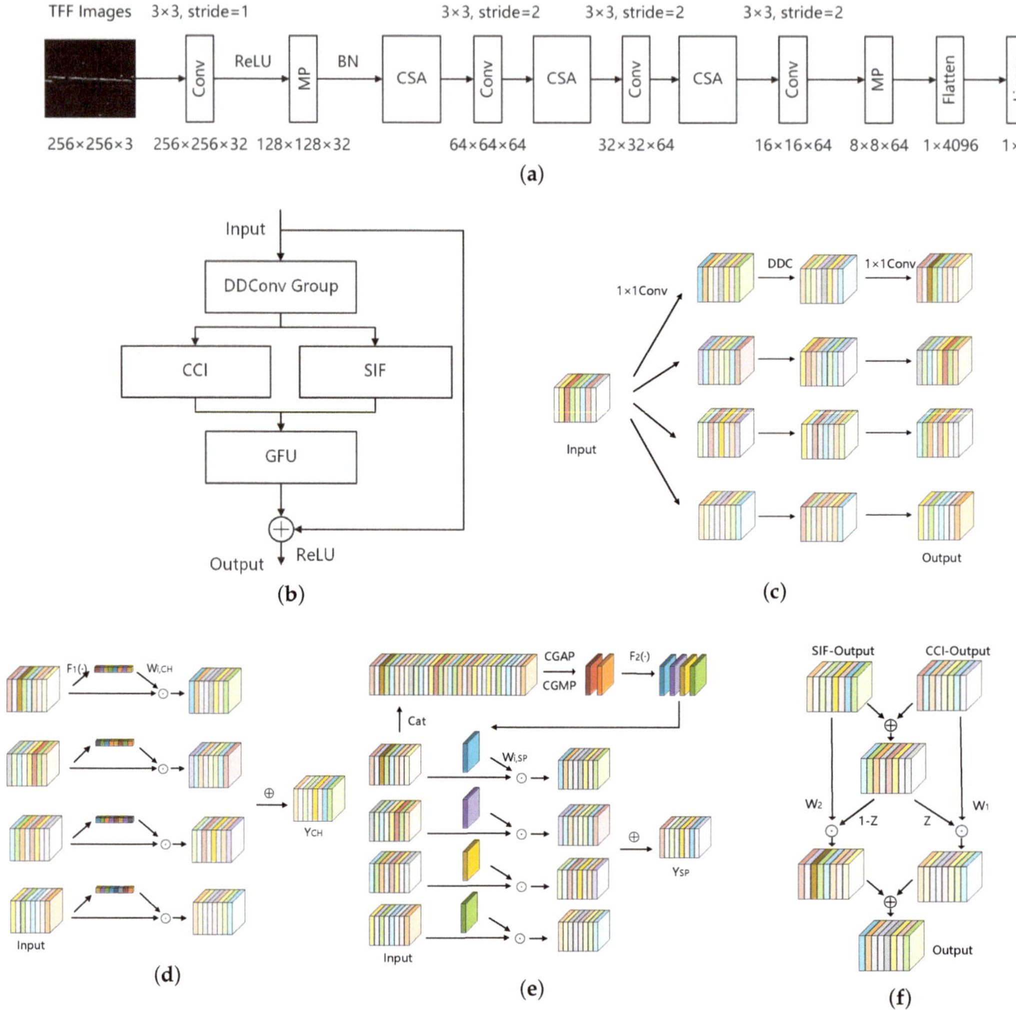

Figure 7. The overall architecture of CSANet: (**a**) CSANet. (**b**) CSA modules. (**c**) DDConv Group module. (**d**) CCI module. (**e**) SIF module. (**f**) GFU module. Multi-scale features are extracted from CSA by DDConv Group, then channel attention is applied by CCI and spatial attention is applied by SIF, and finally, fusion is performed by GFU.

4.1. CSA Module

Figure 7b depicts the operation process of the CSA module. As can be seen, the CSA module employs residual connections and conducts multi-branch feature extraction using the DDConv Group with a parallel structure. It then fuses multiple time–frequency distributions through a cross-channel interaction (CCI) and recognizes the time–frequency structural characteristics of radar signals by spatial information focus (SIF). Finally, it integrates the information through a nonlinear gated fusion unit (GFU).

4.2. Depthwise Dilated Convolution Group (DDConv Group)

Figure 7c shows the structure of the DDConv Group, where DDC represents the depthwise dilated convolution. The expansion rates of the four branches are 1, 3, 5 and 8, respectively. The DDConv Group is used to extract multi-scale features. Instead of the commonly used depthwise separable convolution, we utilize a 3×3 depthwise convolution; each channel operates with an independent convolution kernel, which significantly reduces the computational requirements. The 1×1 convolution in the depthwise separable convolution can reduce the dimensions and carryout channel flow, but dimensionality reduction is not conducive to feature retention. Therefore, we complete the work of channel flow by our designed CCI. Meanwhile, through different receptive fields, SIF can capture the global dependence in the feature information and has advantages over modulated signals with complex time–frequency energy distributions, e.g., polyphase coding and multi-frequency coding signals.

4.3. Cross-Channel Interaction (CCI)

Channel attention is one of the attention mechanism types. An example of channel attention is the squeeze-and-excitation network (SENet) [37]. Generally, channel attention compresses the input feature map from the channel direction, generating weights for each channel to represent the importance of the current channel. In this way, the model can focus on the more important channels, thereby improving performance.

Our designed CCI module aims to strength the integration of TFF characteristics, extract non-overlapping information from three time–frequency distributions, and suppress the noisy channels. CCI applies the channel attention mechanism to each scale of the feature branch. For the DDConv Group, it provides the multi-scale feature $X_i \in \mathbb{R}^{C \times H \times W}$ to the CCI, where $i = [0, 1, 2, 3]$ denote different scale feature blocks.

Figure 7d depicts the operation process of the CCI. First, a global feature representation is obtained through a spatial global average pooling (SGAP), which works by averaging the two-dimensional feature map of each channel. Then, 1×1 convolution (Conv) is used to model the inter-channel relationships. The sigmoid activation function is employed to generate the channel descriptor, and the softmax function is utilized to obtain the representation weight, which is given by

$$W_{i,CH} = \text{softmax}\{\text{sigmoid}[\text{Conv}(\text{SGAP}(X_i))]\} \in \mathbb{R}^{C \times 1 \times 1} \tag{9}$$

where $\text{softmax}\{\text{sigmoid}[\text{Conv}(\text{SGAP}(\cdot))]\}$ is defined as $F_1(\cdot)$. By performing element-wise multiplication of the weights and descriptors, we have the multi-scale fusion feature $Y_{CH} \in \mathbb{R}^{C \times H \times W}$ as

$$Y_{CH} = \sum_{i=0}^{3} W_{i,CH} \odot X_i \tag{10}$$

where $\odot$ denotes the Hadamard product.

4.4. Spatial Information Focus (SIF)

Channel attention focuses on the differences in features in different channels, while spatial attention emphasizes the information in different locations of the image [38]. Basically, spatial attention learns a spatial transformation matrix that is used to transform the input feature map into a new feature map wherein key information is highlighted and irrelevant information is suppressed. This mechanism helps the model to focus more on important spatial locations within the image, thereby enhancing the performance of the network model.

In this paper, SIF is designed as a parallel branch to CCI. While CCI pools spatial information at different scales to compute the channel attention descriptor, SIF requires pooling in the channel dimension. Therefore, to avoid losing information, SIF fuses the multi-scale features $X_i \in \mathbb{R}^{C \times H \times W}$ in the channel dimension and then outputs $[X_F \in \mathbb{R}^{4C \times H \times W}]$. Figure 7e shows the operation process of SIF.

Considering that the channel dimension is a one-dimensional feature and has fewer parameters during global pooling, we design SIF with CGAP (channel global average pooling) and CGMP (channel global maximum pooling) to obtain global features and then fuse them in the channel dimension, given by

$$
\begin{aligned}
X_A &= \mathrm{CGAP}(X_F) \in \mathbb{R}^{1 \times H \times W}, \\
X_M &= \mathrm{CGMP}(X_F) \in \mathbb{R}^{1 \times H \times W}, \\
X_{AM} &= \mathrm{Cat}(X_A, X_M) \in \mathbb{R}^{2 \times H \times W}
\end{aligned}
\tag{11}
$$

where $\mathrm{Cat}(\cdot)$ denotes the feature map fusion in the channel dimension. A 7×7 convolution is used to map the dual-channel X_{AM} into four channels, corresponding to the four scales of the input. The sigmoid activation function is employed to generate the spatial weight representation:

$$
W_{SP} = \mathrm{sigmoid}(\mathrm{Conv}(X_{AM})) \in \mathbb{R}^{4 \times H \times W}
\tag{12}
$$

where $\mathrm{sigmoid}(\mathrm{Conv}(\cdot))$ is defined as $F_2(\cdot)$. Then, the weights $W_{i,SP} \in \mathbb{R}^{1 \times H \times W}$ are from $W_{SP} \in \mathbb{R}^{4 \times H \times W}$ and are used to compute the multi-scale spatially fused feature $Y_{SP} \in \mathbb{R}^{C \times H \times W}$, given by

$$
Y_{SP} = \sum_{i=0}^{3} W_{i,SP} \odot X_i.
\tag{13}
$$

4.5. Gated Fusion Unit (GFU)

The gated fusion unit (GFU), as depicted in Figure 7f, generates adaptive weights to fuse the outputs of the CCI and SIF branches by the Sigmoid activation function in order to restore the original scale size and improve the feature representation. Given $Y_{CH} \in \mathbb{R}^{C \times H \times W}$ and $Y_{SP} \in \mathbb{R}^{C \times H \times W}$, the representative weights $Z \in \mathbb{R}^{C \times H \times W}$ are calculated as

$$
Z = \mathrm{sigmoid}(Y_{CH} W_1 + Y_{SP} W_2)
\tag{14}
$$

where $W_1, W_2 \in \mathbb{R}^{C \times C}$ are the learnable parameters during CSANet training. Then, the cross-scale aware feature is achieved by

$$
Y_{CSA} = Z \odot Y_{CH} + (1 - Z) \odot Y_{SP}.
\tag{15}
$$

5. Experimental Results

The dataset simulates 12 modulation types of radar signals, as shown in Table 1. The parameters of the signals are set as in Table 2. The sampling frequency is 200 MHz, and the signal length is 10 μs. The SNR is set as $[-14, -12, \cdots, 8]$ dB. Therefore, for every type of modulation at each SNR point, we construct 350 training samples, 150 validation samples, and 150 test samples. Moreover, our dataset contains 54,600, 23,400, and 23,400 samples for training, validating, and testing, respectively. The experiments are performed using the PyTorch 2.2.1 framework and an NVIDIA GeForce RTX 4060 laptop GPU.

Table 2. Simulation parameters for radar signals in Table 1.

Modulation	Parameters and Their Ranges
CW	$f_c \in [45, 55]$ MHz
LFM	$f_c \in [45, 55]$ MHz, $B \in [15, 25]$ MHz
NLFM	$f_c \in [45, 55]$ MHz, $B \in [15, 25]$ MHz
BPSK	$f_c \in [45, 55]$ MHz, $N = 13$
QPSK	$f_c \in [45, 55]$ MHz, $N = 28$
FSK	$f_c \in [45, 55]$ MHz, $\Delta f \in [10, 20]$, $N = 13$
4FSK	$f_c \in [45, 55]$ MHz, $\Delta f \in [5, 15]$, $N = 16$
FRANK	$f_c \in [45, 55]$ MHz, $N = 50$, $M = 7$
P1	$f_c \in [45, 55]$ MHz, $N = 50$, $M = 7$
P2	$f_c \in [45, 55]$ MHz, $N = 50$, $M = 7$
P3	$f_c \in [45, 55]$ MHz, $N = 50$, $M = 50$
P4	$f_c \in [45, 55]$ MHz, $N = 50$, $M = 50$

In addition, in order to ensure the comparability and statistical significance of the experimental results, the experimental parameters are uniformly set as follows: the initial learning rate is 0.01, the batch size is 50, the optimization algorithm is stochastic gradient descent, the number of epochs is 50, and the loss function is cross-entropy loss. Through careful adjustment of datasets and parameters, the loss value of each network can reach the convergence state after 50 rounds of training so as to ensure the correct evaluation of the performance of the proposed algorithm under different SNR and parameter conditions.

5.1. Accuracy Analysis of CSANet and Other Networks

First, we evaluate the recognition performance of CSANet. For comparison, we simulate five algorithms that are based on the TFIs-CNN methodology, including CNNQu [39], CNNHuang [32], ResNet50 [36], MobileNetV3 [40], and ShuffleNetV2 [41]. To demonstrate the effectiveness of the TFF, we also generate TFIs, i.e., *SPWVD*, FSST, or HHT, as the input of networks. The experimental results are shown in Figure 8.

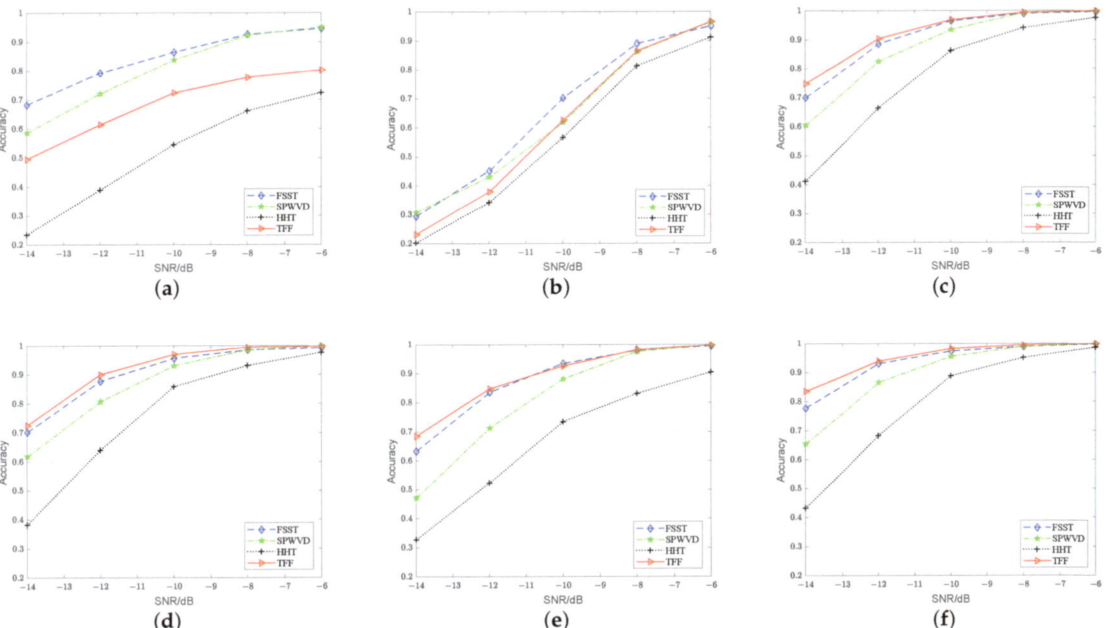

Figure 8. Accuracy evaluation: (**a**) CNNQu, (**b**) CNNHuang, (**c**) ResNet50, (**d**) MobileNetV3, (**e**) ShuffleNetV2 (**b**), and (**f**) CSANet.

As can be seen, the accuracy of ResNet50, MobileNetV3, ShuffleNetV2, and CSANet is higher when TFF is used as the learning object, while for CNNQu and CNNHuang, the accuracy is higher by inputting FSST. This demonstrates that TFF is superior to the time–frequency enhancement methods [25–27].

Overall, CSANet-TFF consistently achieves the highest accuracy across all SNR levels. Notably, at a low SNR of −14 dB, CSANet-TFF attains the highest accuracy of 83.62%, which exceeds the second-highest accuracy of ResNet50-TFF by 8.74%. At SNR = −12 dB, the accuracy of CSANet-TFF is 93.27%, outperforming other networks. Generally, CSANet-TFF excels in low-SNR scenarios. Moreover, CSANet consistently achieves the highest accuracy with *SPWVD*, FSST, HHT, or TFF as inputs, demonstrating that it outperforms existing advanced methods [32,36,39–41]. This superiority is attributed to CSANet's ability to perceive the time–frequency structures of radar signals through the CSA module. The DDConv Group and SIF mechanisms within the module are specifically designed to identify time–frequency ridge features, while CCI is designed to suppress redundant channels caused by noise, particularly in low-SNR conditions.

Furthermore, we compare the network across several metrics, including spatial complexity, computational complexity, and actual running time. These comparisons are based on parameters (Params), floating-point operations (FLOPs), and network inference time (Runtime), with the results presented in Table 3. As can be seen, CNNQu and CNNHuang exhibit lower FLOPs but higher Params compared to CSANet. However, as illustrated in Figure 8, their accuracy is significantly lower than that of CSANet, particularly at low SNRs. CSANet has a more lightweight architecture, with reduced FLOPs and Params compared to ResNet50. Moreover, CSANet's Params are approximately 1/10 of those of MobileNetV3 and ShuffleNetV2. Therefore, our proposed CSANet is proven to be a lightweight network with high accuracy. The inference time of the network is often affected by the hardware resources. Here, we calculate the time for different networks to classify a single sample, which provides a reference for the actual deployment of the network. Experimental results show that CSANet can achieve a shorter time than [36,40,41], which has practical application significance.

Table 3. Computational complexity analysis.

Network	CNNQu	CNNHuang	ResNet50	MobileNetV3	ShuffleNetV2	CSANet
FLOPs/G	0.0334	0.0371	4.1317	0.3118	0.4034	0.3386
Params/M	2.5539	21.0068	23.5326	4.2200	2.4930	0.2152
Runtime/ms	4.8750	5.0065	35.6344	18.1344	10.1563	7.6219

5.2. Ablation Study

This paper designs the CSA module that enables the CSANet to recognize intra-pulse modulation of radar signals. Here, we conduct an ablation study by adding the CSA modules one by one into the CSANet. When one CSA module is added, we employ 3×3 convolutions with a stride of two for further feature mapping and downsampling so as to minimize the data redundancy and forge an efficient architecture for CSANet. When the number of CSA modules grows, CSANet keeps its overall structure and adjusts the parameter number of the "Linear" layer accordingly. Furthermore, ablation experiments are carried out on the internal modules. SIF and GFU are removed, and the remaining modules are named D-CCI. CCI and GFU are removed, and the remaining modules are named D-SIF. We use D-CCI and D-SIF to replace three CSA modules in CSANet. In addition, we contrast CBAM [42], which has a dual-channel and spatial attention mechanism, with some similarities to CSA. In the experiment, we use a CBAM module to replace three CSA modules in CSANet. Using the TFF features as learning objects, the results of the ablation experiments are illustrated in Table 4.

Table 4. Accuracies of different CSANet architectures.

Network	SNR = −14 dB	SNR = −12 dB	SNR = −10 dB	FLOPs
D-SIF	60.79%	84.53%	94.21%	0.2142 G
D-CCI	76.91%	90.96%	95.40%	0.2546 G
CBAM	57.73%	83.48%	92.91%	0.1823 G
CSA = 1	66.90%	87.11%	97.25%	0.2127 G
CSA = 2	77.14%	92.04%	97.24%	0.3134 G
CSA = 3	83.62%	93.99%	98.23%	0.3386 G
CSA = 4	75.91%	92.38%	96.94%	0.3449 G

CBAM is less effective although it exerts both spatial and channel attention, reflecting the importance of multi-scale features. At the same time, D-SIF and D-CCI also have lower recognition accuracies than the CSA = 3 architectures, but they require slightly less computational effort. D-CCI demonstrates superior performance due to the utilization of integrated features, which encompass comprehensive and even redundant information. Therefore, the suppression of channel redundancy is critical. CSANet's accuracy increases when the CSA number grows from one to three, but it declines when the CSA number reaches four. Due to the deployment of convolutional layers for downsampling following each CSA to enhance network efficiency, an increased number of CSA modules may lead to greater information loss. When the number of CSA modules is less than three, the information loss predominantly consists of redundant data caused by noise. Conversely, when the number of CSA modules exceeds four, there is a consequential loss of useful information.

5.3. Signal Confusion Analysis

This section analyzes the confusion matrix of the CSANet results in order to provide insights into the model's performance on different modulation types. Figure 9 depicts the confusion matrices of CSANet based on three methods: *SPWVD*, FSST, and TFF at SNR = −12 dB. HHT is ignored since its performance is the worst.

In Figure 9, the vertical axis lists the true labels, while the horizontal axis lists the predicted labels. The diagonal elements represent the number of samples correctly predicted, the other elements represent the number of samples incorrectly classified, and the darker the color, the more samples. In Figure 9, three time–frequency features and twelve modulation types show excellent discrimination. It is relatively easy to confuse BPSK versus CW and FRANK versus P3. In addition, at very low SNRs, most of the signals are misinterpreted as 4FSK signals because their time–frequency graphs are irregular and scattered.

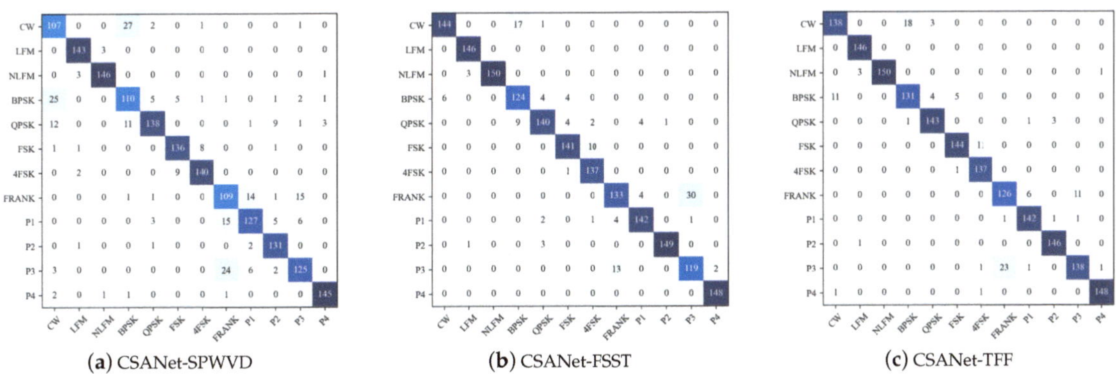

(a) CSANet-SPWVD   (b) CSANet-FSST   (c) CSANet-TFF

Figure 9. CSANet confusion matrix for SNR = −12 dB: (a) *SPWVD*. (b) FSST. (c) TFF.

Figure 10 shows the accuracy of CSANet for the recognition of twelve modulations based on *SPWVD*, FSST, and TFF. As can be seen, the FSST's accuracy varies significantly across different signal types. The FSST performs well in recognizing LFM, NLFM, FSK, and P4, but it performs poorly in recognizing FRANK, P3, and FSK4, even for high SNRs. Hence, the FSST's performance is sensitive to the modulation type. *SPWVD* and TFF are robust in recognition of various signals. The recognition accuracies of NLFM, LFM, P4, and P2 are generally higher, and the time–frequency ridges of NLFM and LFM are simpler. P2 and P4 have two ridges at the edges of their time–frequency maps, as shown in Figure 2. Moreover, P2 has phase mutations and P4 does not, and they are also distinguished from each other. Specifically, CSANet-TFF achieves over 94.73% accuracy at SNR = −10 dB for every signal.

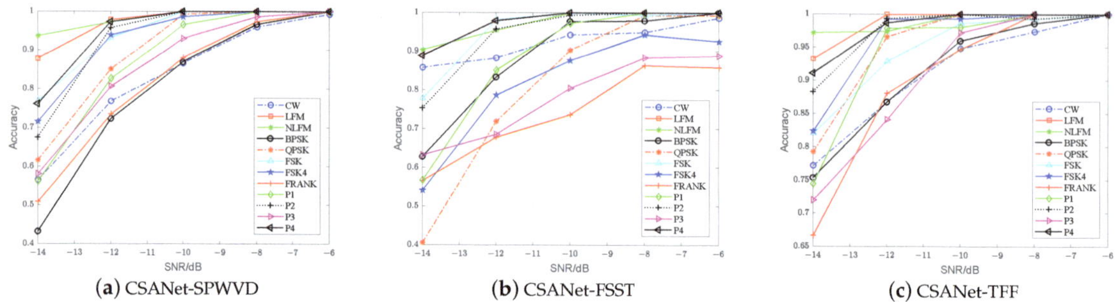

(a) CSANet-SPWVD   (b) CSANet-FSST   (c) CSANet-TFF

Figure 10. Accuracy of each signal: (**a**) *SPWVD*. (**b**) FSST. (**c**) TFF. The figure illustrates the accuracy of CSANet in identifying signal types using three different sets of time–frequency features to compare the classification effect.

5.4. Class Activation Mapping (CAM) Analysis

CAM is widely used for explaining the predictions of DL models [43]. CAM helps researchers understand how a DL model can choose the predicted class by mapping the class activation back to the significant region of the image. Figure 11 shows the CAM analysis results of CSANet and ResNet50 with TFF features, where the brighter regions are more important. The CAM analysis aids in understanding the adaptive receptive field of CSANet. We extract the feature maps before the Linear layer for visualization analysis.

As can be seen from Figure 11, the receptive field of CSANet shows a higher degree of concentration than that of ResNet50. For CSANet, the focus area adaptively adjusts in size, shape, and position, corresponding to the characteristics of varying signals, which is due to the DDConv Group and the SIF modules. Combined with the DDConv Group, which employs convolution kernels of varying dilation ratios, SIF can extract features with different ranges and precision.

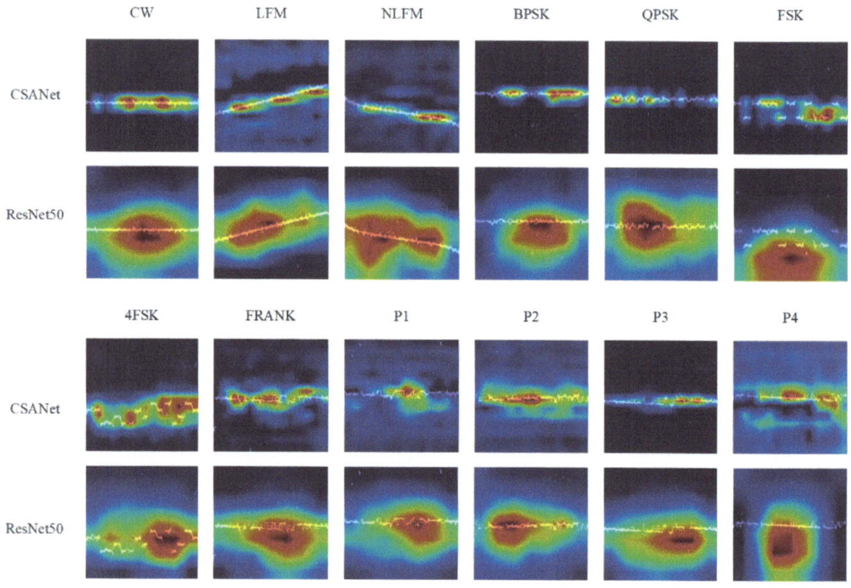

Figure 11. CAMs of CSANet and ResNet50 (SNR = −6 dB).

6. Conclusions

In this paper, we propose CSANet, a lightweight and accurate model for recognizing intra-pulse modulation in radar signals. We design TFF using three types of TFIs, i.e., *SPWVD*, FSST, and HHT. In our experiments with 12 radar signal types, CSANet using TFF achieves accuracies of 83.62%, 93.99%, and 98.23% at SNR levels of -14, -12, and -10 dB, respectively.

CSANet's high precision is mainly attributed to the CSA module, which is specifically designed to effectively address the characteristics of time–frequency ridges, including large spans, narrow curves, and sharp changes. Our solution is to develop a cross-scale strategy that correlates information across different scales and benefits the identification of key features. In the CSA module, the DDConv Group employs multiple dilated convolutions to extract multi-scale feature blocks. Two parallel branches are developed by jointly employing the channel and spatial attention mechanisms to highlight discriminative features and mitigate channel redundancies across various scales.

In terms of network complexity, we employ depthwise dilated convolutions to make the CSA lightweight. Compared to [32] with 21.01 M Params and [39] with 2.55 M Params, CSANet has only 0.22 M Params. Therefore, CSANet is a promising tool for accurate recognition of radar signals, especially in low-SNR conditions.

Author Contributions: Conceptualization, J.L. and Z.L.; methodology, J.L. and Z.L.; software, J.L. and R.L.; validation, J.L., Z.L. and R.L.; formal analysis, J.L.; investigation, R.L.; resources, J.L.; data curation, Z.L.; writing—original draft preparation, J.L., Z.L. and R.L.; writing—review and editing, J.L.; visualization, Z.L.; supervision, J.L.; project administration, J.L.; funding acquisition, J.L. All authors have read and agreed to the published version of the manuscript.

Funding: This research was funded by the National Natural Science Foundation of China (No. 61701067) and the Science and Technology Research Program of the Chongqing Municipal Education Commission (No. KJQN202300633).

Institutional Review Board Statement: Not applicable.

Informed Consent Statement: Not applicable.

Data Availability Statement: The original contributions presented in the study are included in the article, further inquiries can be directed to the corresponding author.

Conflicts of Interest: Author Jingyue Liang was employed by the company Hunan Nanoradar Science and Technology Co., Ltd. The remaining authors declare that the research was conducted in the absence of any commercial or financial relationships that could be construed as a potential conflict of interest.

Abbreviations

The following abbreviations are used in this manuscript:

CNN	Convolutional neural network
TFIs	Ttime–frequency images
SNR	Signal-to-noise ratio
CSANet	Cross-scale aware network
CSA	Cross-scale aware
TFF	Ttime–frequency fusion
WVD	Wigner–Ville distribution
MFRMF	Multi-feature random matching fusion
OD-CNN	Omni-dimensional dynamic convolution
CW	Continuous wave
LFM	Linear frequency modulation
NLFM	Nonlinear frequency modulation
BPSK	Binary phase shift keying
QPSK	Quadrature phase shift keying
FSK	Frequency shift keying
4FSK	Four-frequency shift keying

FSST	Fourier synchrosqueezed transform
SPWVD	Smoothed pseudo-Wigner–Ville distribution
HHT	Hilbert–Huang transform
CWD	Choi–Williams distribution
STFT	Short-time Fourier transform
EMD	Empirical mode decomposition
IMFs	Intrinsic mode functions
VMD	Variational mode decomposition
ResNet	Residual network
DDConv Group	Depthwise dilated convolution group
CCI	Cross-channel interaction
SIF	Spatial information focus
GFU	Gated fusion unit
Conv	Convolution
MP	Maximum pooling
AP	Average pooling
GMP	Global maximum pooling
GAP	Global average pooling
CAM	Class activation mapping

References

1. Rao, G.N.; Sastry, C.V.S.; Divakar, N. Trends in electronic warfare. *IETE Tech. Rev.* **2003**, *20*, 139–150. [CrossRef]
2. Matuszewski, J. The radar signature in recognition system database. In Proceedings of the 19th International Conference on Microwaves, Radar and Wireless Communications, Warsaw, Poland, 21–23 May 2012; pp. 617–622.
3. Zheng, Z.; Qi, C.; Duan, X. Sorting algorithm for pulse radar based on wavelet transform. In Proceedings of the IEEE 2nd Information Technology, Networking, Electronic and Automation Control Conference (ITNEC), Chengdu, China, 15–17 December 2017; pp. 1166–1169.
4. Granger, E.; Rubin, M.A.; Grossberg, S.; Lavoie, P. A What-and-Where fusion neural network for recognition and tracking of multiple radar emitters. *Neural Netw.* **2001**, *14*, 325–344. [CrossRef] [PubMed]
5. Nedyalko, P.; Jordanov, I.; Roe, J. Radar Emitter Signals Recognition and Classification with Feedforward Networks. *Procedia Comput. Sci.* **2013**, *22*, 1192–1200.
6. Dash, D.; Sa, K.D.; Jayaraman, V. Time Frequency Analysis of OFDM-LFM Waveforms for Multistatic Airborne Radar. In Proceedings of the 2018 Second International Conference on Inventive Communication and Computational Technologies (ICICCT), Coimbatore, India, 20–21 April 2018; pp. 1166–1169.
7. Wang, Y.; Huang, G.; Li, W. Waveform design for radar and extended target in the environment of electronic warfare. *J. Syst. Eng. Electron.* **2018**, *29*, 48–57.
8. Shi, Q.; Zhang, J. Radar Emitter Signal Identification Based on Intra-pulse Features. In Proceedings of the IEEE 6th Information Technology and Mechatronics Engineering Conference (ITOEC), Chongqing, China, 15–17 September 2022; pp. 256–260.
9. Xu, S.; Liu, L.; Zhao, Z. DTFTCNet: Radar Modulation Recognition with Deep Time-Frequency Transformation. *IEEE Trans. Cogn. Commun. Netw.* **2023**, *9*, 1200–1210. [CrossRef]
10. Zhang, Z.; Li, Y.; Zhu, M. JDMR-Net: Joint Detection and Modulation Recognition Networks for LPI Radar Signals. *IEEE Trans. Aerosp. Electron. Syst.* **2023**, *59*, 7575–7589. [CrossRef]
11. Ren, B.; Teh, K.C.; An, H. Automatic Modulation Recognition of Dual-Component Radar Signals Using ResSwinT—SwinT Network. *IEEE Trans. Aerosp. Electron. Syst.* **2023**, *59*, 6405–6418. [CrossRef]
12. Walenczykowska, M.; Kawalec, A.; Krenc, K. An application of analytic wavelet transform and convolutional neural network for radar intrapulse modulation recognition. *Sensors* **2023**, *23*, 1986. [CrossRef]
13. Yuan, S.; Li, P.; Wu, B. Radar Emitter Signal Intra-Pulse Modulation Open Set Recognition Based on Deep Neural Network. *Remote Sens.* **2023**, *16*, 108. [CrossRef]
14. Wu, B.; Yuan, S.; Li, P. Radar emitter signal recognition based on one-dimensional convolutional neural network with attention mechanism. *Sensors* **2020**, *20*, 6350. [CrossRef] [PubMed]
15. Wei, S.; Qu, Q.; Su, H. Intra-pulse modulation radar signal recognition based on CLDN network. *IET Radar Sonar Navig.* **2020**, *14*, 803–810. [CrossRef]
16. Gan, F.; Cai, J.; Li, P. Radar Intra-Pulse Signal Modulation Classification Based on Omni-Dimensional Dynamic Convolution. In Proceedings of the 2023 8th International Conference on Signal and Image Processing (ICSIP), Wuxi, China, 8–10 July 2023.
17. Jin, X.; Ma, J.; Ye, F. Radar signal recognition based on deep residual network with attention mechanism. In Proceedings of the 2021 IEEE 4th International Conference on Electronic Information and Communication Technology (ICEICT), Xi'an, China, 18–20 August 2021; pp. 428–432.
18. Wu, Z.L.; Huang, X.X.; Du, M. Intrapulse Recognition of Radar Signals via Bicubic Interpolation WVD. *IEEE Trans. Aerosp. Electron. Syst.* **2023**, *59*, 8668–8680. [CrossRef]

19. Chen, K.Y.; Zhang, J.Y.; Chen, S. Automatic modulation classification of radar signals utilizing X-net. *Digit. Signal Process.* **2022**, *123*, 103396. [CrossRef]
20. Zhang, T.; Shen, H. Improved Radar Signal Recognition by Combining ResNet with Transformer Learning. In Proceedings of the 2024 International Conference on Green Energy, Computing and Sustainable Technology (GECOST), Miri Sarawak, Malaysia, 17–19 January 2024; pp. 94–100.
21. Zhao, Z.; Bai, H.; Zhang, J. Cddfuse: Correlation-driven dual-branch feature decomposition for multi-modality image fusion. In Proceedings of the 2023 IEEE/CVF Conference on Computer Vision and Pattern Recognition (CVPR), Vancouver, BC, Canada, 17–24 June 2023; pp. 5906–5916.
22. Liang, B.; Tang, C.; Zhang, W.; Xu, M. N-Net: An UNet architecture with dual encoder for medical image segmentation. *Signal Image Video Process.* **2023**, *17*, 3073–3081. [CrossRef] [PubMed]
23. Themyr, L.; Rambour, C.; Thome, N. Full contextual attention for multi-resolution transformers in semantic segmentation. In Proceedings of the IEEE/CVF Winter Conference on Applications of Computer Vision (WACV), Waikoloa, HI, USA, 2–7 January 2023; pp. 3223–3232.
24. Pieniężny, A.; Konatowski, S. Intrapulse analysis of radar signal. *Comput. Methods Exp. Meas. XIV* **2009**, *48*, 259.
25. Oberlin, T.; Meignen, S.; Perrier, V. The Fourier-based synchrosqueezing transform. In Proceedings of the 2014 IEEE International Conference on Acoustics, Speech and Signal Processing (ICASSP), Florence, Italy, 4–9 May 2014; pp. 315–319.
26. Toole, J.M.O.; Boashash, B. Fast and memory-efficient algorithms for computing quadratic time–frequency distributions. *Appl. Comput. Harmon. Anal.* **2013**, *35*, 350–358. [CrossRef]
27. Huang, N.E.; Wu, Z.; Long, S.R. On instantaneous frequency. *Adv. Adapt. Data Anal.* **2009**, *1*, 177–229. [CrossRef]
28. Cohen, L. Time-frequency distributions-a review. *Proc. IEEE* **1989**, *77*, 941–981. [CrossRef]
29. Faisal, K.N.; Mir, H.S.; Sharma, R.R. Human Activity Recognition from FMCW Radar Signals Utilizing Cross-Terms Free WVD. *IEEE Internet Things J.* **2024**, *11*, 14383–14394. [CrossRef]
30. Huang, N.E.; Shen, Z.; Long, S.R. The empirical mode decomposition and the Hilbert spectrum for nonlinear and non-stationary time series analysis. *Proc. R. Soc. Lond. Ser. A Math. Phys. Eng. Sci.* **1998**, *454*, 903–995. [CrossRef]
31. Dragomiretskiy, K.; Zosso, D. Variational mode decomposition. *IEEE Trans. Signal Process.* **2013**, *62*, 531–544. [CrossRef]
32. Huang, H.; Li, Y.; Liu, J. LPI waveform recognition using adaptive feature construction and convolutional neural networks. *IEEE Aerosp. Electron. Syst. Mag.* **2023**, *38*, 14–26. [CrossRef]
33. Tang, P. A digitalization-based image edge detection algorithm in intelligent recognition of 5G smart grid. *Expert Syst. Appl.* **2023**, *233*, 120919. [CrossRef]
34. Yu, Z.Y.; Tang, J.L. Radar signal intra-pulse modulation recognition based on contour extraction. In Proceedings of the IGARSS 2020—2020 IEEE International Geoscience and Remote Sensing Symposium, Las Vegas, NV, USA, 26 September 2020; pp. 2783–2786.
35. Lecun, Y.; Bottou, L. Gradient-based learning applied to document recognition. *Proc. IEEE* **1998**, *86*, 2278–2324. [CrossRef]
36. He, K.; Zhang, X.; Ren, S. Deep residual learning for image recognition. In Proceedings of the IEEE Conference on Computer Vision and Pattern Recognition (CVPR), Las Vegas, NV, USA, 26 June 2016; pp. 770–778.
37. Hu, J.; Shen, L.; Sun, G. Squeeze-and-excitation networks. In Proceedings of the IEEE Conference on Computer Vision and Pattern Recognition (CVPR), Salt Lake City, UT, USA, 19–21 June 2018; pp. 7132–7141.
38. Li, Y.; Hou, Q.; Zheng, Z. Large selective kernel network for remote sensing object detection. In Proceedings of the IEEE/CVF International Conference on Computer Vision (ICCV), Paris, France, 2–6 October 2023; pp. 16794–16805.
39. Qu, Z.; Mao, X.; Deng, Z. Radar signal intra-pulse modulation recognition based on convolutional neural network. *IEEE Access* **2018**, *6*, 43874–43884. [CrossRef]
40. Howard, A.; Sandler, M.; Chu, G. Searching for MobileNetV3. In Proceedings of the IEEE/CVF International Conference on Computer Vision (ICCV), Seoul, Republic of Korea, 27 October 2019; pp. 1314–1324.
41. Ma, N.; Zhang, X.; Zheng, H.T. ShuffleNet v2: Practical guidelines for efficient CNN architecture design. In Proceedings of the European Conference on Computer Vision (ECCV), Munich, Germany, 8–14 September 2018; pp. 116–131.
42. Woo, S.; Park, J.; Lee, J.Y. Cbam: Convolutional block attention module. In Proceedings of the European Conference on Computer Vision (ECCV), Munich, Germany, 8–14 September 2018; pp. 3–19.
43. Baehrens, D.; Schroeter, T.; Harmeling, S. How to explain individual classification decisions. *J. Mach. Learn. Res.* **2010**, *11*, 1803–1831.

Article

On a Closer Look of a Doppler Tolerant Noise Radar Waveform in Surveillance Applications [†]

Maximiliano Barbosa [1,*], Leandro Pralon [2], Antonio L. L. Ramos [3] and José Antonio Apolinário, Jr. [4]

[1] Brazilian Navy Weapons Systems Directorate, Rio de Janeiro 20010-000, Brazil
[2] Brazilian Army Technological Center, Rio de Janeiro 23020-470, Brazil; pralon.leandro@eb.mil.br
[3] Department of Science and Industry Systems, University of South-Eastern Norway (USN), 3616 Kongsberg, Norway; antonior@usn.no
[4] Military Institute of Engineering, Rio de Janeiro 22290-270, Brazil; apolin@ime.eb.br
* Correspondence: maximiliano.barbosa@marinha.mil.br
[†] This paper is an extended version of our paper published in Barbosa, M.; Pralon, L.; Apolinário, J.A. Slow-Moving Target Detection Performance of an LPI APCN Waveform in Surveillance Applications. In Proceedings of the 23rd International Radar Symposium (IRS), Gdansk, Poland, 12–14 September 2022; pp. 147–152.

Abstract: The prevalence of Low Probability of Interception (LPI) and Low Probability of Exploitation (LPE) radars in contemporary Electronic Warfare (EW) presents an ongoing challenge to defense mechanisms, compelling constant advances in protective strategies. Noise radars are examples of LPI and LPE systems that gained substantial prominence in the past decade despite exhibiting a common drawback of limited Doppler tolerance. The Advanced Pulse Compression Noise (APCN) waveform is a stochastic radar signal proposed to amalgamate the LPI and LPE attributes of a random waveform with the Doppler tolerance feature inherent to a linear frequency modulation. In the present work, we derive closed-form expressions describing the APCN signal's ambiguity function and spectral containment that allow for a proper analysis of its detection performance and ability to remove range ambiguities as a function of its stochastic parameters. This paper also presents a more detailed address of the LPI/LPE characteristic of APCN signals claimed in previous works. We show that sophisticated Electronic Intelligence (ELINT) systems that employ Time Frequency Analysis (TFA) and image processing methods may intercept APCN and estimate important parameters of APCN waveforms, such as bandwidth, operating frequency, time duration, and pulse repetition interval. We also present a method designed to intercept and exploit the unique characteristics of the APCN waveform. Its performance is evaluated based on the probability of such an ELINT system detecting an APCN radar signal as a function of the Signal-to-Noise Ratio (SNR) in the ELINT system. We evaluated the accuracy and precision of the random variables characterizing the proposed estimators as a function of the SNR. Results indicate a probability of detection close to 1 and show good performance, even for scenarios with a SNR slightly less than -10 dB. The contributions in this work offer enhancements to noise radar capabilities while facilitating improvements in ESM systems.

Keywords: noise radar; APCN; electronic support measures; time-frequency analysis

Citation: Barbosa, M.; Pralon, L.; Ramos, A.L.L.; Apolinário, J.A., Jr. On a Closer Look of a Doppler Tolerant Noise Radar Waveform in Surveillance Applications. *Sensors* **2024**, *24*, 2532. https://doi.org/10.3390/s24082532

Academic Editors: Janusz Dudczyk and Piotr Samczyński

Received: 4 March 2024
Revised: 9 April 2024
Accepted: 11 April 2024
Published: 15 April 2024

1. Introduction

Electronic Intelligence systems for Electronic Warfare rely on the detection, identification, and further processing of radar signals [1]. The advances in the semiconductor industry over the last decade enabled huge breakthroughs in ELINT systems. Indeed, modern equipment can implement high-frequency, high-bandwidth digital receivers with complex signal processing algorithms. An example is jammers capable of reproducing radar signal characteristics after extracting them. The design of Low Probability of Interception and Low Probability of Exploitation radars has attracted significant attention in the current technological race. As a result, the transmission of random or pseudo-random signals has gained considerable notoriety within the radar system community over the last few years [2–10]. Commonly known as "noise radars", they manage to achieve LPI by employing pulse compression in the reception, along with a proper choice of transmitted waveform stochastic properties [2,11], LPE [9,12,13], and low sidelobe levels, while suppressing range ambiguity.

Many different approaches have been proposed in the literature to generate waveforms characterized by stochastic processes that better fulfill these radar system requirements [5,14–18]. Still, they all suffer from low Doppler tolerance [19], an inherent characteristic of traditional noise radars that hinders their use in several applications, including surveillance. Researchers have proposed alternative approaches to bypass such a weakness while preserving the LPI and the LPE properties of random waveforms. Included in this group is the so-called Advanced Pulse Compression Noise (APCN) waveform [3,20,21], which combines a random signal with a Doppler-tolerant Linear Frequency Modulation (LFM) waveform. The random signal may consist of amplitude terms, phase terms, or both.

In [22], we expanded the discussion over the performance of APCN waveforms. We derived closed-form expressions to characterize the APCN's narrowband ambiguity function, the power spectral density, and the spectral containment, thus accounting for random components in the amplitude and phase of the transmitted signal. Furthermore, we identified and addressed a significant drawback in employing the APCN signal for detecting slow-moving targets. In the present work, we extend the results and findings of [22] by investigating the same properties for APCN signals with Phase-Only random components. Furthermore, we examine APCN's exploitation of power transmission through the Peak-to-Average Power Ratio, its ability to mitigate range ambiguities, detect uncooperative targets, and determine their range and radial velocity in real-time. These factors are crucial in radar waveform design, particularly in surveillance applications.

From the perspective of an ESM system, as discussed in [20], employing an APCN signal with a particular level of phase randomness and constant amplitude introduces a challenge in accurately discerning the waveform characteristics for an intercept-receiver system. However, owing to significant recent advancements in emitter detection and exploitation techniques, modern ESM systems and traditional spectral analysis methods can generate Time-Frequency Analysis maps, as mentioned in [6,11,23]. As a result, it has become feasible to ascertain the waveform characteristics of deterministic LPI/LPE radar signal modulations.

References [24–28] introduce techniques for conducting TFA and extracting modulation parameters from deterministic signals assumed LPI/LPE. However, these techniques have a drawback: the extraction process relies exclusively on visual analysis. This dependency on human interpretation of TFA results hampers the effectiveness of non-real-time EW receivers. In contrast, the works in [29–31] propose autonomous extraction methods, albeit still primarily designed for deterministic signals. Nonetheless, a significant contribution of these works is in developing a technique that leverages image processing methodologies to autonomously extract characteristics of a noise radar that employs APCN, allegedly LPI, as discussed in [3,20]. Our analysis, however, considers varied signal-to-noise ratio levels in an ESM receiver chain to ensure robust and satisfactory results. In addition, it examines APCN waveforms featuring phase and amplitude random components, in contrast to the approach in [3] that focuses exclusively on phase randomness. Indeed, the extended configuration augments the level of randomness in APCN waveforms and enhances their LPI and LPE characteristics, as corroborated by the authors of the present work in [22].

The rest of this paper unfolds as follows. Section 2 explores noise radar operations focusing on the APCN waveform as the transmitting signal in surveillance applications. Section 3 delves into the modeling of an ESM digital system, offering an in-depth analysis of the APCN radar waveform from the perspective of an intercept–receiver system. Moreover, the section introduces a methodology designed to detect and extract the distinctive characteristics of APCN signals. It also includes a comprehensive discussion of the performance evaluations of this methodology, laying out the grounds for conclusions regarding the

LPI and LPE potential of the APCN waveform in an EW scenario. Section 4 presents the conclusions and summarizes the key findings and insights in this paper.

2. Advanced Pulse Compression Noise Radars

In traditional noise radar systems, the transmit signal is characterized by a stochastic process, $s(t)$ [2]. Consequently, the corresponding matched filter outputs, relative to the pulse compression architectures, are all characterized by complex random processes. Therefore, a proper analysis of noise radars must rely on probabilistic tools. That is the case of the narrowband ambiguity function of a random signal, given as [32]

$$\chi_{\tilde{s}}(\tau, f_{\mathrm{D}}) = \int_{-\infty}^{\infty} \tilde{s}(t)\, \tilde{s}^*(t - \tau)\, \mathrm{e}^{-\mathrm{j}2\pi f_{\mathrm{D}} t} dt, \tag{1}$$

where $\tilde{s}(t)$ is the complex envelope of $s(t)$, f_{D} is the Doppler frequency, and τ is the time delay. Assuming $\tilde{s}(t)$ to be Wide-Sense Stationary (WSS) and time-limited with duration τ_s, and that f_{D} is deterministic, the expected value of the ambiguity function of Equation (1) simplifies to [5,19]

$$\mathrm{E}[\chi_{\tilde{s}}(\tau, f_{\mathrm{D}})] = \tau_s\, R_{\tilde{s}}(\tau)\, \mathrm{sinc}(f_{\mathrm{D}}\tau_s); \quad -\tau_s \leq \tau \leq \tau_s, \tag{2}$$

where $R_{\tilde{s}}(\tau) = \mathrm{E}[\tilde{s}(t)\,\tilde{s}^*(t - \tau)]$ is the autocorrelation function of the complex stochastic process $\tilde{s}(t)$ [33].

The near thumbtack format depicted in Equation (2) shows that the expected range and Doppler profiles in noise radar systems are independent functions; therefore, no range-Doppler coupling is expected in noise radars, inhibiting their usage in surveillance applications, for example [34].

To overcome the no range-Doppler coupling limitation while preserving the random nature of the transmit signal, the authors in [3,20,21] proposed the Advanced Pulse Compression Noise (APCN) radar architecture. APCN waveforms combine a deterministic Linear Frequency Modulation signal with a stochastic component. Its complex envelope is given by [20,22]

$$\tilde{s}(t) = \sqrt{P}\, a(t)\, \mathrm{e}^{\mathrm{j}[\theta(t) + \kappa\,\phi(t)]}, \tag{3}$$

where the samples of the random process $a(t)$ have a Rayleigh distribution, i.e., $p(a) = (a/\alpha^2)\,\mathrm{e}^{-a^2/\alpha^2}$, with $a \geq 0$ and α being the scale parameter. Moreover, $\phi(t)$ is uniformly distributed in the interval $(0, 2\pi]$ and $\theta(t)$ represents the LFM deterministic component. The signal's complex envelope mean power, P, is assumed, with no loss of generality, to be unitary throughout the present work. We can rewrite Equation (3) as $\tilde{s}(t) = \tilde{s}_r(t)\,\tilde{s}_c(t)$, where $\tilde{s}_r(t) = a(t)\,\mathrm{e}^{\mathrm{j}\kappa\phi(t)}$ is the transmit signal random component, with $0 \leq \kappa \leq 1$, and $\tilde{s}_c(t) = \mathrm{e}^{\mathrm{j}\theta(t)}$ is the LFM waveform envelope, with bandwidth $\beta_{\tilde{s}_c}$.

In modern radar systems, the matched filter is implemented digitally. Therefore, we consider the discrete-time case in our analysis and define

$$\tilde{s}(n) = \tilde{s}(t)\big|_{t=nT},$$

where n is the discrete-time index and $T = 1/f_s$, with f_s being the sampling frequency. We also assume that $\tau = \bar{\tau}T$, where $\bar{\tau}$ corresponds to the discrete-time delay in samples. Moreover, $\tau_s = \bar{\tau}_s T$, with $\bar{\tau}_s$ being the number of samples in a pulse signal with duration τ_s.

We show in [22] that the APCN narrowband ambiguity function is such that

$$\mathrm{E}[\chi_{\tilde{s}}(\bar{\tau}, f_{\mathrm{D}})] = R_{\tilde{s}_r}(\bar{\tau})\, \chi_{\tilde{s}_c}(\bar{\tau}, f_{\mathrm{D}}); \quad \bar{\tau} \leq |\bar{\tau}_s|, \tag{4}$$

where $R_{\tilde{s}_r}(\bar{\tau})$ is the autocorrelation sequence of the transmit signal random component given as

$$R_{\tilde{s}_r}(\bar{\tau}) = \alpha^2 \left(\frac{1 - \cos 2\kappa\pi}{4\kappa^2\pi} \right) + \left[2\alpha^2 - \alpha^2 \left(\frac{1 - \cos 2\kappa\pi}{4\kappa^2\pi} \right) \right] \delta(\bar{\tau}), \tag{5}$$

where $\delta(\bar{\tau})$ is the Dirac delta sequence and $\chi_{\tilde{s}_c}(\bar{\tau}, f_{\mathrm{D}})$ in (4) is the ambiguity function of $\tilde{s}_c(t)$ defined as

$$\chi_{\tilde{s}_c}(\bar{\tau}, f_{\mathrm{D}}) = \left(1 - \frac{|\bar{\tau}|}{\bar{\tau}_s}\right)\left|\frac{\sin\left[\pi\left(\frac{f_{\mathrm{D}}}{f_s} + \frac{\mu}{f_s^2}\bar{\tau}\right)(\bar{\tau}_s - |\bar{\tau}|)\right]}{\pi\left(\frac{f_{\mathrm{D}}}{f_s} + \frac{\mu}{f_s^2}\bar{\tau}\right)(\bar{\tau}_s - |\bar{\tau}|)}\right|, \ \bar{\tau} \le |\bar{\tau}_s|, \tag{6}$$

wherein $\mu = \beta_{\tilde{s}_c}/\tau_s$.

Figure 1a illustrates the range profile for different values of Doppler shifts, namely $f_{\mathrm{D}} = 0$, $f_{\mathrm{D}} = 0.55/\tau_s$, $f_{\mathrm{D}} = 0.7/\tau_s$, and $f_{\mathrm{D}} = 1.2/\tau_s$, of an APCN normalized ambiguity function, with $\kappa = 0.5$ and $\alpha = 1$, and considering $\beta_{\tilde{s}_c} = 30$ MHz and $\tau_s = 50$ μs. The sampling frequency was assumed to be $f_s = 1$ GHz. Figure 1b shows the zero-delay cut behavior (Doppler profile).

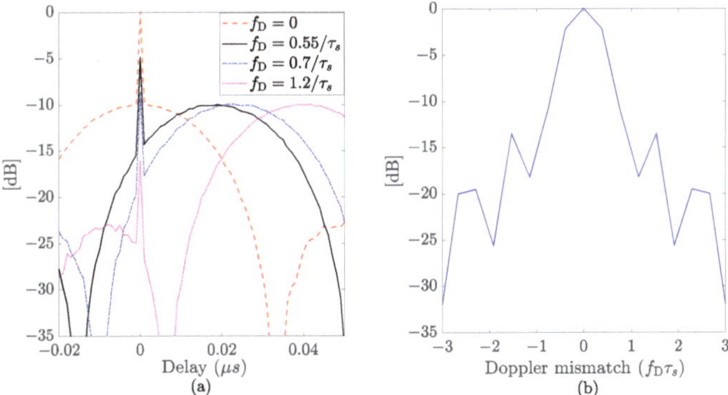

Figure 1. Normalized ambiguity function (in dB) considering a single pulse realization of an APCN waveform ($\kappa = 0.5$ and $\alpha = 1$): (**a**) range profile (Doppler cuts); (**b**) Doppler profile (zero-delay cut).

Despite being Doppler tolerant, APCN waveforms exhibit an anomaly at zero delays, potentially impairing radar performance in the presence of slow-moving targets. A closer inspection of Equations (4)–(6) evidences the presence of a sinc function at the pulse compression output whose maximum peak exists at $f_{\mathrm{D}}/fs + \mu\bar{\tau}/f_s^2 = 0$. The sinc function is attenuated by a factor $L = H\left(1 - f_{\mathrm{D}}/\beta_{\tilde{s}_c}\right)$, for $\bar{\tau} \ne 0$, and amplified by $A = 2\alpha^2$, for $\bar{\tau} = 0$, with $H = \alpha^2[1 - \cos(2\kappa\pi)]/4\kappa^2\pi$, as a result of Equation (5). Additionally, and analogously to the LFM case, it presents a range measurement error $\epsilon_R = -cf_{\mathrm{D}}\tau_s/2\beta_{\tilde{s}_c}$, that can be eliminated [34].

Figure 2 illustrates one realization of the APCN random component autocorrelation sequence (see Equation (5)), considering a signal $\tilde{s}_r(n)$ composed of a random amplitude with scale parameter $\alpha = 1$ and a random phase uniformly distributed in the interval $(0, 2\kappa\pi]$, for different values of κ. Here, we can see that the attenuation on the APCN range profile increases with κ for $\bar{\tau} \ne 0$. The implication is that when a target exhibits a Doppler shift the peak of the range profile shifts away from the origin because of the Doppler-tolerant behavior of the LFM component, making it susceptible to attenuation. Indeed, this behavior reduces the SNR, diminishing the system's Probability of Detection.

The Dirac delta component in Equation (5) is always present for $\bar{\tau} = 0$. For slow-moving targets, i.e., those that give rise to a Doppler shift above a threshold greater than $0.55/\tau_s$, in this specific case, the spike falls outside the range resolution, given by the sinc function main lobe's 3 dB width, but still exhibits enough power to pose as another target's response. Thus, the contribution of the noisy component to the matched filter output can lead to target misdetection or false alarms due to misinterpretations.

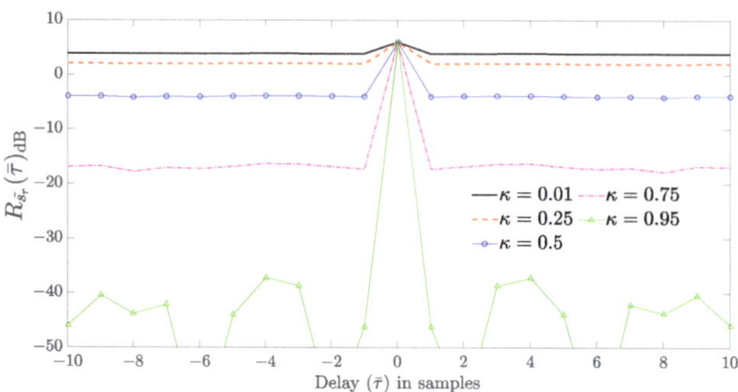

Figure 2. Autocorrelation sequence of the APCN complex random component for different values of κ, considering $\alpha = 1$.

In [22], the authors proposed a method to eliminate the spike at $\bar{\tau} = 0$ by correlating the received signal with the deterministic component of the APCN waveform instead of using the transmit signal's replica in a pulse compression architecture. Figure 3 presents a simplified block diagram of the proposed system implemented in a digital receiver signal processing chain. This modification makes it possible to eliminate the impulse anomaly at $\bar{\tau} = 0$, leaving only the deterministic component attenuation $(1 - f_D/\beta_{\tilde{s}_c})$ due to the Doppler mismatch. In this diagram, index "i" represents the i-th generated sample function of a stochastic process in the transmission of the i-th pulse, and $\omega_D = 2\pi f_D/f_s$ is the digital Doppler frequency. We adopted the Hann window to weigh the replica due to its demonstrated advantages in minimizing the integrated side-lobe ratio, which favors range resolution [21].

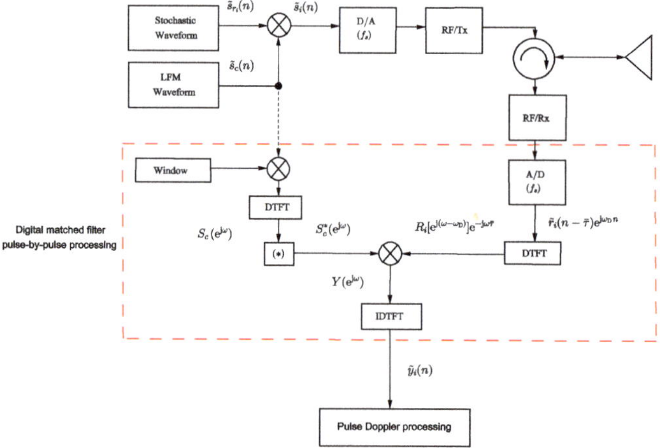

Figure 3. Proposed simplified radar block diagram to obtain the matched filter output pulse-by-pulse, where (*) is the complex conjugate operation.

In light of the physics governing the phenomenon, for simplification and without sacrificing generality, let us consider a received signal $\tilde{r}(t) = \tilde{s}(t - T_0)e^{-j2\pi f_D t}$ from a scatter located at $R_0 = c/2T_0$, with c being the vacuum light speed. The expected value of the pulse compression output resulting from the proposed architecture [22] (see Figure 3) is thus given by

$$\mathrm{E}[\tilde{y}(\tau)] = \mathrm{E}\left[\int_{-\infty}^{\infty} \left[\tilde{s}_r(t - T_0)\tilde{s}_c(t - T_0)\,e^{-j2\pi f_D t}\right] \tilde{s}_c^*(t - \tau)dt\right]. \tag{7}$$

Since $a(t)$ and $\phi(t)$ are WSS and independent process [20], Equation (7) can then be written as [33]

$$E[\tilde{y}(\tau)] = E[a(t)]E\left[e^{j\kappa\phi(t)}\right]\chi_{\tilde{s}_c}(\tau - T_0, f_D); \quad T_0 - \tau_s \le \tau \le T_0 + \tau_s, \tag{8}$$

which, in turn, can be shown to reduce to

$$E[\tilde{y}(\tau)] = \alpha\sqrt{\frac{\pi}{2}}\left(\frac{\sin\kappa\pi}{\kappa\pi}\right)\chi_{\tilde{s}_c}(\tau - T_0, f_D); \quad T_0 - \tau_s \le \tau \le T_0 + \tau_s. \tag{9}$$

As stated in Equation (9), the output of our proposed architecture [22] presents the shape of the sinc function relative to the LFM matched filter component, attenuated by a factor $B = \alpha\sqrt{\frac{\pi}{2}}\left(\frac{\sin\kappa\pi}{\kappa\pi}\right)$. Therefore, the pulse compression gain is compromised, and the proposed technique becomes effective only when the chosen APCN random component parameters are such that B is not so high as to start jeopardizing the SNR at the detector. Figure 4a, illustrates the normalized matched filter output of a conventional pulse compression radar architecture based on filtering the received APCN signal ($\alpha = 1$ and $\kappa = 0.5$) with a transmit replica (proposed in [21]). Note the attenuation of the normalized sinc main lobe, given by $L + A$, for $\bar{\tau} \ne 0$, as derived previously. Note further that, since the attenuation is near the same for all $\bar{\tau} \ne 0$, the peak side-lobe (PSL) ratio achieved with the LFM signal is preserved and remains close to 13 dB [34]. These observations demonstrate that APCN signals do not present a PSL ratio close to the well-known time-bandwidth product inherent to traditional noise radar waveforms.

Figure 4b illustrates the normalized pulse compression output of our proposed architecture, considering targets with the same Doppler shifts as in Figure 4a. Note that the anomaly was eliminated for all analyzed Doppler shifts, minimizing the possibility of false targets. The PSL ratio also remained close to 13 dB.

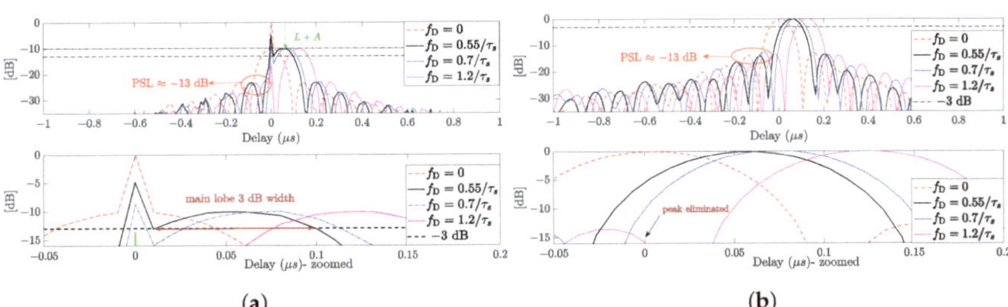

(a)   **(b)**

Figure 4. Pulse compression output for targets with different Doppler shifts: (**a**) considering the receiver system configuration based on filtering the received APCN signal with a transmit replica; (**b**) considering the receiver system using our proposal configuration.

The stochastic component $a(t)$ in Equation (3) increases the transmit signal's degree of randomness, potentially enhancing its LPI and LPE properties. Nevertheless, it introduces a significant drawback to the system related to the limited exploitation of the power transmitter. Such a drawback is evident in the reduction in the the Peak-to-Average Power Ratio (PAPR) of the system, given by [6,35]

$$\text{PAPR} = \frac{\max_{\{n\}}|\tilde{s}(n)|^2}{\frac{1}{\tau_s}\sum_{n=1}^{\tau_s}|\tilde{s}(n)|^2}. \tag{10}$$

Signals characterized by a consistent amplitude envelope exhibit a PAPR of 1, rendering them highly power-efficient. These signals enable the driving of power amplifiers close to saturation, maximizing power utilization during transmission. Any departure from the unity amplitude level results in an energy loss within the correlation main lobe, ultimately reducing the peak response level. The energy loss can be interpreted as an SNR degradation, leading to a decline in detection performance. Define the unavoidable decrease in performance as [6,13]

$$\text{SNR}_{Loss} = -10\log_{10}(\text{PAPR}). \tag{11}$$

As an alternative to amplitude modulation, the community has devoted significant effort to deriving closed-form expressions of random frequency-modulated signals for radar applications better suited for systems requiring high power efficiency. Assuming $\tilde{s}_r^{PO}(t) = e^{j\kappa\phi(t)}$, the autocorrelation sequence of the random component of the transmit Phase-Only (PO) APCN signal is given by [22]

$$R_{\tilde{s}_r^{PO}}(\bar{\tau}) = \frac{1 - \cos 2\kappa\pi}{2\kappa^2\pi^2} + \left(1 - \frac{1 - \cos 2\kappa\pi}{2\kappa^2\pi^2}\right)\delta(\bar{\tau}). \tag{12}$$

Analogously to Figure 1a, Figure 5a illustrates the Phase-Only APCN normalized range profile, considering different values of Doppler shifts ($f_D = 0$, $f_D = 0.55/\tau_s$, $f_D = 0.7/\tau_s$, and $f_D = 1.2/\tau_s$), $\kappa = 0.5$, $\beta_{\tilde{s}_c} = 30$ MHz, and $\tau_s = 50\,\mu$s. The sampling frequency was also assumed to be $f_s = 1$ GHz. Figure 5b shows the zero-delay cut behavior, nearly the same as in Figure 1b.

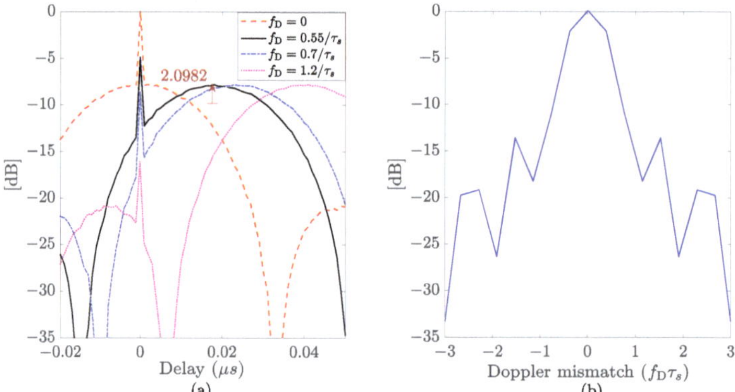

Figure 5. Normalized ambiguity (in dB) function considering a single pulse realization of a Phase-Only APCN waveform ($\kappa = 0.5$): (**a**) range profile (Doppler cuts); (**b**) Doppler profile (zero-delay cut).

In Figure 5a, it is noteworthy that the power level difference between the zero-delay ($\tau = 0\,\mu$s) spike and the main lobe of the sync function is reduced by a factor of $\pi/4$ compared to Figure 1a. When converted to dB, this difference becomes $20\log_{10}(\pi/4) = 2.0982$ dB. The reduction helps mitigate the adverse effects of such anomalies on the system's detection performance. Additionally, Figure 6 illustrates the relationship between SNR_{Loss} and various PAPR values, considering different scale parameters α associated with randomness in amplitude. A value of $\kappa = 0.5$ is assumed. Notably, the Phase-Only (constant amplitude) APCN waveform achieves a PAPR < 2, which is acceptable in noise radar applications without substantial degradation in detection performance ($\text{SNR}_{\text{Loss}} < 3$ dB) [6]. Furthermore, according to [13], when the random amplitude is employed, the resultant noisy waveform has a PAPR of around 10 (or greater), with reduced transmitted energy.

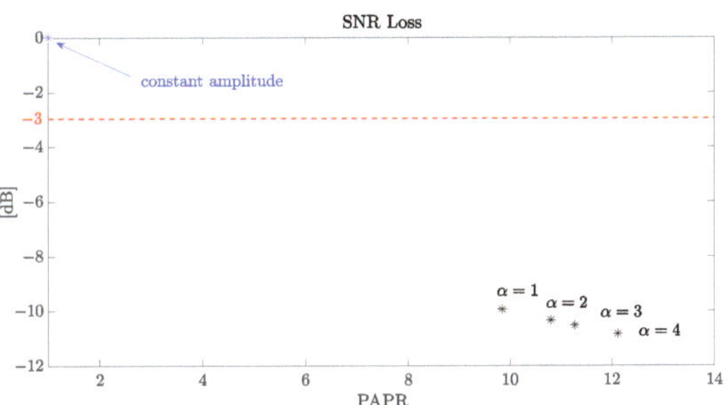

Figure 6. SNR loss versus PAPR assuming $\kappa = 0.5$ and different scale parameters α.

Another property that we evaluated in dealing with random transmit signals is related to its spectral containment, which usually is not efficient. The analysis of the transmit signal's Power Spectral Density (PSD), $\Gamma_{\tilde{s}}(\omega)$, given by the Discrete-Time Fourier Transform (DTFT) of its autocorrelation sequence, $R_{\tilde{s}}(\bar{\tau}) = \mathrm{E}[\chi_{\tilde{s}}(\bar{\tau}, 0)]$ [33,36], allows for the evaluation of its spectral containment and the derivation of closed-form expressions describing its bandwidth. Considering Equation (5), the APCN PSD is given by

$$\Gamma_{\tilde{s}}(\omega) = \mathcal{F}\{H\,\chi_{\tilde{s}_c}(\bar{\tau}, 0)\} + (2\alpha^2 - H)\,\mathcal{F}\{\delta(\bar{\tau})\} * \Gamma_{\tilde{s}_c}(\omega), \tag{13}$$

where $*$ and $\mathcal{F}\{\cdot\}$ are, respectively, the convolution and DTFT operators, $H = \alpha^2[1 - \cos{(2\kappa\pi)}]/4\kappa^2\pi$, as previously defined, and $\Gamma_{\tilde{s}_c}(\omega)$ is the PSD of $R_{\tilde{s}_c}(\bar{\tau}) = \chi_{\tilde{s}_c}(\bar{\tau}, 0)$. Expanding Equation (13) and knowing that $\mathcal{F}\{\delta(n)\} = 1$ yields

$$\Gamma_{\tilde{s}}(\omega) = H\,\Gamma_{\tilde{s}_c}(\omega) + (2\alpha^2 - H)\,\underbrace{\frac{1}{2\pi}\int_{-\pi}^{\pi}\Gamma_{\tilde{s}_c}(\varsigma)\,d\varsigma}_{1}. \tag{14}$$

Note that the integral on the right-hand side represents, in the digital frequency domain $(-\pi \leq \omega \leq \pi)$, the periodic sum of the samples of a complex envelope linear chirp signal with unitary power. Consequently, according to Parseval's Theorem [36], it evaluates to one. Finally, Equation (14) becomes

$$\Gamma_{\tilde{s}}(\omega) = H\,\Gamma_{\tilde{s}_c}(\omega) + (2\alpha^2 - H), \quad |\omega| \leq \pi. \tag{15}$$

From the closed-form transmits signal's PSD in (15), there are two widespread approaches to define a signal's bandwidth: the 3 dB bandwidth ($\beta_{3\mathrm{dB}}$) and the portion of the spectrum where $p\%$ of the total power is concentrated ($\beta_{p\%}$) [5]. Concerning the former and according to Equation (15), $\beta_{3\mathrm{dB}}$ can be considered the same $\beta_{\tilde{s}_c}$ of the APCN signal deterministic LFM component [3]. Nevertheless, the percentage of the total power within $\beta_{\tilde{s}_c}$ reduces as κ increases, leading to the rise of $\beta_{p\%}$, which can compromise the APCN waveform's performance in practical applications.

The percentage of total power within $\beta_{\tilde{s}}$ is given by the integral of $\Gamma_{\tilde{s}}(\omega)$ in Equation (15) over the interval $[-\pi\beta_{\tilde{s}_c}/f_s, \pi\beta_{\tilde{s}_c}/f_s]$; hence

$$\%P_{\beta_{\tilde{s}}} = \int_{-\pi\beta_{\tilde{s}_c}/f_s}^{\pi\beta_{\tilde{s}_c}/f_s} \left[H\,\Gamma_{\tilde{s}_c}(\omega) + (2\alpha^2 - H)\right]d\omega$$

$$= H\left[\int_{-\pi\beta_{\tilde{s}_c}/f_s}^{\pi\beta_{\tilde{s}_c}/f_s}\Gamma_{\tilde{s}_c}(\omega)\,d\omega\right] + (2\alpha^2 - H)\frac{2\pi\beta_{\tilde{s}_c}}{f_s}. \tag{16}$$

Practical situations require a higher $\beta_{\tilde{s}_c}\tau_s$ product to well-define the spectrum of an LFM waveform [34]. Owing to the maximum total power contained within a rectangular bandwidth shape, there is minimal spreading of the LFM spectrum. Consequently, the integral in Equation (16) can be approximated as one, simplifying to

$$\%P_{\beta_{\tilde{s}}} = H + 2\pi(2\alpha^2 - H)\,\frac{\beta_{\tilde{s}_c}}{f_s}. \tag{17}$$

In turn, $\beta_{p\%}$ can be shown as in

$$\beta_{p\%} = \frac{(p-H)f_s}{2\pi(2\alpha^2 - H)}, \tag{18}$$

where p is the desired percentage power within $\beta_{\tilde{s}}$.

A similar derivation applies to the Phase-Only APCN waveform. Analogously to the previous formulation, one can show that the percentage of total power within $\beta_{\tilde{s}}$ when the APCN signal's amplitude is constant is given by

$$\%P^{PO}_{\beta_{\tilde{s}}} = D + 2\pi(1-D)\,\frac{\beta_{\tilde{s}_c}}{f_s}, \tag{19}$$

where $D = [1 - \cos(2\kappa\pi)]/2\kappa^2\pi^2$. Thus, it be shown that $\beta_{p\%}$ becomes

$$\beta^{PO}_{p\%} = \frac{(p-D)f_s}{2\pi(1-D)}, \tag{20}$$

where p is the desired percentage power within $\beta_{\tilde{s}}$.

Figure 7 presents the behavior of $\%P_{\beta_{\tilde{s}}}$ for different values of κ, assuming constant and random discrete amplitude $a(n)$. Note that the variability in phase randomness directly impacts the mean power of the APCN transmitted signal within the desired bandwidth. Moreover, introducing randomness in amplitude creates a greater challenge in maintaining the transmit waveform spectral confinement. Consequently, the RF transmitter, receiver, and the entire signal processing chain must account for these effects to prevent signal distortions or a misformulation of the radar range equation, ultimately leading to a degradation in detection performance.

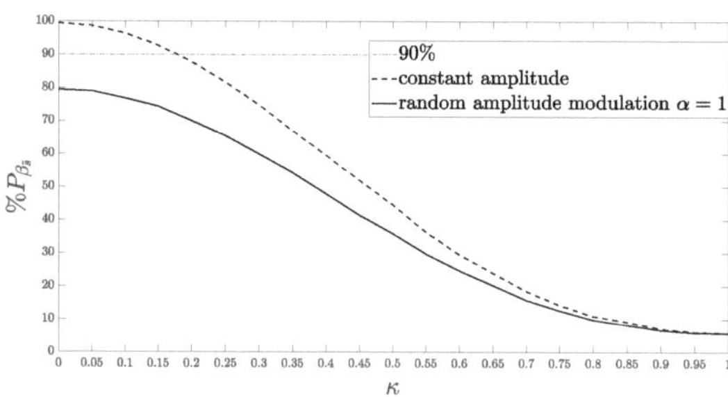

Figure 7. Relationship between % of total power in bandwidth and κ, assuming constant and random amplitude ($\alpha = 1$) modulation.

As a rule of thumb, radar system design good practice recommends that the percentage of power within the designated bandwidth should be close to 90% [5]. Nevertheless, this criterion is met only for low values of κ regardless of whether or not one considers random amplitude. On the other hand, the value of κ directly influences the APCN signal's degree of randomness and, consequently, its LPI/LPE performance [20] as well. Specifically, the greater the value of κ, the more random the resultant APCN signal should be. Therefore, one must evaluate the LPI/LPE performance as a function of the stochastic component parameters of APCN waveforms to establish proper trade-offs during design.

The noise radar system LPI/LPE characteristics are twofold. Firstly, these systems can effectively mimic thermal noise, rendering them indistinguishable from less sophisticated ELINT systems and ensuring covert operation. Secondly, they can produce varied sample functions from the same stochastic process. This capability diminishes the effectiveness of deception systems that do not operate in real time. Many works in the literature, e.g., [4–6,9,13], attempt to define proper metrics to evaluate a system's LPI/LPE characteristic properly. Despite the different approaches, a common sense within the community is that these features relate to the transmitted waveform degree of randomness.

In [4,5,9,13], for instance, the authors propose the analysis of the transmitted signal's Spectral Flatness Measure (SFM) and Mutual Information Rate (MIR) as measures of its information increase with time. For Gaussian processes [37], the SFM is directly related to the MIR and is defined as the ratio of the geometric mean to the arithmetic mean of the signal's PSD, given by [37]

$$
\text{SFM} = \frac{\exp\left[\frac{1}{2\pi} \int\limits_{-\pi}^{\pi} \ln\left(\Gamma_{\tilde{s}}(\omega)\right) d\omega\right]}{\frac{1}{2\pi} \int\limits_{-\pi}^{\pi} \Gamma_{\tilde{s}}(\omega) \, d\omega}.
\tag{21}
$$

From Figure 8, it is possible to observe that the SFM (calculated as the average of 100 independent trials) increases as κ increases, reaching an upper bound value when the phase factor is $\kappa = 0.8$. Note also that the SFM average exhibits a similar behavior, assuming either random or constant amplitude. This observation suggests that the random phase contributes more significantly to the LPI/LPE characteristics.

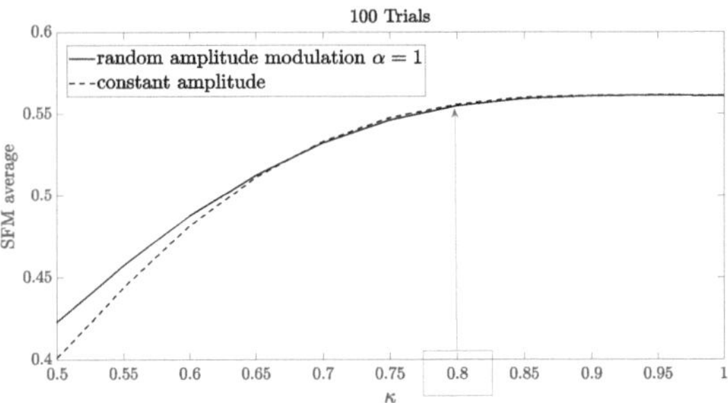

Figure 8. SFM average of APCN ($\kappa \geq 0.5$) waveforms: assuming Phase-Only randomness (constant amplitude) and random amplitude modulation ($\alpha = 1$).

The degree of randomness in a given stochastic waveform introduces an additional advantage to pulsed noise radars. Random signals are expected to present a low cross-correlation between pulses transmitted at different times. This feature enables the elimination of range ambiguities in pulse compression architectures.

A short-integration-time pulsed noise radar emits a train of noise-modulated electromagnetic pulses toward the target [2], specifically N_p time-limited signals, separated in time by the Pulse Repetition Interval (P_{RI}). When we assume the presence of a single non-fluctuating scattering point moving at a range of R_0, at the time it starts being illuminated by the radar, we can express the complex envelope of the i-th received signal as [38]

$$
\tilde{r}_i(t) = G e^{-j4\pi\left(\frac{R_0 + vT_i}{\lambda}\right)} e^{-j2\pi f_D t} \tilde{s}_i\left(t - T_0 - \frac{2v}{c} T_i\right),
\tag{22}
$$

where G is a term that reflects the backscattering effects, channel fading, and the gains and distortions introduced by the receiver RF chain (assumed to be constant over the coherent processing interval), λ is the operating wavelength, and v is the target's radial velocity, also assumed constant. Finally, T_0 in Equation (22) represents the time spent by the echo signal to return to the radar, which is given by $T_0 = 2R_0/c$ and $T_i = (i-1)P_{RI}$.

Range ambiguity arises when the scatter is far enough so that the i-th echo, relative to the reflection of the i-th transmitted pulse, arrives at the receptor after the transmission of the subsequent signal, $s_{i+1}(t)$, and, considering the pulse compression gain, has enough power to be detected. Let us again consider the simplified received signal model [21] for simplification and no loss of generality $\tilde{r}_i(t) = \tilde{s}(t - T_0)e^{-j2\pi f_D t}$, then, the expected value of the pulse compression output when range ambiguity is present is given by

$$E[\tilde{y}(\tau)_{RA}] = E\left[\int_{-\infty}^{\infty} \left[\tilde{s}_{r_i}(t - T_0)\tilde{s}_{c_i}(t - T_0)\,e^{-j2\pi f_D t}\right] \tilde{s}_{r_{i+1}}^*(t - \tau)\tilde{s}_{c_{i+1}}^*(t - \tau)dt\right]. \tag{23}$$

The stochastic nature of pure noise radar waveforms may contribute to the suppression of range ambiguities in target detection [39]. The cross-correlation process between the i-th replica and the j-th received pulse, where $i \neq j$, plays a crucial role in radar systems employing random signals. The elimination of range ambiguity is directly contingent upon this process. It is no different for APCN signals. One can also evaluate range ambiguity suppression using the cross-correlation between the i-th and the j-th, transmitted pulses, $i \neq j$, since it is possible to rewrite Equation (23) as

$$E[\tilde{y}_{RA}(\tau)] = E[a_i(t)a_{i+1}(t - \tau)]E\left[e^{j\kappa\phi_i(t)}e^{j\kappa\phi_{i+1}(t-\tau)}\right]\chi_{\tilde{s}_c}(\tau - T_0, f_D), \tag{24}$$

with $(T_0 - \tau_s \leq \tau \leq T_0 + \tau_s)$ and considering that $\tilde{s}_{c_i}(t) = \tilde{s}_{c_{i+1}}(t)$.

Finally, after some mathematical manipulations, it can be shown that

$$E[\tilde{y}_{RA}(\tau)] = \left(\alpha^2\frac{\pi}{2}\right)\left(\frac{1 - \cos 2\kappa\pi}{2\kappa^2\pi^2}\right)\chi_{\tilde{s}_c}(\tau - T_0, f_D); \quad T_0 - \tau_s \leq \tau \leq T_0 + \tau_s. \tag{25}$$

The attenuation in the pulse compression output of APCN signals introduced by the presence of the random component $\tilde{s}_r(t)$, when range ambiguous targets are present, is given by $H = \alpha^2[1 - \cos(2\kappa\pi)]/4\kappa^2\pi$, the same introduced as an effect of the Doppler mismatch of moving targets (see Equation (5)). The same analysis can be performed for the Phase-Only APCN waveform, leading to

$$E\left[\tilde{y}_{RA}^{PO}(\tau)\right] = \left(\frac{1 - \cos 2\kappa\pi}{2\kappa^2\pi^2}\right)\chi_{\tilde{s}_c}(\tau - T_0, f_D); \quad T_0 - \tau_s \leq \tau \leq T_0 + \tau_s, \tag{26}$$

with an attenuation given by $D = [1 - \cos(2\kappa\pi)]/2\kappa^2\pi^2$.

The primary objective of the radar system in surveillance applications is to provide the user with situational awareness of the operational scenario. That entails detecting uncooperative targets and determining their range and radial velocity in real-time. Contemporary digital systems utilize the Pulse Doppler technique, separable two-dimensional processing in the fast and slow-time dimensions, for that purpose [34].

In the fast-time dimension, the i-th single-pulse matched filter, denoted as $\tilde{y}_i(n) = \tilde{y}_i(t)|_{t=n/f_s}$ (see Figure 3) is performed. Next, the k-points Discrete Fourier Transform (DFT) of the slow-time data sequence is employed in each range bin [38]. The result is the well-known range-Doppler matrix, used as the input for the subsequent step of the detection process.

In the present analysis, we chose the Cell-Averaging (CA) CFAR detection technique to investigate the performance of APCN signals. We assume a single Signal of Interest (SOI) scenario and consider the predominant interference in the radar reception chain from thermal noise origin. Furthermore, employing a quadratic law detector, we set the probability of false alarm to 10^{-5}. This detection process is first applied to the echo of the APCN waveform using the receiver system configuration detailed in [21]. Therefore, one must derive the matched filter output by correlating the receiver echo with the transmitted stochastic signal. Subsequently, the detection process assesses the corresponding filter output, as determined by Figure 3. Given the time and frequency sampling, we incorporate two "guard cells" on each side of the "cell under test". We take this precaution to account for the potential occurrence of "straddle range-Doppler cells" [34].

Next, we discuss an experiment assuming a digital radar receiver. Below is an outline of the main parameters employed in the simulation for this experiment:

- Waveform wavelength: 0.0321 m;
- Waveform bandwidth: $\beta_s = 30$ MHz;
- Waveform pulse width: $\tau_s = 50\ \mu s$;
- Waveform pulse repetition interval: $\tau_s = 500\ \mu s$;
- Fast-time sampling frequency : $f_s = 125$ MHz;
- Pulse Repetition Frequency: $PRF = 2$ kHz;
- Slow-time sampling frequency: 2 kHz;
- Number of received pulses: $N_p = 27$;
- Number of DFT Doppler domain points: $N = 64$;
- Target's Doppler shift: $f_D = 700$ Hz;
- Target's range: $R_0 = 50$ km;
- Maximum radar unambiguous range: $Runam_{max} = 75$ km.

Figure 9 displays the results of the detection process with the receiver system configuration detailed in [21]. In addition to successfully detecting the SOI highlighted in green, the five false targets depicted in red were also identified in a scenario with an SNR as low as -15 dB.

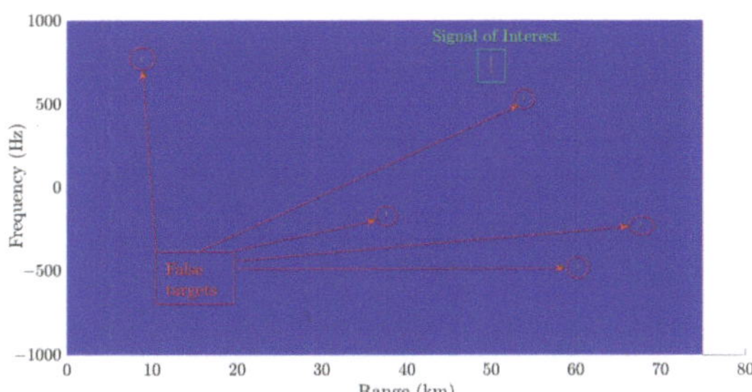

Figure 9. Detection process output considering the CA-CFAR technique implemented on the digital radar receiver proposed in [21].

Figure 10 shows the detected SOI when the configuration outlined in Figure 3 is employed. Note that this detection procedure eliminates false targets.

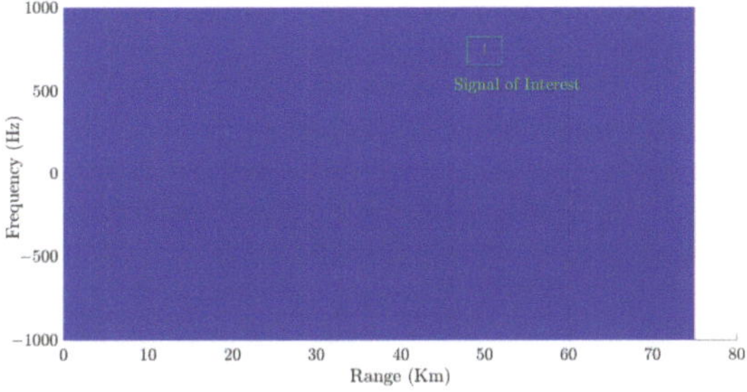

Figure 10. Detection process output considering the CA-CFAR technique applied to our proposed architecture.

To demonstrate and validate the improvement in performance in detecting and estimating range and radial velocity, we conducted an experiment that consists of transmitting different APCN signals with randomness in amplitude (scale parameter $\alpha = 1$) and varied randomness in phase (scale parameter κ). The results illustrated in Figure 8 corroborate that phase randomness impacts the increase in APCN LPI/LPE characteristics most significantly. After applying the Pulse-Doppler processing, we generated a range-Doppler matrix, assuming a low SNR of -15 dB on the receiver's end of the radar chain. Next, we applied the CA-CFAR detection technique to this dataset and calculated the average number of false targets. We conducted this analysis using Monte Carlo simulations comprising 100 independent trials. The procedure adheres to the recommended receiver configuration of [21] and the receiver configuration specifications provided in Figure 3.

Figure 11 illustrates the results of an experiment in which we increased the waveform randomness while maintaining the SNR level low in the radar receiver chain. Notably, using the configuration detailed in [21] resulted in a substantial increase in the average count of false targets. Therefore, this configuration significantly compromises the radar's detection and estimation capabilities despite enhancing its LPI/LPE characteristics. Specifically, when the variable κ reaches a value of 0.8, the average count of false alarms escalates to eight. This elevated count remains consistent regardless of increments in the value of κ. This behavior relates to the maximum value of the SFM average observed in Figure 8, which reaches an upper bound when the phase factor is $\kappa = 0.8$.

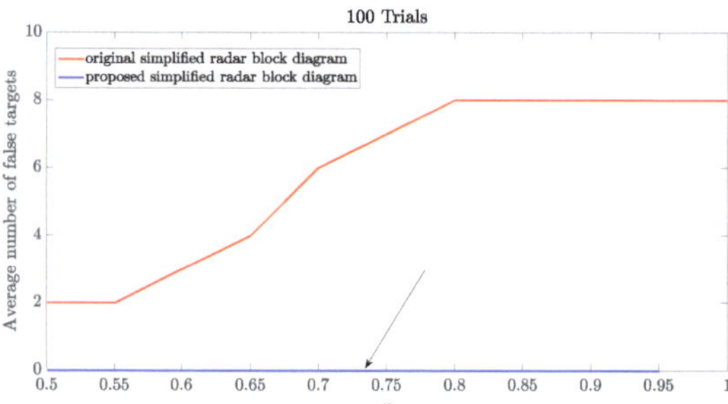

Figure 11. APCN waveform ($\kappa \geq 0.5$ and $\alpha = 1$) detection performance: average number of false targets.

However, when we applied our proposed architecture detailed in Figure 3, it maintained the crucial LPI/LPE capability by preserving the transmission waveform. Moreover, as we explored phase scale factors up to $\kappa = 0.95$, we eliminated false target occurrences, reducing them notably to zero, as visually indicated by the black arrow in Figure 11. However, the waveform lost its Doppler tolerance capacity at the uppermost limit of randomness, namely $\kappa = 1$. Consequently, correlating the received echo from the APCN waveform with its deterministic component became unfeasible.

It is clear from the previous derivations and analysis that proper selection of the parameters governing the random components of the APCN transmitted signal is relevant for the overall system performance. The greater the value of κ, the more random the waveform. That leads to a higher Spectral Flatness Measure, improving its range ambiguity suppression. On the other hand, high values of κ lead to a waveform with less spectral efficiency (low power within the desired bandwidth) and less Doppler tolerance. Using a stochastic signal to modulate the transmit waveform in amplitude also increases its degree of randomness, which enhances the trade-offs discussed above. Additionally, the random component in amplitude also deteriorates the system performance concerning power efficiency, increasing the transmit signal's PAPR. That needs to be considered, especially in long-range applications.

In Section 3, we examine APCN waveforms from the point of view of a passive intercept-receiver system and propose, analyze, and discuss a methodology to detect and extract the characteristics of this noisy waveform automatically. These waveforms are recognized for their LPI/LPE attributes as detailed in [3,20].

3. The Proposed Metodology for Identifying APCN Signals

In this section, we investigate the performance of the APCN waveform in an electronic warfare context. We first address the modeling of a digital superheterodyne ESM receiver system. We consider interception, A/D conversion, and digital processing operations performed on the analog SOI. We introduce a specific time-frequency transform technique to analyze radar signals deemed as LPI/LPE [11,23]. Then, we outline techniques used to extract signal characteristics and evaluate the performance of the proposed methodology in estimating the radar parameters of the SOI in an EW scenario.

The primary objective of an ESM system is to identify emitting sources. The ESM system's digital processing chain extracts intrapulse and interpulse parameters from the intercepted waveform. Examples of intrapulse parameters are pulse width (τ_s), operating frequency (f_0), and bandwidth (β_s), whereas for interpulse we have the Pulse Repetition Interval (P_{RI}). The way to determine the Pulse Repetition Interval is by estimating the arrival time (T_e) between successive intercepted pulses [40]. This information is intrinsic to the signal's identity. As for the amplitude of the received signal, it relies, in part, on the distance between the radar and the ESM system since measuring the amplitude is challenging due to deleterious effects, such as multipath that may lead to constructive interference within pulses [40].

Commonly, an ESM system has a listening time (Δt) longer than the radar's pulse repetition interval, enabling parameter estimation based on multiple intercepted pulses. Thus, the intercepted RF signal can be expressed as

$$r_e^{RF}(t) = \sum_{i=1}^{Np_e} Q\, s_i^{RF}(t - T_e - (i-1)P_{RI}), \tag{27}$$

where $s_i^{RF}(t)$ is the i-th transmitted pulse one-way Doppler shifted by the radial relative velocity. Moreover, Q accounts for gains and attenuation, and N_{p_e} is the number of intercepted pulses.

Typically, six pulses are needed to allow parameter extraction [40]. The received signal is routed to the RF chain and contaminated with thermal noise $w(t)$. Therefore, the actual signal at the output of the RF module is given by

$$x(t) = r_e(t) + w(t) = \sum_{i=1}^{N_{p_e}} V\, s_i(t - T_e - (i-1)P_{RI}) + w(t), \tag{28}$$

where $s_i(t)$, in noise radars, are sample functions of the stochastic process that characterizes the transmitted random signal assumed to be statistically independent of the thermal noise. Moreover, V accounts for gains and attenuation introduced by the RF chain cascaded to the gains and attenuation of the intercepted signal. We assume real-valued and time-invariant quantities throughout this work.

Since we employ discrete-time analysis, we denote $x(n) = x(t)|_{t=n/f_{s_e}}$, assuming a digital ESM system sampling frequency of f_{s_e}. Additionally, the TFA of the intercepted radar signal adopts its analytical form [11] and is given by

$$\tilde{x}(n) = x(n) + j\mathcal{H}\{x(n)\}, \tag{29}$$

where $\mathcal{H}$ denotes the Hilbert transform. Figure 12 illustrates the simplified diagram of an ESM superheterodyne digital receiver [11], outlining the process that starts with the reception of the analog signal up to the display of the extracted information in a Human–Machine Interface (HMI). The diagram excludes the pulse deinterleaving step, focusing the analysis on individual waveforms.

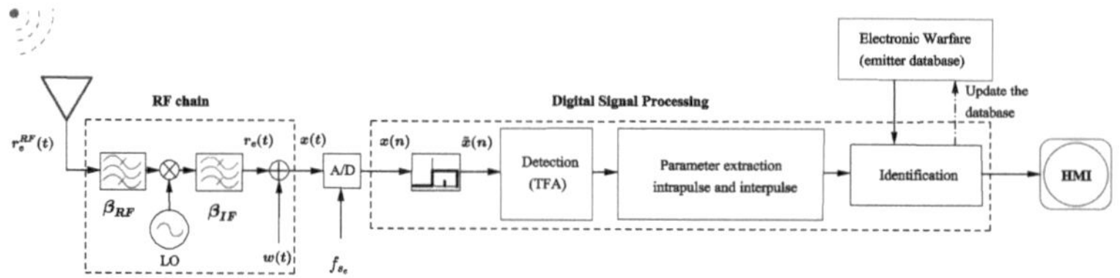

Figure 12. ESM superheterodyne receiver: simplified block diagram incorporating digital technology.

In numerous EW scenarios, the nature of a received radar signal often deals with time and frequency information and, therefore, requires time-frequency analysis techniques to effectively characterize the non-stationary behavior exhibited by the signal [23,41]. Several variations of these techniques are available in the literature. We use the Short-Time Fourier Transform (STFT) in this work, with no loss of generality.

The discrete version of the STFT of signal $\tilde{x}(n)$ is defined as

$$S_F(k, m) = \sum_{n=0}^{N_e-1} \tilde{x}(n + m\,R)\, g(n)\, e^{-j\left(\frac{2\pi k}{N_e}\right)n}; \ 0 \leq k \leq N_e - 1, \tag{30}$$

where $2\pi k / N_e$ is the k-th discrete frequency bin, m represents the m-th discrete tile in time, N_e is the number of points of the FFT, equal to the window size, and R is the hop size (with an overlapping of $N_e - R$ samples in this case). Moreover, $g(n)$ is a window of size N_e, that is, $g(n) = 0$ outside the interval $0 \leq n \leq N_e - 1$. In this work, we chose the Hamming window, widely used in EW systems [11,34,41]. In most applications that involve STFT, the interest is in the magnitude response, with a focus on the short-time quadratic magnitude $|S_F(k, m)|^2$, representing the short-time energy spectral density, and usually displayed as a function of time and frequency in the form of a spectrogram [42].

From the computational implementation of Equation (30), we obtain the $N_e \times M$ matrix $\mathbf{E}$ of time-frequency distribution, given by

$$\mathbf{E} = \begin{bmatrix} \vdots & \vdots & \vdots \\ |S_F(k,1)|^2 & \cdots & |S_F(k,M)|^2 \\ \vdots & \vdots & \vdots \end{bmatrix}_{N_e \times M}, \tag{31}$$

where k corresponds to k-th frequency bin, N_e is the total number of bins, and M is the total number of tiles in time. The determination of overlap, crucial for achieving resolution between fixed tile and frequency bin quantities, is computed according to [43]

$$L = \left\lceil \frac{MN_e - N_s}{M - 1} \right\rceil, \tag{32}$$

where "$\lceil \ \ \rceil$" is the round-to-nearest integer operator and N_s is the number of samples.

Displayed in Figure 13 are the time "(a)" and time-frequency "(b)" representations of an intercepted signal, derived from six pulses of an APCN waveform ($\kappa = 0.5$ and $\alpha = 1$). It is important to note that while representation "(a)" highlights time-related features, the time-frequency display in "(b)" provides valuable insights into the energy carried by the SOI. These depictions assume an SNR of -10 dB at the passive intercept-receiver input.

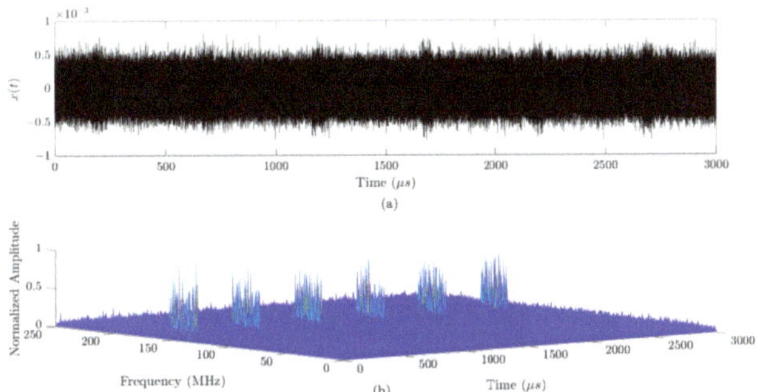

Figure 13. Intercepted signal $x(t)$: (**a**) time representation; (**b**) time-frequency representation. In this figure, the x-axis is represented by "Time (µs)" and the y-axis is represented by "Frequency (MHz)".

The subsequent stage after the TFA in the block diagram depicted in Figure 12 involves parameter extraction. Matrix **E** described by Equation (31) furnishes details regarding the energy the intercepted signal carries, enabling us to visualize the time-frequency plane as a 2D image. Consequently, it is feasible and natural to use image processing techniques to extract parameters related to the SOI. In the present work, we use the Hough transform for detecting geometric shapes such as lines in a binary image [44]. The way to represent a line equation in the Hough space is as follows [45]:

$$\rho = x\cos(\psi) + y\sin(\psi), \tag{33}$$

where ρ is the distance between the line and the origin of the Cartesian system and ψ is the angle between the axis x and the segment perpendicular to the line.

Figure 14 illustrates the process of estimating a line in Cartesian space (a) from the Hough space (b), with the parameters ψ' and ρ' determined by the intersection of the two sinusoids within the Hough space.

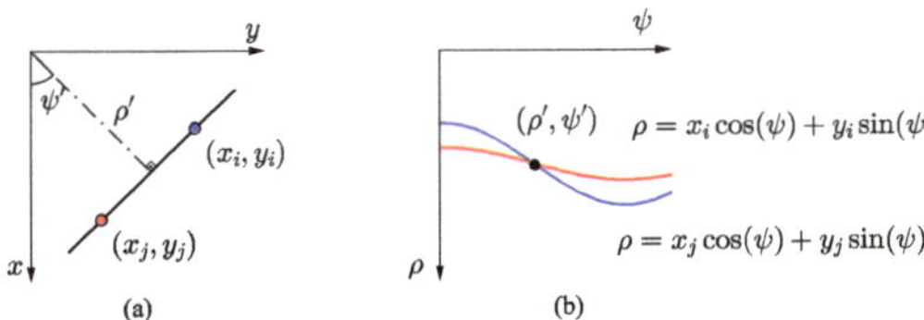

Figure 14. Representation of a line: (**a**) Cartesian space; (**b**) Hough space.

The computational implementation of the Hough transform [46] yields a structured array represented as $[\mathbf{H}; \boldsymbol{\psi}; \boldsymbol{\rho}]$, where $\mathbf{H}$ denotes the histogram amplitude matrix, with each of its elements standing for the number of increments within each cell of the quantized Hough space. Additionally, $\boldsymbol{\psi}$ represents the slopes vector, while $\boldsymbol{\rho}$ is the distance vector.

The peaks in $\mathbf{H}$ are then obtained and stored in a matrix $\mathbf{P}$ of order $n_p \times 2$ whose form is given by

$$\mathbf{P} = [\boldsymbol{\rho}^* \quad \boldsymbol{\psi}^*] = \begin{bmatrix} \rho_1 & \psi_1 \\ \rho_2 & \psi_2 \\ \vdots & \vdots \\ \rho_{n_p} & \psi_{n_p} \end{bmatrix}, \tag{34}$$

where n_p represents the desired number of peak estimates, assumed to be the minimum number of pulses N_{p_e} required to enable the parameter extraction output. Finally, the detection of lines in the image space (matrix **BW**) is performed, leading to **L**, a *restructured array* of the form $[\mathbf{L}_1; \mathbf{L}_2]$ where

$$\mathbf{L}_1 = \begin{bmatrix} x_1|_{P_1} & y_1|_{P_1} \\ \vdots & \vdots \\ x_1|_{P_i} & y_1|_{P_i} \\ \vdots & \vdots \\ x_1|_{P_{N_{pe}}} & y_1|_{P_{N_{pe}}} \end{bmatrix}, \tag{35}$$

with $x_1|_{P_i}$ and $y_1|_{P_i}$ being the i-th ordered pair matrix referring to the beginning of the i-th detect line, and

$$\mathbf{L}_2 = \begin{bmatrix} x_2|_{P_1} & y_2|_{P_1} \\ \vdots & \vdots \\ x_2|_{P_i} & y_2|_{P_i} \\ \vdots & \vdots \\ x_2|_{P_{N_{pe}}} & y_2|_{P_{N_{pe}}} \end{bmatrix}, \tag{36}$$

is the matrix representing the ordered pairs at the end of the i-th detected line.

The intrapulse and interpulse parameters of the SOI are then estimated, with the estimated bandwidth given by

$$\hat{\beta}_s = \frac{1}{N_{pe}} \left[\sum_{i=1}^{N_{pe}} (y_2|_{P_i} - y_1|_{P_i}) \frac{f_{s_e}}{2N_e} \right]. \tag{37}$$

The SOI operating frequency (f_0) can be estimated as

$$\hat{f}_c = \frac{1}{N_{pe}} \left[\sum_{i=1}^{N_{pe}} (y_1|_{P_i} + 0.5 \left(y_2|_{P_i} - y_1|_{P_i} \right)) \frac{f_{s_e}}{2N_e} \right]. \tag{38}$$

Therefore, $\hat{f}_0 = \hat{f}_c + f_{LO}$, where f_{LO} is the ESM system Local-Oscillator (LO) frequency. The estimated intercepted signal time duration is calculated as

$$\hat{\tau}_s = \frac{1}{N_{pe}} \left[\sum_{i=1}^{N_{pe}} (x_2|_{P_i} - x_1|_{P_i}) \frac{\Delta t}{M} \right]. \tag{39}$$

Finally, to estimate the radar P_{RI}, it is necessary to measure the difference between the arrival times of N_{pe} successive pulses in such a way that

$$\hat{P}_{RI} = \frac{P_{RI}|_{P_{(21)}} + P_{RI}|_{P_{(32)}} + \cdots + P_{RI}|_{P_{(i(i-1))}} + \cdots + P_{RI}|_{P_{(N_{pe}(N_{pe}-1))}}}{N_{pe} - 1}, \tag{40}$$

where $P_{RI}|_{P_{(i(i-1))}} = \left(x_1|_{P_i} - x_1|_{P_{(i-1))}} \right) \frac{\Delta t}{M}$ is the i-th measure pulse repetition interval between two successive lines detected.

The block diagram in Figure 15 illustrates the methodology proposed in this study for parameters extraction from the APCN noise radar waveform. In this diagram, solid black lines represent the input and output of the block diagram, while dashed black lines indicate intermediate inputs and outputs necessary for preprocessing.

In the final stage of the architecture outlined in Figure 12, emphasis is placed on identifying the source emitter. In the digital processing realm of an ESM system, a predefined set of mean parameters $[\hat{\beta}_s \ \hat{f}_0 \ \hat{\tau}_s \ \hat{P}_{RI}]^T$, referred to as a *fingerprint*, can aid in the identification phase of the radar model [1,11,40]. These parameters offer a degree of tolerance and facilitate the identification of the emitting source. In cases where the intercepted signal fails to correlate with an existing emitter in the EW database, we add the new signal to the database for future recognition.

Subsequent sections delve into the proposed methodology, utilizing a numerical example to detail the process of extracting signal information to construct a waveform *fingerprint*.

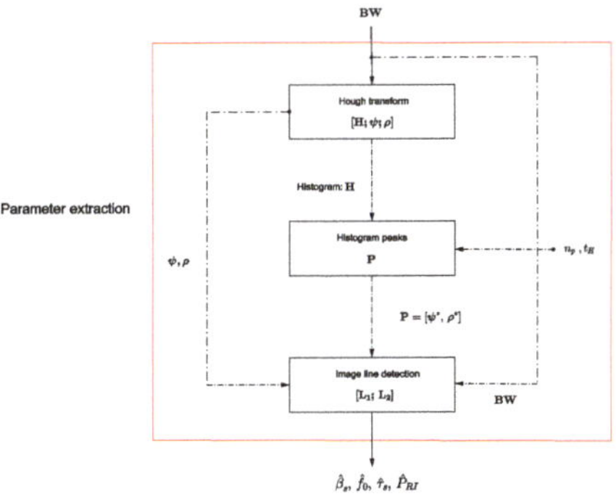

Figure 15. Block diagram showing the approach for extracting parameters from the APCN noise radar waveform.

3.1. Numerical Example

The block diagram of Figure 16 outlines the proposed methodology. It is designed to aid in identifying APCN signals through image processing, thereby enabling the identification of emitting sources transmitting this type of waveform in an EW scenario. The synthesized scenario is based on Figure 12 and the main parameters used in the simulations are:

- Waveform bandwidth: $\beta_s = 30$ MHz;
- Waveform pulse width: $\tau_s = 50$ μs;
- Waveform pulse repetition interval: $P_{RI} = 500$ μs;
- ESM system sampling Frequency: $f_{s_e} = 500$ MHz;
- ESM system listening time: $\Delta t = 3000$ μs;
- ESM system local oscillator frequency: $f_{LO} = 9.2$ GHz;
- Number of pulses intercepted by the ESM system: $N_{p_e} = 6$;
- Radar center frequency: $f_c = 160$ MHz.

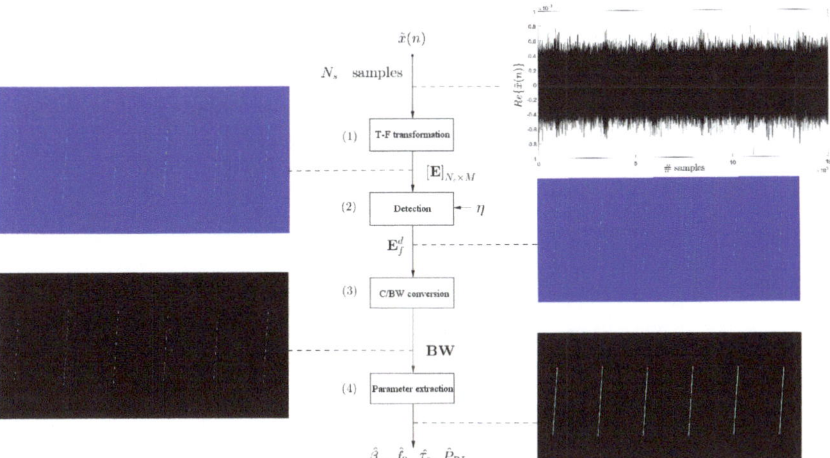

Figure 16. Block diagram illustrating a proposed methodology utilizing image processing techniques. Parameters to be extracted include bandwidth, operating frequency, pulse width, and pulse repetition interval.

It is noteworthy to emphasize that, in [20], the utilization of an APCN waveform with specific parameters ($\kappa = 0.5$ and $\alpha = 1$) and a designed bandwidth of 30 MHz was identified as potentially posing challenges to the intercept–receiver system in a communication channel where thermal noise was present. Acknowledging this insight from the referenced work, our investigation aligns with this perspective, thus adopting a bandwidth of $\beta_s = 30$ MHz for our analysis.

The proposed methodology to analyze this noise radar signal comprises four steps (as seen in Figure 16) and are described in the following.

1. In Step 1, time-frequency (T-F) transformation, we obtain matrix **E** using Equation (31). To address the uncertainty principle [42], we use a window $g(n)$ of size $N_e = 1024$ samples, corresponding to 2.048 µs. This choice aims to balance local signal analysis and stationary conditions, facilitating FFT applications while ensuring an adequate balance of time and frequency resolutions. The analytical nature of signal $\tilde{x}(n)$ allows for 512 frequency bins within the range $0 \leq f \leq f_{s_e}/2$, corresponding to the positive half of the spectrum $0 \leq \omega \leq \pi$, given the window size. Simultaneously, we set the number of STFT tiles to $M = 1024$, with an overlap of 536 samples at each hop. As illustrated in Figure 16, this transformation showcases the signal's shift from the time domain to its time-frequency representation. Although the radar signal's intent is discernible amidst system thermal noise, preprocessing remains necessary for automatic and accurate characteristic extraction.

2. Step 2 performs detection in the T-F plane. For this purpose, we define a threshold η as

$$\eta = -\left[\bar{e}\,\ln(P_{fa})\right], \tag{41}$$

with

$$\mathbf{e} = \underbrace{\left[\ \overline{\mathbf{E}_f(:,1)}\quad \overline{\mathbf{E}_f(:,2)}\quad \cdots \quad \overline{\mathbf{E}_f(:,M)}\ \right]^{\mathrm{T}}}_{\text{Mean vector in matrix rows}} \quad \text{and} \quad \bar{e} = \frac{1}{M}\sum_{m=1}^{M}\overline{\mathbf{E}_f(:,m)}, \tag{42}$$

where the desired probability of false alarm is $P_{fa} = 10^{-5}$. After the detection process, matrix $\mathbf{E}_f^d$ is obtained as the output, as illustrated in Figure 16.

3. Conversion from grayscale (C) to black-white (BW): $\mathbf{E}_f^d$ is converted from grayscale to black-and-white [47] to obtain the matrix **BW**, and the signal's amplitude information is encoded into binary values.

4. Parameter extraction step: as previously mentioned, the proposed approach for extracting information from the APCN waveform relies on the Hough transform. Due to its deterministic component, and according to [3,22], the bandwidth β_s can be considered the same as that of its linear chirp component, i.e., β_{s_c}. The number of input peaks, assumed to be the minimum number of pulses to allow for parameter extraction, was considered $n_p = 6$ [40]. Moreover, we fixed the threshold t_H at $0.5\max[\mathbf{H}]$, which is the default minimum value for identifying a peak.

By defining a threshold t_H and applying it to the Hough space matrix **H**, along with knowing the desired minimum number of peaks n_p, we can obtain the matrix **P** as denoted by Equation (34). Figure 17 illustrates the detected peaks stored in **P**. Subsequently, using $\mathbf{P} = [\rho^* \quad \psi^*]$, the conversion from Hough space to Cartesian space [45] was performed based on the parameter relationship in Equation (33), resulting in the detection of lines $[\mathbf{L}_1; \mathbf{L}_2]$. Figure 18b illustrates some of the lines detected in Cartesian space.

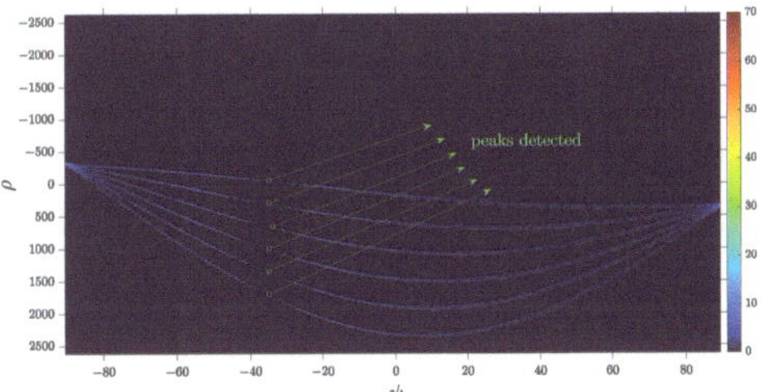

Figure 17. Peaks detected from Hough's histogram $[\mathbf{H}]$.

Finally, the intrapulse and interpulse parameters are estimated as depicted in the detailed extraction block diagram (Figure 15). This process establishes a connection between the desired information illustrated in Figure 18a and the information obtained in the Cartesian space through the Hough space (Figure 18b).

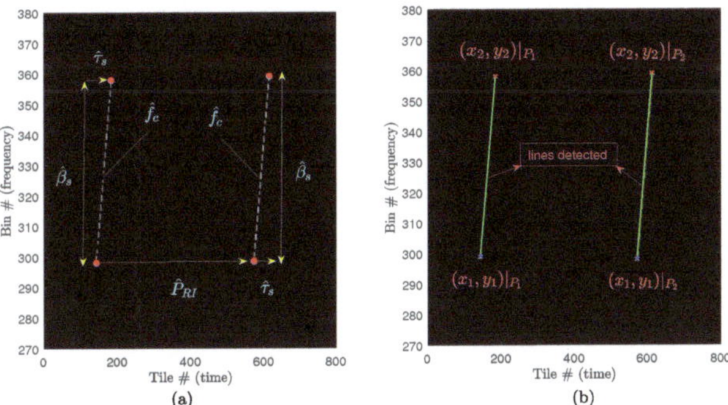

Figure 18. Detected lines. For better observation, the image has been zoomed in on the first two intercepted pulses (**a**) desired parameters estimation; (**b**) Cartesian lines detected. In both figures "#" stands for the number of the Bin or Tile, respectively.

3.2. Performance Assessment

Several studies in the literature aim to establish performance benchmarks for ESM systems [29,48,49], yet a universally accepted standard for ESM development remains elusive [30]. As previously mentioned, tolerances may correlate a particular set of estimated parameters to a specific emitter, and their significance in the overall ESM performance is crucial [40]. For instance, in [29], the assumed tolerances are ± 1 MHz for operating frequency and bandwidth and ± 1 µs for the modulation period. From this perspective, one presumes that the probability of an ESM detecting a radar signal (P_{d_e}) is directly linked to the accuracy/precision of its parameter estimation by such a system, making it an evaluation metric. As an alternative, the authors in [23,29,31] consider the percentage relative error to evaluate the efficiency of their proposed methodologies for radar parameter extraction of deterministic radar signals considered LPI/LPE.

In this assessment, we start with P_{d_e} of the ESM system employing the proposed methodology to identify APCN signals through a Monte Carlo simulation, assuming 100 independent trials. For this purpose, we record a detection when the tolerance between the actual and estimated parameters falls below a certain threshold: ± 2 MHz for β_s, ± 5 MHz for f_0, ± 5 µs for τ_s and ± 25 µs for P_{RI}. These tolerances were derived from the information in [29,40]. Figure 19 presents the obtained ESM probability of detection, P_{d_e}, of an APCN signal. One can see that the detection performance remains above 99% for SNR levels considered low [23,29–31], i.e., less than -10 dB, for both intrapulse as well as interpulse radar parameters. Performance is degraded for SNR less than -11 dB.

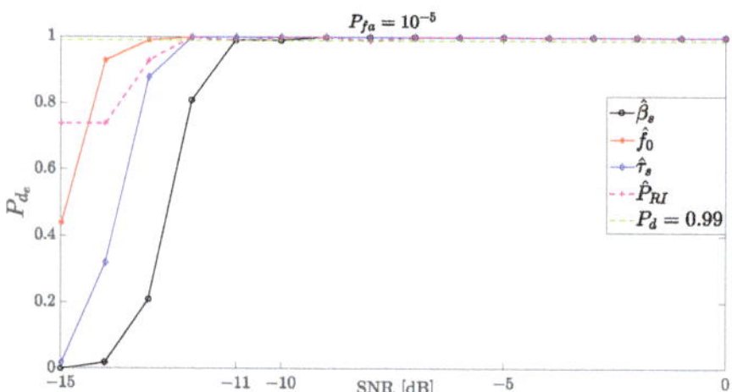

Figure 19. Probability of detection APCN waveform parameters ($\kappa = 0.5$ and $\alpha = 1$) considering 100 independents trials.

We also evaluated the accuracy and precision of the random variables characterizing the proposed estimators. Accuracy, represented by the bias of an estimator $\hat{\Theta}$ for a parameter Θ [11,34], is defined as the expected value of the difference between the mean of the estimate and the actual parameter value

$$B_{\hat{\Theta}}(\Theta) = \mathrm{E}[\hat{\Theta}] - \Theta. \tag{43}$$

Precision, on the other hand, is the standard deviation of the estimate

$$\sigma_{\hat{\Theta}}(\Theta) = \mathrm{E}\left[\sqrt{(\hat{\Theta} - \mathrm{E}[\hat{\Theta}])^2}\right]. \tag{44}$$

Figure 20a and Figure 20b, respectively, depict the precision and accuracy of the proposed estimators across varying SNR levels in an ESM system employing the methodology to identify APCN signals. The estimators exhibit high precision, implied by the low standard deviation of the random variables, up to an SNR of -12 dB. However, beyond this threshold, a notable decline in precision is observed. Additionally, while the proposed f_0 estimator maintains high accuracy up to an SNR of -12 dB before exhibiting bias, the proposed bandwidth estimator displays a slight bias (approximately 1 MHz) independent of the SNR in the passive intercept–receiver chain. Nonetheless, the estimators for intrapulse and interpulse temporal parameters demonstrate high accuracy up to an SNR of -12 dB before showing signs of bias.

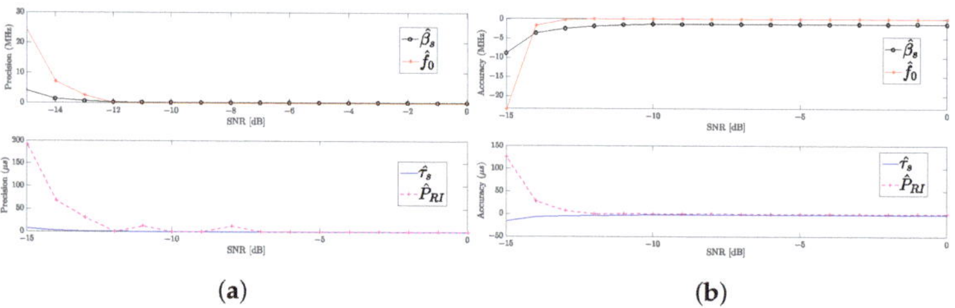

(a)   (b)

Figure 20. APCN ($\kappa = 0.5$ and $\alpha = 1$) waveform parameters estimation performance: (**a**) Precision as a function of the SNR; (**b**) Accuracy as a function of the SNR.

Thus, from the analyzed perspectives, a digital intercept receiver that employs the automatic parameter extraction approach proposed in the present work can detect the APCN noise radar signal, with $\kappa = 0.5$ and $\alpha = 1$, and explore it, inhibiting this waveform from being claimed as either LPI or LPE.

Lastly, Figure 21 presents the assessment metric based on percentage error, defined as

$$e(\%) = \left| \frac{\text{actual value} - \text{estimated value}}{\text{actual value}} \right| \times 100, \tag{45}$$

wherein the mean percentage relative error $e(\%)_{ensemble}$ is derived for observation in the experiment ensemble. As per [1,31,50], a margin of up to 10% in parameter estimation error can be deemed acceptable in the context of ESM equipment.

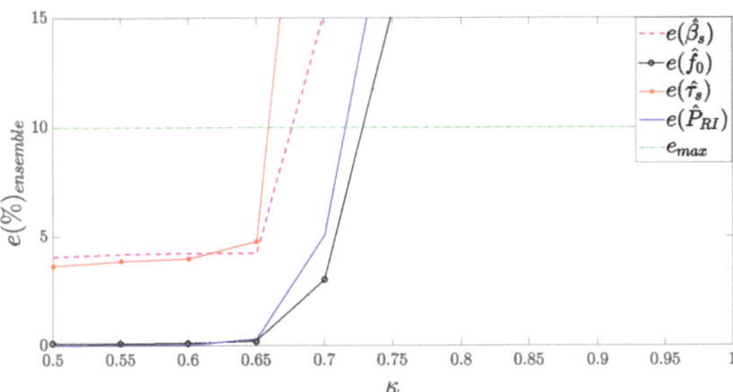

Figure 21. APCN parameter intrapulse and interpulse estimation with different values of κ, considering $\alpha = 1$.

We have used that metric to evaluate the performance of the proposed methodology for APCN signals generated with different values of κ, assuming $\alpha = 1$. Moreover, we have considered a fixed SNR of -3 dB in the receiver chain. As we can see, the proposed methodology managed to estimate all parameters within an acceptable error limit for APCN waveforms that employ $\kappa < 0.7$ when the signal becomes too noisy, at the expense of spectral confinement and Doppler tolerance, as previously mentioned.

4. Conclusions

This paper investigated the performance of the Doppler-tolerant Advanced Pulse Compression Noise waveform radar in surveillance applications. We analyzed its performance as an LPI/LPE signal under the framework of a proposed detection/information extraction method. From the perspective of a radar system, we showed an expression of the narrowband ambiguity function to assess its Doppler tolerance capacity. The analysis revealed an anomaly inherent to the waveform that can jeopardize detecting slow-moving targets in surveillance applications. Thus, we proposed a novel configuration for a digital radar receiver to address this issue. The proposed solution involves correlating the received signal with the deterministic component of the APCN waveform instead of relying on the transmit signal's replica in a pulse compression architecture. This approach eliminates the anomaly and improves the accuracy and reliability of slow-moving target detection within noisy environments at the expense of an additional attenuation of the resulting signal. Closed-form expressions characterizing the pulse compression output in such architecture were also derived.

Moreover, we showed that the meticulous selection of parameters governing the random components in the APCN transmit signal emerges as a pivotal factor influencing overall system performance. In particular, the scale parameter associated with phase randomness assumes a critical role: larger values yield a more random waveform characterized by a higher Spectral Flatness Measure, thereby enhancing range ambiguity suppression. However, this improvement is offset by reduced spectral efficiency, as higher scale parameters lead to lower power within the desired bandwidth and decreased Doppler tolerance. Furthermore, introducing a stochastic signal to modulate the transmit waveform's amplitude intensifies its randomness, exacerbating these trade-offs. Additionally, the inclusion of a random component in amplitude results in a deterioration in system performance concerning power efficiency, as evidenced by the increased Peak-to-Average Power Ratio in the transmit signal. These considerations are particularly pertinent in long-range applications and necessitate careful deliberation in designing and optimizing APCN-based systems.

Regarding an intercept–receiver system point of view, a system with digital processing was modeled assuming a plausible number of intercepted pulses, according to the recent literature. We then proposed a candidate method to use in an ESM system for APCN noise waveform detection and parameter extraction. We employed a time-frequency transform to accurately extract the radar parameters since interception and exploitation of signals considered LPI/LPE requires sophisticated receivers that use time-frequency signal processing. The transformation made it possible to detect the radar signal immersed in thermal noise. We assumed the nonexistence of any replica of the intercepted signal, as the sample functions of noise radar are theoretically uncorrelated with each other, and the incoming signal parameters are unknown.

The proposed methodology parameter extraction was based on image processing techniques generated by the time-frequency transform. We described each step of the developed methodology to finally generate a *fingerprint* to assist in identifying the emitting source. We evaluated the intercept receiver performance based on the probability of such an ELINT system detecting an APCN radar signal, considering LPI/LPE as a function of the signal-to-noise ratio of the ELINT system. Results showed a probability of detection close to 1 for SNRs less than -10 dB. We also evaluated the accuracy and precision of the random variables characterizing the APCN estimated parameters (bandwidth, operating frequency, time duration, and pulse repetition interval) as a function of the SNR. Results also showed that the proposed ELINT system performed well in estimating such parameters in scenarios with SNR less than -10 dB. Finally, we concluded that defining a radar as LPI and LPE, or either, necessarily involves defining the corresponding intercept–receiver system.

Author Contributions: Conceptualization, M.B., L.P., A.L.L.R. and J.A.A.J.; software, M.B. and L.P.; validation, M.B. and L.P.; formal analysis, M.B., L.P., A.L.L.R. and J.A.A.J.; investigation, M.B., L.P., A.L.L.R. and J.A.A.J.; writing—original draft preparation, M.B. and L.P.; writing—review and editing, M.B., L.P., A.L.L.R. and J.A.A.J.; visualization, J.A.A.J.; project administration, L.P. All authors have read and agreed to the published version of the manuscript.

Funding: This study is partially financed by the Brazilian Navy, the Brazilian Army, the National Council for Scientific and Technological Development—CNPq, the Brazilian Development Bank (BNDES), the Studies and Projects Financing Agency (Finep) and the Coordenação de Aperfeiçoamento de Pessoal de Nível Superior–Brazil (CAPES)—Finance Code 001.

Institutional Review Board Statement: Not applicable.

Informed Consent Statement: Not applicable.

Data Availability Statement: Data are contained within the article.

Conflicts of Interest: The authors declare no conflict of interest.

References

1. Neri, F. *Introduction to Electronic Defense Systems*, 3rd ed.; Artech House: London, UK, 2018.
2. Kulpa, K. *Signal Processing in Noise Waveform Radar*; Artech House: London, UK, 2013.
3. Govoni, M.A.; Li, H.; Kosinski, J.A. Range-Doppler resolution of the linear-FM noise radar waveform. *IEEE Trans. Aerosp. Electron. Syst.* **2013**, *49*, 658–664. [CrossRef]
4. Galati, G.; Pavan, G.; De Palo, F. Noise radar technology: Pseudorandom waveforms and their information rate. In Proceedings of the 15th International Radar Symposium (IRS), Gdansk, Poland, 16–18 June 2014. [CrossRef]
5. Pralon, L.; Beltrão, G.; Barreto, A.; Cosenza, B. On the analysis of PM/FM noise radar waveforms considering modulating signals with varied stochastic properties. *Sensors* **2021**, *21*, 1727. [CrossRef] [PubMed]
6. Savci, K.; Galati, G.; Pavan, G. Low-PAPR waveforms with shaped spectrum for enhanced low probability of intercept noise radars. *Remote Sens.* **2021**, *13*, 2372. [CrossRef]
7. Palo, F.D.; Galati, G.; Pavan, G.; Wasserzier, C.; Savci, K. Introduction to noise radar and its waveforms. *Sensors* **2020**, *20*, 5187. [CrossRef]
8. Savci, K.; Stove, A.G.; De Palo, F.; Erdogan, A.Y.; Galati, G.; Lukin, K.A.; Lukin, S.; Marques, P.; Pavan, G.; Wasserzier, C. Noise Radar—Overview and Recent Developments. *IEEE Aerosp. Electron. Syst. Mag.* **2020**, *35*, 8–20. [CrossRef]
9. Stove, A.G.; Lukin, K.A.; Orlenko, V.M. Analysis of Partially Deterministic Waveforms in Noise Radar Applications. In Proceedings of the 23rd International Radar Symposium (IRS), Gdansk, Poland, 12–14 September 2022; pp. 159–163.
10. Galati, G.; Pavan, G.; Wasserzier, C. Interception of Continuous-Emission Noise Radars Transmitting Different Waveform Configurations. In Proceedings of the 23rd International Radar Symposium (IRS), Gdansk, Poland, 12–14 September 2022; pp. 153–158.
11. Martino, A.D. *Introduction to Modern EW Systems*, 2nd ed.; Artech House: London, UK, 2018.

12. Galati, G.; Pavan, G.; De Palo, F.; Stove, A. Potential applications of noise radar technology and related waveform diversity. In Proceedings of the 2016 17th International Radar Symposium (IRS), Krakow, Poland, 10–12 May 2016; pp. 1–5. [CrossRef]
13. Galati, G.; Pavan, G.; Savci, K.; Wasserzier, C. Counter-interception and counter-exploitation features of noise radar technology. *Remote Sens.* **2021**, *13*, 4509. [CrossRef]
14. Lukin, K.; Kulyk, V.; Zemlyaniy, O. Application of dynamical chaos for design of random waveform generators. In Proceedings of the Noise Radar Technology Workshop, Yalta, Ukraine, 18–20 September 2002; pp. 127–135.
15. Thayaparan, T.; Wernik, C. *Noise Radar Technology Basics*; Technical Report; Defense Research and Development: Ottawa, ON, Canada, 2006.
16. Axelsson, S.R. Noise radar using random phase and frequency modulation. *IEEE Trans. Geosci. Remote Sens.* **2004**, *42*, 2370–2384. [CrossRef]
17. Dawood, M.; Narayanan, R.M. Generalised wideband ambiguity function of a coherent ultrawideband random noise radar. *IEE Proc.-Radar Sonar Navig.* **2003**, *150*, 379–386. [CrossRef]
18. Pralon, L.; Pompeo, B.; Beltrão, G.; Cioqueta, H.; Cosenza, B.; Fortes, J.M. Random phase/frequency modulated waveforms for noise radar systems considering phase shift. In Proceedings of the 2012 9th European Radar Conference, Amsterdam, The Netherlands, 31 October–2 November 2012.
19. Pralon, L.; Beltrão, G.; Pompeo, B.; Pralon, M.; Fortes, J.M. Near-thumbtack ambiguity function of random frequency modulated signals. In Proceedings of the 2017 IEEE Radar Conference (RadarConf), Seattle, WA, USA, 8–12 May 2017; pp. 352–355.
20. Govoni, M.A.; Li, H.; Kosinski, J.A. Low Probability of Interception of an Advanced Noise Radar Waveform with Linear-FM. *IEEE Trans. Aerosp. Electron. Syst.* **2013**, *49*, 1351–1356. [CrossRef]
21. Govoni, M.A.; Elwell, R.A. Radar spectrum spreading using Advanced Pulse Compression Noise (APCN). In Proceedings of the IEEE Radar Conference, Cincinnati, OH, USA, 19–23 May 2014; pp. 1471–1475. [CrossRef]
22. Barbosa, M.; Pralon, L.; Apolinário, J. Slow-Moving Target Detection Performance of an LPI APCN Waveform in Surveillance Applications. In Proceedings of the 23rd International Radar Symposium (IRS), Gdansk, Poland, 12–14 September 2022; pp. 147–152. [CrossRef]
23. Pace, P. *Detecting and Classifying Low Probability of Intercept Radar*, 2nd ed.; Artech House: London, UK, 2009.
24. Gupta, A.; Bazil Rai, A.A. Feature Extraction of Intra-Pulse Modulated LPI Waveforms Using STFT. In Proceedings of the 4th International Conference on Recent Trends on Electronics, Information, Communication Technology (RTEICT), Bangalore, India, 17–18 May 2019; pp. 742–746. [CrossRef]
25. Shyamsunder, M.; Subbarao, K.; Regimanu, B.; Teja, C.K. Estimation of modulation parameters for LPI radar using Quadrature Mirror Filter Bank. In Proceedings of the IEEE Uttar Pradesh Section International Conference on Electrical, Computer and Electronics Engineering (UPCON), Varanasi, India, 9–11 December 2016; pp. 239–244. [CrossRef]
26. Stevens, D.L.; Schuckers, S.A. Detection and Parameter Extraction of Low Probability of Intercept Radar Signals using the Hough Transform. *Glob. J. Res. Eng.* **2016**, *15*, 9–25. [CrossRef]
27. Yu Gau, J. Analysis of Low Probability of Intercept (LPI) Radar Signals Using the Wigner Distribution. Master's Thesis, Naval Postgraduate School, Monterey, CA, USA, 2002. [CrossRef]
28. Jarpa, P. Quantifying the Differences in Low Probability of Intercept Radar Waveforms Using Quadrature Mirror Filtering. Master's Thesis, Naval Postgraduate School, Monterey, CA, USA, 2002.
29. Guner, K.K.; Gulum, T.O.; Erkmen, B. FPGA-Based Wigner–Hough Transform System for Detection and Parameter Extraction of LPI Radar LFMCW Signals. *IEEE Trans. Instrum. Meas.* **2021**, *70*, 2003515. [CrossRef]
30. Erdogan, A.Y.; Gulum, T.O.; Durak-Ata, L.; Yildirim, T.; Pace, P.E. FMCW Signal Detection and Parameter Extraction by Cross Wigner–Hough Transform. *IEEE Trans. Aerosp. Electron. Syst.* **2017**, *53*, 334–344. [CrossRef]
31. Gulum, T.O.; Pace, P.E.; Cristi, R. Extraction of polyphase radar modulation parameters using a Wigner-Ville distribution—Radon transform. In Proceedings of the IEEE International Conference on Acoustics, Speech and Signal Processing (ICASSP), Las Vegas, NV, USA, 31 March–4 April 2008; pp. 1505–1508. [CrossRef]
32. Boashash, B. *Time-Frequency Signal Analysis and Processing: A Comprehensive Reference*, 2nd ed.; Elsevier: Amsterdam, The Netherlands, 2016.
33. Popoulis, A.; Pillai, S.U. *Probability, Random Variables, and Stochastic Processes*; McGraw-Hill: Boston, MA, USA, 1991.
34. Richards, M.A. *Fundamentals of Radar Signal Processing*, 3rd ed.; McGraw-Hill: Boston, MA, USA, 2022.
35. Galati, G.; Pavan, G.; Savci, K.; Wasserzier, C. Noise radar technology: Waveforms design and field trials. *Sensors* **2021**, *21*, 3216. [CrossRef] [PubMed]
36. Oppenheim, A.V.; Schafer, W.R. *Discrete-Time Signal Processing*, 3rd ed.; Pearson: London, UK, 2014.
37. Dubnov, S. Generalization of spectral flatness measure for non-Gaussian linear processes. *IEEE Signal Process. Lett.* **2004**, *11*, 698–701. [CrossRef]
38. Beltrão, G.; Pralon, L.; Barreto, A.; Alaee-Kerahroodi, M.; Bhavani Shankar, M.R. Subpulse Processing for Unambiguous Doppler Estimation in Pulse-Doppler Noise Radars. *IEEE TRansactions Aerosp. Electron. Syst.* **2021**, *57*, 3813–3826. [CrossRef]
39. Pralon, L.; Pompeo, B.; Fortes, J.M. Stochastic analysis of random frequency modulated waveforms for noise radar systems. *IEEE Trans. Aerosp. Electron. Syst.* **2015**, *51*, 1447–1461. [CrossRef]
40. Robertson, S. *Practical ESM Analysis*; Artech House: London, UK, 2019.
41. Tsui, J.B. *Digital Techniques for Wideband Receivers*; SciTech Publishing: Raleigh, NC, USA, 2015; Volume 3.

42. Cohen, L. *Time-Frequency Analysis*; Prentice Hall: Upper Saddle River, NJ, USA, 1995.
43. Smith, J.O. *Spectral Audio Signal Processing*; W3K: São Leopoldo, Brazil, 2011.
44. Gonzalez, R.C.; Woods, R.E. *Digital Image Processing*, 2nd ed.; Pearson Prentice Hall: Upper Saddle River, NJ, USA, 2008.
45. Duda, R.O.; Hart, P.E. Use of the Hough transformation to detect lines and curves in pictures. *Commun. ACM* **1972**, *15*, 11–15 [CrossRef]
46. Parker, J.R. *Algorithms for Image Processing and Computer Vision*; John Wiley & Sons: Hoboken, NJ, USA, 2010.
47. Otsu, N. A Threshold Selection Method from Gray-Level Histograms. *IEEE Trans. Syst. Man Cybern.* **1979**, *9*, 62–66. [CrossRef]
48. Tsui, J.B.Y.; Shaw, R.L.; Davis, R.L. Performance standards for wideband EW receivers. *Microw. J.* **1989**, *32*, 46.
49. Watson, R. Receiver dynamic range. II—Use one figure of merit to compare all receivers. *Microwaves* **1987**, *26*, 99.
50. Figueirêdo, R. Approaches for Analysis and Extraction of LPI Radar Features. Master's Thesis, Postgraduate Program in Electrical Engineering, COPPE, UFRJ, Rio de Janeiro, Brazil, 2019.

Article

Efficient FPGA Implementation of Convolutional Neural Networks and Long Short-Term Memory for Radar Emitter Signal Recognition

Bin Wu [1], Xinyu Wu [1,*], Peng Li [1], Youbing Gao [2], Jiangbo Si [3,4] and Naofal Al-Dhahir [4]

[1] School of Electronic Engineering, Xidian University, Xi'an 710071, China; bwu@xidian.edu.cn (B.W.); penglixd@xidian.edu.cn (P.L.)

[2] Science and Technology on Electronic Information Control Laboratory, Chengdu 610036, China; gaoyb@cetc.com

[3] Integrated Service Networks Laboratory, Xidian University, Xi'an 710071, China; jbsi@xidian.edu.cn

[4] Department of Electrical and Computer Engineering, The University of Texas at Dallas, Richardson, TX 75080, USA; aldhahir@utdallas.edu

* Correspondence: wuxy2021@stu.xidian.edu.cn

Abstract: In recent years, radar emitter signal recognition has enjoyed a wide range of applications in electronic support measure systems and communication security. More and more deep learning algorithms have been used to improve the recognition accuracy of radar emitter signals. However, complex deep learning algorithms and data preprocessing operations have a huge demand for computing power, which cannot meet the requirements of low power consumption and high real-time processing scenarios. Therefore, many research works have remained in the experimental stage and cannot be actually implemented. To tackle this problem, this paper proposes a resource reuse computing acceleration platform based on field programmable gate arrays (FPGA), and implements a one-dimensional (1D) convolutional neural network (CNN) and long short-term memory (LSTM) neural network (NN) model for radar emitter signal recognition, directly targeting the intermediate frequency (IF) data of radar emitter signal for classification and recognition. The implementation of the 1D-CNN-LSTM neural network on FPGA is realized by multiplexing the same systolic array to accomplish the parallel acceleration of 1D convolution and matrix vector multiplication operations. We implemented our network on Xilinx XCKU040 to evaluate the effectiveness of our proposed solution. Our experiments show that the system can achieve 7.34 giga operations per second (GOPS) data throughput with only 5.022 W power consumption when the radar emitter signal recognition rate is 96.53%, which greatly improves the energy efficiency ratio and real-time performance of the radar emitter recognition system.

Keywords: convolutional neural network (CNN); field programmable gate array (FPGA); hardware accelerators; long short-term memory (LSTM); radar emitter signal recognition

Citation: Wu, B.; Wu, X.; Li, P.; Gao, Y.; Si, J.; Al-Dhahir, N. Efficient FPGA Implementation of Convolutional Neural Networks and Long Short-Term Memory for Radar Emitter Signal Recognition. *Sensors* **2024**, 24, 889. https://doi.org/10.3390/s24030889

Academic Editors: Janusz Dudczyk and Piotr Samczyński

Received: 5 January 2024
Revised: 24 January 2024
Accepted: 25 January 2024
Published: 30 January 2024

1. Introduction

Most of the traditional radar emitter signal (RES) identification methods are based on the RES parameters, such as carrier frequency, signal pulse width, signal amplitude, direction of arrival (DOA), and signal time of arrival (TOA). However, with the RES system and the modern electromagnetic environment becoming more complicated, the traditional identification method based on parameters cannot meet the requirements of RES identification requirements. In addition, whether in the field of electronic support measure systems or communications, RESs always need to be intentionally or unintentionally modulated before being radiated into space [1]. In recent years, according to this RES characteristic, there has been significant research on modulation type recognition, behavior recognition, and even specific RES identification using the methods of artificial intelligence and machine learning [2]. Compared with the traditional methods, recognition algorithms based on deep learning are widely used because of their powerful feature extraction ability, which greatly improves the RES recognition rate.

However, many recognition algorithms based on deep learning require complex RES preprocessing to complete the recognition task. For example, the RES is transformed into the frequency domain, time–frequency domain, or other transform domains, and then these features or images are classified and recognized after obtaining the transform domain features. In [3], the Hilbert transform and bispectrum of the RES are combined to form a signal image, which is fed into a convolutional neural network (CNN) to realize individual RES recognition. There are also other approaches to directly classify and identify RES intermediate frequency (IF) data using deep learning networks with complex structures, such as residual networks, attention mechanisms or hybrid neural network (NN) [4–7]. In [8], a combination of NN with an inception mechanism and long short-term memory (LSTM) neural networks was used to achieve individual identification of five universal software radio peripheral (USRP) communication emitter signal devices. Although such methods have high RES identification accuracy, the complex preprocessing process and algorithm model will inevitably lead to an increase in computing complexity and system latency, especially for power-limited computing devices, such as spaceborne devices or portable terminals, where most of the complex deep learning algorithms cannot be implemented. Therefore, in the design process of RES identification algorithms, more attention should be paid to the balance between signal recognition rate and algorithm complexity, and the implementation ability of an RES identification algorithm in engineering applications should not be ignored.

The rapid development of deep learning cannot be achieved without its powerful computing power, which is usually accelerated by parallel processors such as a graphics processing unit (GPU). Although a GPU can provide powerful computational support for deep learning algorithms, the huge power consumption of GPUs limits their application in low-power scenarios. An application-specific integrated circuit (ASIC) [9] can provide high-performance and low-power computational support for deep learning algorithms, but its high cost and long-term development process have been major deterrents. Hence, programmable logic gate arrays (FPGAs) have received increasing attention due to their reprogrammable, low power consumption, and abundant computational resources, and many FPGA-based computational acceleration schemes for deep learning algorithms have been proposed in recent years. A method to automatically deploy CNNs on on-board FPGAs was proposed in [10], which achieved 23.06 GOPS and 22.17 GOPS throughput rates for the simplified VGG16 network and YOLOv2 network deployed on a Xilinx AC701. In [11], the FPGA implementation of CNNs for radar signal processing was carefully optimized for better performance and energy efficiency. The authors of [12] achieved substantial improvements in computational speed and energy efficiency ratio of the LSTM network acceleration engine implemented on an FPGA compared to CPU and GPU using fixed-point parameters, systolic arrays, and nonlinear function lookup tables. In [13], an architecture for CNN implementation in FPGAs using the Winograd algorithm was proposed to reduce the complexity of convolutional operations and accelerate the computational process. In [14], a CNN acceleration was implemented specifically using deep separable convolution.

However, most of the current research on FPGA-based deep learning computational acceleration only focuses on the computational acceleration of one of the NN models, CNNs or RNNs [15], and cannot support the computation of C-RNN models combining a CNN and an RNN, and most of the related research focuses on the processing of two-dimensional image data.

Based on the above discussion, we constructed a one-dimensional (1D) CNN-LSTM model for RES classification and identification, considering the characteristics that the actual signal duration varies randomly and the information among sampling points is correlated in the time dimension. We exploited the different advantages of CNNs for reducing frequency variation, LSTM for temporal modeling, and deep NNs for mapping features to a more separable space [16], and designed an FPGA-based resource multiplexing computing acceleration platform as shown in Figure 1. The deployment of the CNN-LSTM algorithm model on FPGAs was realized, which enables the deep learning-based RES identification algorithm to be truly implemented in low-power and stringent real-time scenarios. The main contributions of this paper are summarized as follows:

1. We constructed a 1D-CNN-LSTM model for RES recognition, which can directly process IF data and guarantee high recognition accuracy with a simple structure that is more convenient for FPGA hardware implementation. Compared with a single CNN, the network model has no uniform requirement on the length of the input signal and is more suitable for processing pulse RESs with randomly varying length.
2. We designed an FPGA-based resource multiplexing computational acceleration platform for the 1D-CNN-LSTM model constructed in this paper, which achieves parallel acceleration of both 1D convolution and matrix multiplication operations by multiplexing the same systolic array, reducing the processing delay while greatly improving the utilization of FPGA computational resources.
3. For the different operation characteristics of CNNs and LSTM, a special instruction set of the FPGA acceleration platform was developed, which can realize rapid redeployment by adjusting instructions during the change of NN model structure or algorithm iteration.
4. On the Xilinx XKU040 FPGA development board, we have implemented a 1D-CNN-LSTM RES recognition system. The experimental results show that the system achieves a data throughput rate of up to 7.34 GOPS with a power consumption of only 5.022 W with a recognition rate of 97.53% for RES recognition, which is suitable for the scenario of low-power requirements for RES recognition while guaranteeing high computing performance. This ensures both efficient resource utilization and optimal system performance for RES recognition.

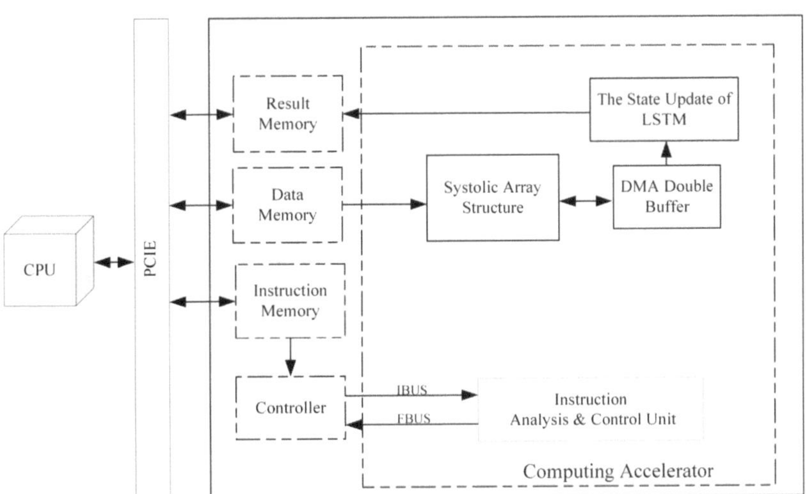

Figure 1. Resource multiplexing computational accelerator architecture.

The rest of this paper is organized as follows: Section 2 introduces RESs, the operational properties of CNNs and RNNs, and the 1D-CNN-LSTM algorithm constructed in this paper for RES identification. Section 3 introduces the design ideas and methods of the FPGA-based resource multiplexed computing acceleration platform. Section 4 presents the related experimental results. Section 5 is the discussion of the results, and Section 6 concludes this paper.

2. Related Works

2.1. Radar Emitter Signal Recognition

Early radar signals were single-carrier pulses without intra-pulse modulation, but with the advancement of radar technology, various modulations of radar signals began to be performed to improve the radar range of action and Doppler resolution. The five parameters (carrier frequency, amplitude, pulse width, direction of arrival, and signal time of arrival) of the conventional pulse description word (PDW) can no longer fully characterize the state information of the radar pulse. The RES intra-pulse modulation characteristics have consequently become an important parameter to describe the characteristics of the radar pulse. The main radar signal intra-pulse modulation methods are frequency modulation and phase modulation, which can be divided into continuous frequency modulation and discrete frequency modulation. Phase modulation mainly has two-phase coding, four-phase coding, and multi-phase coding [17].

Regardless of the modulation method, the modulation information is contained in the time domain signal or transform domain characteristics of the pulse. Traditional CNNs require uniformity in the dimensionality of the input time domain signal or transform domain features for RES identification. However, the pulse width, carrier frequency, amplitude, and signal bandwidth of the RESs are parameters that can change randomly; this requires that the algorithm model used for recognition must set redundancy according to the range of parameters required by the system and, thus, improve the applicability of the algorithm [18]. For example, the radar IF signal pulse width varies from a few microseconds to several tens of microseconds, and for some special functions of the radar signal, the signal pulse width may even reach the millisecond level. The carrier frequency ranges are generally in the 30 to 500 MHz range. Even without considering special requirements, if the signal is sampled for signal processing with a sampling rate that meets the bandwidth, the data length of the signal varies from several hundred to several tens of thousands of points. Whether the signal dimension is unified by using time domain or transform domain redundancy, it will inevitably result in wasted computational resources.

2.2. Convolutional Neural Networks

A typical 1D-CNN algorithm model [19] is shown in Figure 2, which generally consists of an input layer, convolutional layer, pooling layer, fully connected layer, and output layer. In the actual algorithmic model, the convolutional and pooling layers are generally used alternately several times to form the depth structure.

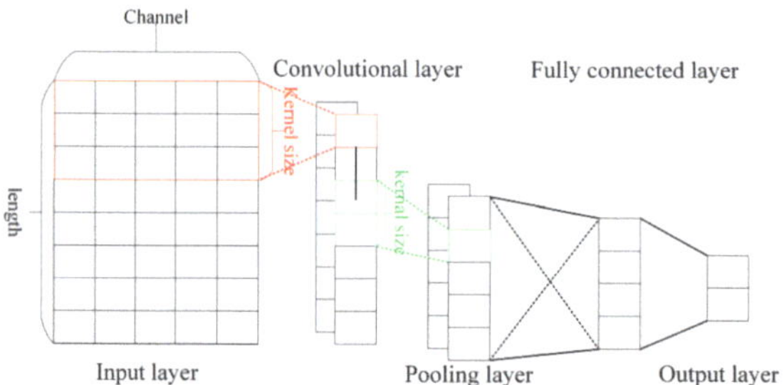

Figure 2. A typical 1D-CNN algorithm model.

Among them, the convolutional layer extracts the abstract features of the input data by convolutional operations, which are defined as follows:

$$x_j^l = f\left(\sum_{i \in M} x_i^{l-1} \otimes k_{ji}^l + b_j^l \right), \tag{1}$$

where x_j^l denotes the channel output of the j channel of the l convolutional layer, $f(\bullet)$ denotes the activation function, M denotes the number of channels of the input data, x_i^{l-1} denotes the i channel of the input data, $\otimes$ denotes the sequence convolution, k_{ji}^l denotes the i channel of the j convolutional kernel of the l layer, and b_j^l denotes the bias parameter of the j convolutional kernel of the l layer.

For the RES, if the quadrature dual-channel sampled IF data is used as the input of the convolution operation, the number of channels of the input data is fixed to 2, but the pulse length of the RES generally varies randomly. Therefore, using a fixed number of convolution kernels, the abstract features of the RES with randomly varying length and fixed number of output channels can be extracted by the convolution operation [20].

The main role of the pooling layer in a CNN is to obtain key features and achieve information dimensionality reduction. The pooling methods used (down-sampling methods) are generally average pooling and maximum pooling, where average pooling takes the average value of the features in the pooling window as the output, and maximum pooling takes the maximum value of the features in the pooling window as the output. The pooling operation is defined as follows:

$$\hat{x}_j^l = D(x_j^l), \tag{2}$$

where $\hat{x}_j^l$ denotes the pooled output of the j channel of the l layer, $D(\bullet)$ denotes the down-sampling function, and x_j^l denotes the pooled input of the j channel of the l layer. From the definition of the pooling operation, it can be seen that the number of channels and the length of the output features of the pooling layer depend on the dimensionality of the input features. Therefore, the abstract features of the RES extracted by the convolutional layer are still abstract feature data with randomly varying lengths and a fixed number of channels after down-sampling by the pooling layer.

The fully connected layer is used to map the feature information extracted from the convolutional and pooling layers to a more separable space for the final classification output [21], and the fully connected operation is defined as follows:

$$\begin{aligned}
(y_1, y_2, \ldots, y_m)^T =& W_{mj}\left(\hat{x}_1^l, \hat{x}_2^l, \ldots, \hat{x}_j^l\right)^T \\
&+ (b_1, b_2, \ldots, b_m)^T,
\end{aligned} \tag{3}$$

where $(y_1, y_2, \ldots, y_m)^T$ is the output of the fully connected operation, W_{mj} is the weight matrix, $W_{mj}\left(\hat{x}_1^l, \hat{x}_2^l, \ldots, \hat{x}_j^l\right)^T$ is the expanded feature vector as the input to the fully connected operation, and $(b_1, b_2, \ldots, b_m)^T$ is the bias vector. From the definition of the fully connected operation, it is clear that once the dimensionality of the weight matrix is determined, the dimensionality of the input features must be determined, which obviously cannot be adapted to the operation of data with random length radiation source signal features.

The above analysis leads to the following conclusion: the convolution and pooling operation layers of the 1D-CNN have no requirement on the length of the input signal, under the condition that the number of input data channels is determined. Since the convolutional computation uses weight sharing to extract data features, the change in the input data length does not affect the properties of the output features. Although the fully-connected layer can map the feature data to a separable space for classification and recognition, it cannot adapt to changes in the dimensionality of the input features. Therefore, when using a CNN to solve the problem of inconsistent input data dimensions, a redundancy mechanism must be used to unify the signal dimensions before they are fed into the algorithm model for processing. This will inevitably lead to a waste of computational resources. To solve this problem, we discuss the LSTM in the next section.

2.3. Long Short-Term Memory

An RNN generally consists of an input layer, hidden layer and output layer [21], whose operations are expanded in the time dimension in Figure 3. RNN operations are defined as follows:

$$s_t = [W, U](s_{t-1}, x_t)^T, \tag{4}$$

$$o_t = V(s_t), \tag{5}$$

where W and U are the weight matrices of s_{t-1} and x_t, respectively, and $V(\bullet)$ is the activation function. The output o_t of the RNN at time t depends on the state s_t of the network at the current time, which is related not only to the current input x_t, but also to the network state s_{t-1} at the time $t-1$ of the network. The network records the hidden state s_t at time t and passes it to time $t+1$ until the last iteration, and the output of the network contains the state information of all historical moments, which enables the RNN to capture the order-dependent features such as location or time in the input data. Meanwhile, the cyclic structure of the RNN network itself determines its ability to adapt to random variations in the input sequence length.

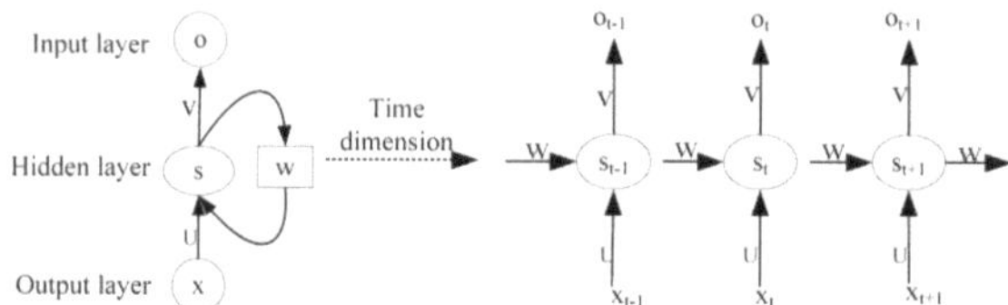

Figure 3. Recurrent neural network.

$$f_t = sigmoid(w_f \cdot [h_{t-1}, x_t] + b_f); \tag{6a}$$

$$i_t = sigmoid(w_i \cdot [h_{t-1}, x_t] + b_i); \tag{6b}$$

$$o_t = sigmoid(w_o \cdot [h_{t-1}, x_t] + b_o); \tag{6c}$$

$$\hat{c}_t = tanh(w_c \cdot [h_{t-1}, x_t] + b_c); \tag{6d}$$

$$C_t = f_t * C_{t-1} + i_t * \hat{c}_t; \tag{6e}$$

$$h_t = o_t * C_t \tag{6f}$$

However, the simple RNN network structure often suffers from the problem of gradient dispersion and gradient explosion during the training process [22], which makes the network parameters fail to converge for a long time during the training process. The author proposed the LSTM [23], which solves this problem to a certain extent. The structure of the LSTM NN is shown in Figure 4, which introduces the concept of gate in the traditional RNN to control the opening of an information flow in the network cycle by simulating the characteristics of human memory to achieve the goal of local key information filtering and long-time-span feature synthesis. In (6), the relevant definitions of Figure 4 are explained. The three gates in the LSTM are the forget gate f_t, the input gate i_t, and the output gate o_t. All three gates are composed of *sigmoid* cells controlled by the current input x_t and the previous time output h_{t-1}. The forget gate controls the opening of the historical state C_{t-1} into the current state C_t, the input gate controls the opening of x_t and h_{t-1} into C_t, and the output gate controls the opening of C_t into the current output h_t.

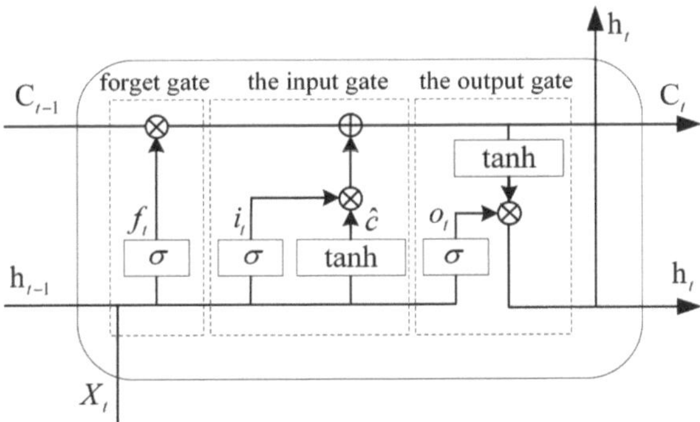

Figure 4. Structure of an LSTM neural network.

The LSTM not only increases the gating structure compared with the simple RNN, which improves the network trainability and sequence feature synthesis ability, but also inherits the characteristics of the simple RNN to adapt to the indefinitely long input. This makes the LSTM more suitable for synthesizing the information from the local radiation source signal features extracted by the convolution operation. The reason why the LSTM network is not used directly for signal feature extraction is that its structure determines that it can only be executed sequentially in the time dimension and is not suitable for parallel computing, and the parallelism of the algorithm is especially important for processing RESs with sequence lengths in the thousands [23].

2.4. 1D-CNN-LSTM

We combined the characteristics of RESs with a large variation range of parameters, strong randomness, and 1D time domain sequence, using the good local feature extraction ability of CNNs. The LSTM is good at capturing time series information and can adapt to the random variation of the input data length and the fully connected network can map the features to the separable space. The 1D-CNN-LSTM model is constructed as shown in Figure 5; "n" represents the length of the RES with variable length. The network model directly processes the IF data of the RES, and completes the classification and identification of six different modulation types of RES, which include continuous wave (CW), binary frequency shift keying (BFSK) signals, binary phase-shift keying (BPSK) signals, quadrature phase-shift keying (QPSK) signals, linear frequency modulation (LFM) signals, and nonlinear frequency modulation (NLFM) signals [24].

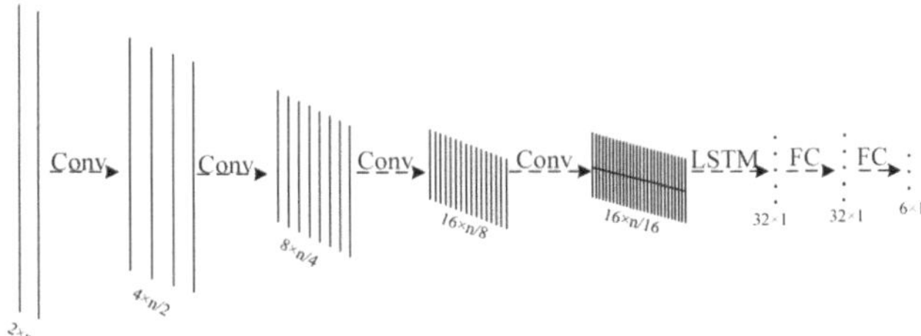

Figure 5. 1D-CNN- LSTM neural network model.

The NN model consists of four convolutional layers, one LSTM layer, and two fully connected layers. The number of input channels of each convolutional layer is 2, 4, 8, and 16, the number of output channels is 4, 8, 16, and 32, and the length of the convolutional kernel is 15. Each convolutional layer uses a maximum pooling method with a pooling kernel length of 2 to reduce the dimensionality of the feature data, and the linear rectification function *ReLU* is used as the activation function. The input dimension of the LSTM layer is

32, and the size of the hidden layer of the LSTM is 32. The input of the first fully connected layer is the last updated hidden layer state vector of the LSTM, and the output dimension is 32×1. The input dimension of the second fully connected layer is the same as the output dimension of the first convolutional layer, and the output dimension is 6×1, which corresponds to six different modulation types of source signals. However, this process can be omitted when deploying the inference network on FPGAs, and the classification results can be derived from the numerical magnitude of the output of the fully-connected layer alone.

3. System Design and Structure

In this section, the design ideas and methods of the FPGA-based resource multiplexing computing acceleration platform will be presented, and the 1D-CNN-LSTM algorithm constructed in this paper for RES identification will be implemented, as shown in Figure 1.

3.1. One-Dimensional Discrete Convolution and Matrix–Vector Multiplication

The 1D discrete convolution operation is an operation that computes the output feature sequence $y(k)$ by sliding multiplication and accumulation of a fixed-length weight kernel $w(m)$ with the input sequence $x(n)$, as shown in Figure 6.

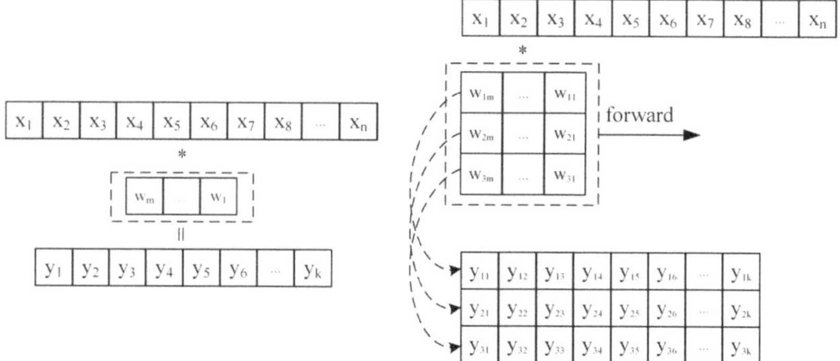

Figure 6. One-dimensional discrete convolution operation process. $*$ represents the inner product operation.

The calculation process shows that each result $y(k)$ of the convolution calculation can be viewed as the inner product of vector $(w_m, w_{m-1}, \cdots, w_1)$ and vector $(x_k, x_{k+1}, \cdots, x_{k+m})$. Similarly, when multiple convolution kernels are convolved with the input sequence at the same time, multiple output sequences can be obtained. For example, if the kernel size is 3, the operation process is shown in Figure 6. It can be seen that the result (y_1, y_2, y_3) obtained from each sliding calculation of the convolution kernel is actually the result of multiplying the weight matrix and the vector $(x_k, x_{k+1}, \cdots, x_{k+m})$ formed by w.

Through the above analysis, it is easy to find that the convolution operation of a multicore is actually composed of multiple matrix–vector multiplication operations. This makes it theoretically feasible to reuse FPGA hardware computing resources to achieve parallel acceleration of both convolutions in 1D-CNN and matrix–vector multiplications in LSTM [22]. However, the input data for the convolution operation in 1D-CNN is not a single channel, but the number of channels of input data increases as the number of convolution layers increases. Corresponding to the number of channels of input data, the dimension of convolution kernel increases, and the convolution result becomes the sum of the convolution results of each channel. The operation process in the actual 1D-CNN is shown in Figure 7.

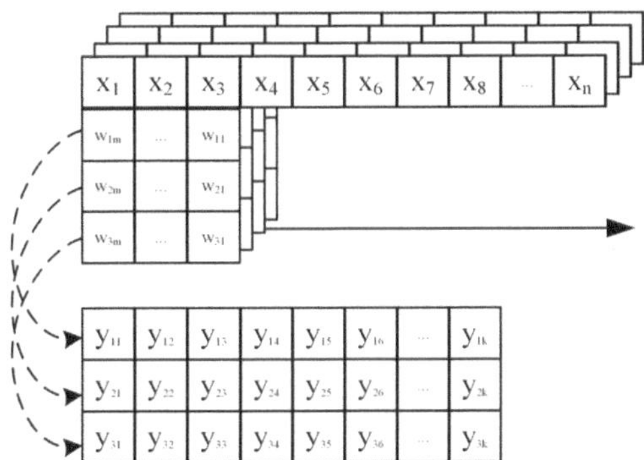

Figure 7. The actual operation process of 1D-CNN. Taking the 1D convolution operation with multi-channel input and multi-channel output with the number of input channels of 4 and a convolution kernel size of 3 as an example.

3.2. Design of Systolic Array Structures

A systolic array [25] is a pipeline structure that can do multiple computations per memory access and consists of a set of interconnected elementary operators, each of which is capable of performing some simple operations. Because simple, regular communication and control structures have significant design and implementation advantages over complex ones, they can accelerate the execution of edge computing problems without increasing I/O requests [26]. First, 1D convolution and matrix vector multiplication operations are typically computationally constrained computations because the total number of computation operations is greater than the total number of input and output elements, and are well suited for parallel computation acceleration using systolic arrays. Secondly, because the structure of each processing element (PE) in the systolic array structure is relatively independent, it is less likely to cause wiring congestion when deployed on FPGAs, which helps to increase the operating frequency of the system.

Based on the above advantages of systolic arrays, the systolic array structure shown in Figure 8 is designed to perform parallel acceleration of 1D convolutional and matrix vector multiplication operations in the CNN-LSTM. According to the 1D-CNN-LSTM structure constructed in this paper, the size of the systolic array is designed to be 32×64, which can support the parallel computation of 1D single-channel convolution with the length of the convolution kernel not larger than 64 and the number of kernels not larger than 32, or the parallel computation of matrix vector multiplication with the size of the weight matrix not larger than 32×64.

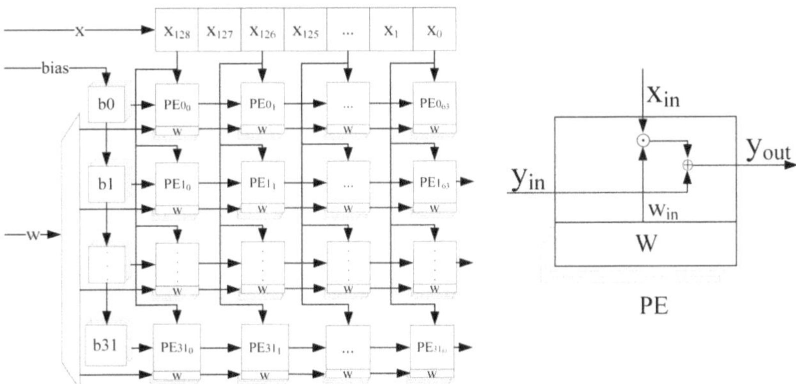

Figure 8. Systolic array structure and PE.

The structure of each PE of the systolic array is shown in Figure 8, including a bias input port y_{in}, a data input port x_{in}, a weight input port w_{in}, a weight cache unit, and a result output port y_{out}. Among them, the weight cache unit is composed of four registers with 16-bit width to store four different weight parameters, and the value in one of the registers is selected as a valid parameter to be used during the calculation as needed. In this paper, all the data involved in the operation are 16-bit fixed-point numbers, including 1-bit sign bits, 4-bit integer bits and 11-bit fractional bits. The basic operations performed by each PE in each clock cycle is:

$$y_{out} = x_{in} * w_{in} + y_{in} \tag{7}$$

The weight matrices of the four gating coefficients are loaded into the weight cache unit at the same time during the iterative process of the LSTM loop, thus avoiding the system time delay caused by the update of the weight matrix at each loop and effectively improving the data throughput rate of the whole systolic array.

To compute multi-channel 1D convolution, the cache structure of the systolic array computation results is designed in this paper as shown in Figure 9.

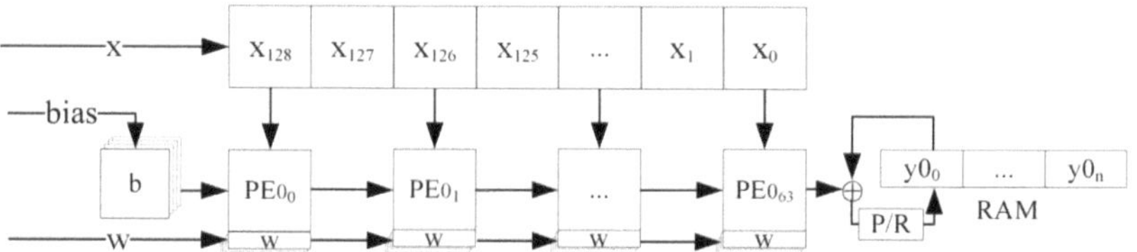

Figure 9. The cache structure of the systolic array computation results.

When computing multi-channel convolution, the result of the previous channel is read from the corresponding address of the result cache RAM, added to the result of the current channel, and then stored back to the original address. The summation result can also be activated by the maximum pooling and linear rectification operations as needed before storing it back to the source address. The operation is defined as follows:

$$P/R(y_k, y_{k+1}, \cdots, y_{k+l-1})$$
$$= \begin{cases} \max(y_k, y_{k+1}, \cdots, y_{k+l-1}) & act_{en} = 0 \\ \max(y_k, y_{k+1}, \cdots, y_{k+l-1}, 0) & act_{en} = 1 \end{cases} \tag{8}$$

where y_k denotes the k pooling result, l denotes the pooling kernel size, and act_{en} denotes whether to activate the result of the operation or not.

For CNNs, the operation input of an intermediate layer is the operation result of the previous layer, and the operation output of that layer is the operation input of the next layer. Therefore, a single structure of systolic result cache is not enough to support multi-layer CNN operations. To solve this problem, we design a direct memory access (DMA) [27] double buffer to achieve the systolic operation result reuse seen in Figure 10.

The result cache array consists of two sets of 32 channels of RAM with the same structure, *BUF0* and *BUF1*. When one of the *BUF* is used as the input source for the systolic operation, the other is used as the result cache. Of course, both of them support both single-channel row outputs and multi-channel column outputs, where the multi-channel column output function is used to realize the last convolutional layer operation result as the input of LSTM and the reuse of matrix vector multiplication results.

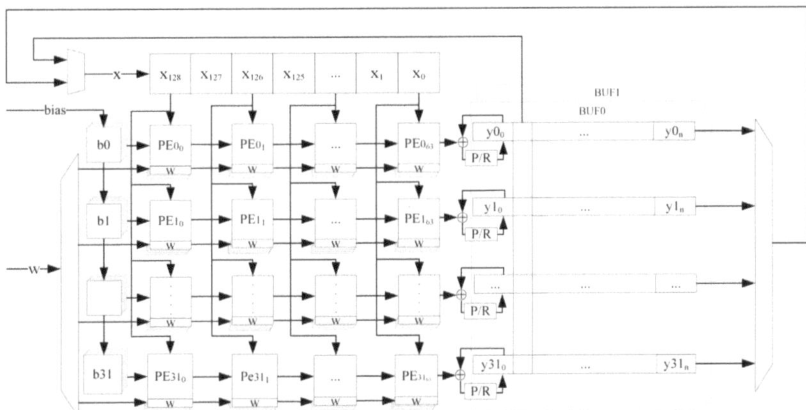

Figure 10. DMA double buffer.

3.3. LSTM Neural Network State Update

To cooperate with the systolic array to complete the whole LSTM operation process, we design the operation module shown in Figure 11 to complete the state update, according to the LSTM operation definition.

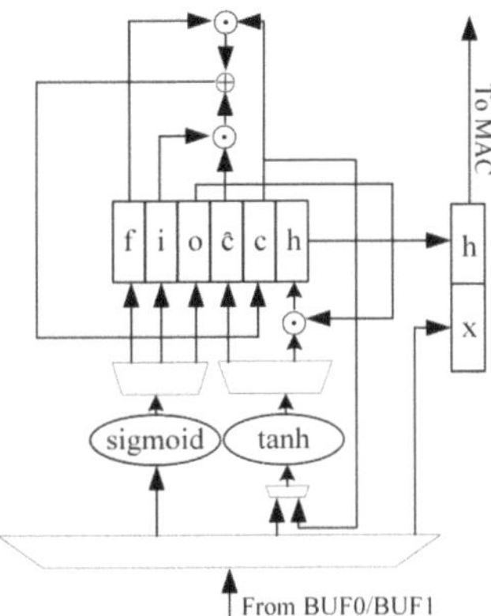

Figure 11. The state update of LSTM neural network.

The module reads the CNN result x_t from the result cache of the pulsating array, and then loads it into the pulsating array for matrix–vector multiplication after forming the $[h_{t-1}, x_t]$ vector with the LSTM output h_{t-1} at the previous time. Then, it reads the result from the result cache of the pulsating array and sends it to the *sigmoid* or *tanh* activation unit.

The two nonlinear activation units, *sigmoid* and *tanh*, are implemented using a table lookup. The values of the two functions are quantized and stored in the ROM in advance, and when the nonlinear activation of the matrix vector multiplication results is required, they are first converted to the corresponding ROM addresses and then read directly from the ROM to obtain the function values. We use 16-bit fixed-points to quantize the nonlinear functions. The quantization process intercepts the part of the independent variables of *sigmoid* and *tanh* in $[-4, 4)$ for sampling, and the part of the independent variables beyond the sampling range is taken as the boundary point in Figure 12.

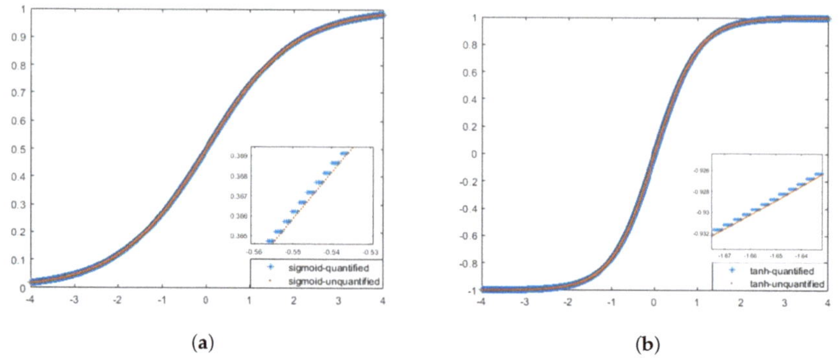

(a)   (b)

Figure 12. Quantization process of two nonlinear activation units. (**a**) *Sigmoid* nonlinear activation units. (**b**) *Tanh* nonlinear activation units.

3.4. Acceleration of the Overall Architecture of the Platform

Combined with the designed systolic array and the LSTM state update circuit, we designed a 1D-CNN-LSTM resource multiplexing computational accelerator using a heterogeneous computational architecture of a Von Neumann-like system [28], as shown in Figure 13.

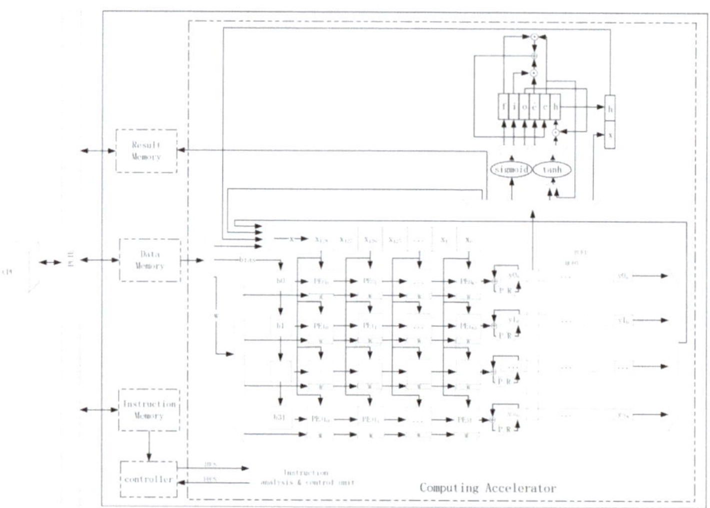

Figure 13. 1D-CNN-LSTM resource multiplexing computational accelerator framework.

The accelerator consists of two parts: a general-purpose CPU and an FPGA, which realize data interaction through a peripheral component interconnect express (PCIE) bus. The general-purpose CPU is used to analyze the NN structure, generate corresponding operation instructions, and load them into the instruction memory on the FPGA side. Then, the RES data to be identified is loaded into the FPGA data memory, and the identification results are read out from the memory after the operation is completed. The FPGA includes five modules: instruction memory, data memory, result memory, controller, and operator. The controller module reads the operation instructions from the instruction memory, which, in turn, controls the operator module to complete the corresponding calculation, and writes the final operation results to the result memory for the CPU to read.

For the computational acceleration platform designed in this paper, a dedicated instruction set with a 64-bit bit width was developed to support 1D convolution, matrix multiplication, LSTM operations, and data read/write functions, as shown in Table 1.

Table 1. Instruction examples from the hardware instruction set.

Instruction ID	Function Code	Operand Length	Source Address
63:60	59:36	35:18	17:01 [1]

[1] Numbers indicate the start and end positions of the instructions.

The instruction content is divided into four parts: instruction ID, function code, operand length, and data source start address. The instruction ID field, as a unique code for each instruction, is used to distinguish different operations. The function code field is used to control some optional functions during the instruction execution. The operand length field is used to specify the total amount of data for an instruction operation. The source address field is used to specify the starting address of the operand. The entire instruction set is divided into four types: systolic array parameter loading instruction, systolic array data loading instruction, LSTM operation instruction, and data moving instruction. Among them, the systolic array parameter loading instructions include two instructions: weight parameter loading and bias parameter loading. The systolic array data load instructions include five instructions: loading from data memory, loading from $BUF0$ row, loading from $BUF1$ row, loading from $BUF0/BUF1$ column, and loading from LSTM output. With LSTM operation instructions and data moving instructions, the entire acceleration platform-specific instruction set consists of nine instructions.

When the hyperparameters of the NN model, such as the number of convolutional layers and size of the convolutional kernel, or the model parameters need to be adjusted, the computational acceleration platform designed in this paper does not need to redesign the accelerator hardware logic. It only needs to make adjustments to the instruction content according to complete the rapid deployment of the new algorithm model, greatly improving the radar emitter source identification system. This is particularly important for deep learning algorithms to adapt to the rapidly changing electromagnetic environment.

4. Experiments

In this section, the 1D-CNN-LSTM model and its experiments based on FPGA deployment are presented. The details of the experiments are explained, the experimental results and performance analysis are shown, and a comparison of different works is performed to evaluate the performance of our approach.

In the experiment, a computer with Intel i7-10700@2.9GHz CPU and the PyTorch deep learning framework was used. We used Vivado 2018.3 development tool to implement and deploy the algorithm on Xilinx XCKU040-ffva1156-2-i for the 1D-CNN-LSTM acceleration designed in this paper. The entire FPGA accelerator system computing unit frequency was 120 MHz. In the comparative experiment, NVIDIA GPU RTX 3090 was used.

4.1. Dataset

To demonstrate the effectiveness of our designed RES identification system, RESs were are generated according to Table 2. We randomly generated six modulation types by simulation, with 3000 sample signals for each type of RES. We divided the 18,000 signals into two parts—the training set and the test set—according to the ratio of 5:1, for the training and generalization performance testing of constructing a 1D-CNN-LSTM model.

Table 2. Specific parameters of the radar emitter signals.

Type	CW	BFSK	BPSK	QPSK	LFM	NLFM
Carrier Frequency			100–400 MHz			
Pulse Width			1–5 µs			
Band Width	–	5–50 MHz	–	–	5–50 MHz	5–50 MHz
Code	–	13-digit Barker code	13-digit Barker code	16-digit frank code	–	–
Frequency of Sample			500 MHz			
Signal to Noise Ratio			5–15 dB			

4.2. Result

The physical object of the designed FPGA acceleration for RES identification is shown in Figure 14. In the Vivado 2018.3 development environment, the detailed FPGA hardware design part of the acceleration is shown in Figure 15.

Figure 14. FPGA Experiment platform, connected to CPU by PCIE bus.

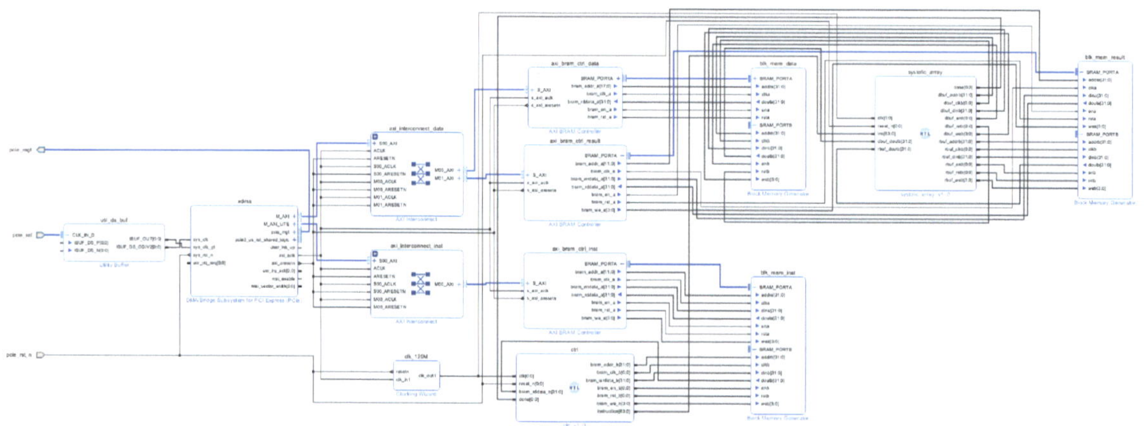

Figure 15. FPGA hardware design schematic.

We used the the Pytorch deep learning framework to train the constructed 1D-CNN-LSTM model. The normalized recognition accuracy and loss values of the algorithm model for the training and test sets during 500 iterations of the training set are shown in Figure 16. At the end of the training, the recognition accuracy of the model for the training and test sets reached 100% and 97.27%, respectively. To demonstrate the identification of the various categories, we present the confusion matrix in Figure 17. We found that between LFM and NLFM, BPSK and QPSK were more difficult to identify. Other types were better identified, and the identification results were relatively evenly distributed.

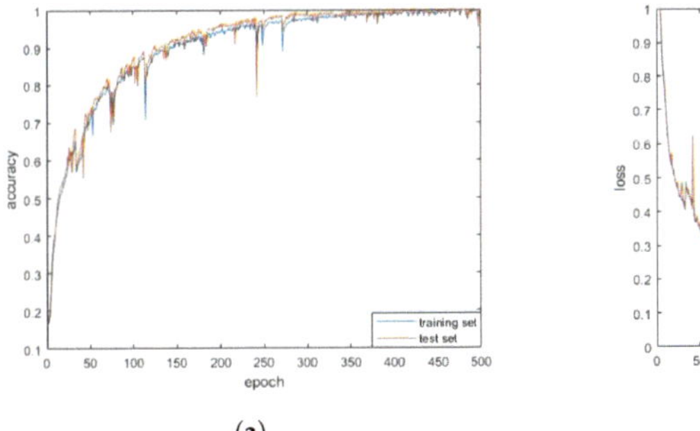 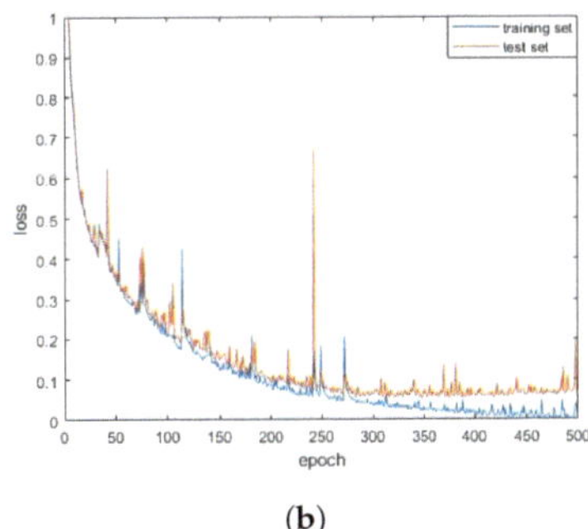

(a)   (b)

Figure 16. (**a**) Normalized recognition accuracy and loss values for training sets; (**b**) normalized recognition accuracy and loss values for testing sets.

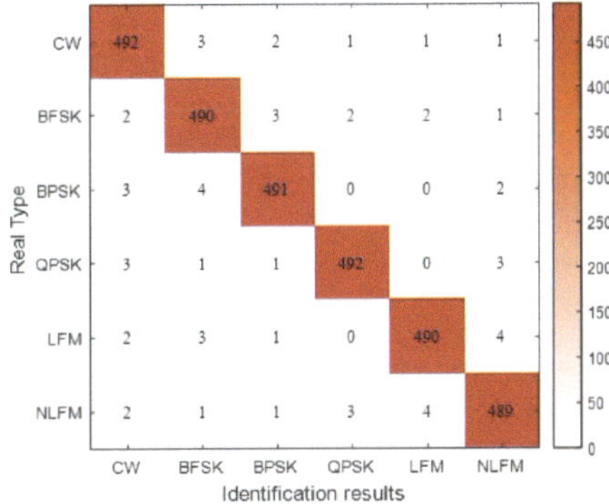

Figure 17. Confusion matrix for each type of recognition result.

The above experiments were all completed under the random SNR of 5–15 dB for recognition. To identify the RESs for different SNR, the SNR of the test set data samples was reset, and the recognition experiments were conducted between 1–15 dB, and the results are shown in Figure 18.

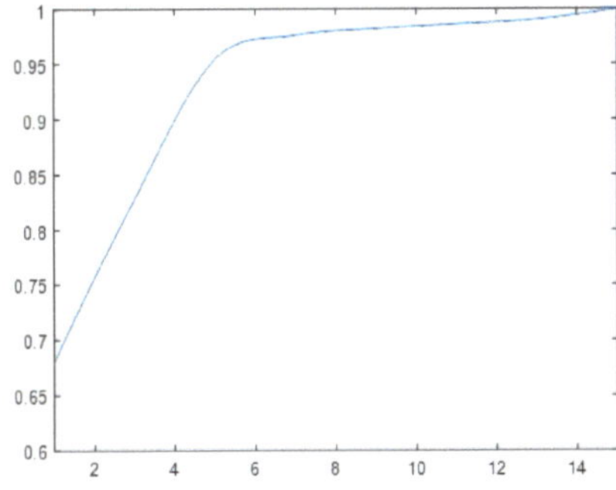

Figure 18. Recognition results with different SNR.

It can be seen that the recognition rate of the algorithm exceeded 80% when the SNR was higher than 5 dB, recovered to the level after the end of training, and the lowest still had an 82.7% accuracy in the case of a low SNR of 3 dB. This indicates that the 1D-CNN-LSTM network has better adaptability in completing the RES recognition under different SNRs.

The trained algorithm model was coded and the parameters quantized, and then deployed on the computational acceleration platform designed in this paper for testing. For the training and test sets constructed in this paper, the processing performance of a total of 18,000 signals was tested with the Pytorch deep learning framework on the CPU and the FPGA computing acceleration platform designed in this paper as acceleration devices to run the same models. The comparison of recognition accuracy is shown in Table 3.

Table 3. Comparison of recognition accuracy in different environments.

Inference Environment	Training Set	Testing Set
Pytorch deep learning framework	100%	97.27%
FPGA acceleration platform	99.54%	96.53%

The experimental results show that the recognition rate of the NN deployed on the FPGA computing platform for the training and test set signals was reduced by 0.46% and 0.74%, respectively, compared to that under the Pytorch deep learning framework. The recognition rate decreased due to the quantization error of both parameters and signals. When implemented on FPGAs, we quantized to 16 bits to reduce the computation and design complexity for hardware implementation. Moreover, the acceleration platform designed in this paper needs to adapt to the inference computation of different structural NN models. Hence, the quantization of NN models was performed with 16-bit fixed-point arithmetic considering the quantization accuracy and model adaptation capability. For our 1D-CNN-LSTM model, the recognition accuracy is considerably high, indicating that the algorithm can be implemented on the FPGA accelerated platform to perform RES recognition.

5. Discussion

5.1. Adaptability for Varying Pulse Width

To demonstrate the adaptability of the 1D-CNN-LSTM model in processing RESs with randomly varying pulse widths, we uses nine sets of RES samples with the same range of other parameters as the training set and pulse widths distributed between 1 μs and 10 μs for recognition accuracy testing.

The test results in Figure 19 show that, regardless of the inference environment, the recognition accuracy of the model started to show a decreasing trend after the signal pulse width exceeded the distribution of the training set parameters. However, the recognition rate of the model remained at a high level when the pulse width reached twice the distribution of the training set. Due to the training set with pulses of 1–5 μs RES, it inevitably led to a decrease in recognition rate when the training set range was exceeded. However, recognition could still be performed with increasing variation in pulse length, and even 9–10 μs could reach more than 55%. It is worth noting that the CNN-LSTM has the advantage of being adaptable to pulse widths beyond the training set distribution, which is not achieved by a single CNN.

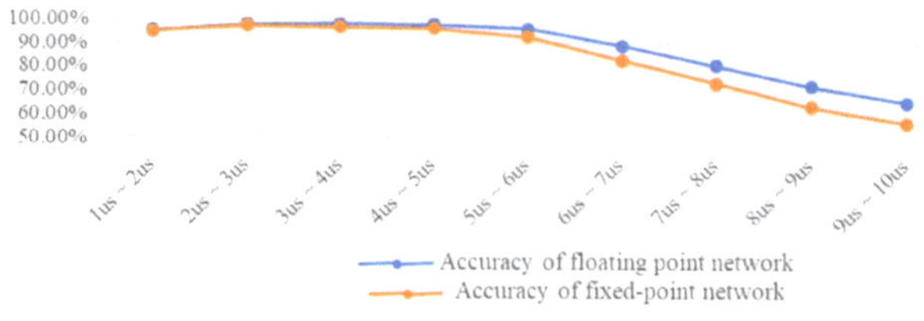

Figure 19. Recognition accuracy for different pulse widths.

5.2. Efficiency Analysis of Acceleration

To evaluate FPGA acceleration performance, the processing performance was tested on the CPU, GPU, and the FPGA computing acceleration platform designed in this paper. On the test set, we ran the same model separately to process the same calculation for inference.

As shown in the test results of Table 4, the processing speed and data throughput of the FPGA acceleration platform were 4.16 times higher than that of the CPU. When in the GPU environment, the Pytorch deep learning framework could take full advantage of the GPU's multi-core parallel processing, which greatly improved the system's data throughput rate to 16.73 GOPS, which was about 2.3 times higher than that of the FPGA compute acceleration platform, but the system power consumption also increased. In terms of energy efficiency, the FPGA computing acceleration platform designed in this paper improved 73 times and 9.125 times compared to the CPU and GPU, respectively. Our FPGA implementation showed much higher energy efficiency when still meeting the data processing requirement.

Table 4. CPU, GPU, and FPGA computing acceleration platform processing performance.

Platform	Calculation	Processing Time (s)	Throughput (GOPS)	Consumption (W)	Energy Efficiency (GOPS/W)
CPU	108.28 G	61.36	1.76	81.19	0.02
GPU	108.28 G	6.47	16.73	102.5	0.16
FPGA	108.28 G	14.75	7.34	5.022	1.46

5.3. Resource Utilization

To represent the advantages of the acceleration performance proposed in this paper, the following resource utilization analysis was performed after using measures such as fixed-point and systolic array structures. The FPGA hardware resource consumption is shown in Table 5.

Table 5. FPGA resource utilization

Logic	LUT	FF	BRAM	DSP
Results	171,497 (71%)	188,405 (39%)	472.0 (79%)	1920 (100%)

The FPGA chip used in this paper has a total of 1920 DSP hardware computing resources, and the systolic array was used in the acceleration optimization with a scale of 32×64. If calculated according to the consumption of one DSP resource per PE, the entire on-chip resources are far from sufficient, but other logic resources can be fully used. For example, through Vivado, we obtained LUT usage of 71%, FF usage of 39%, BRAM usage of 79%, and DPS usage of 100%. These resources can also be integrated to complete the multiplication and addition operations to satisfy the requirements. It also demonstrates one of the major advantages of FPGAs in the design process of large-scale integrated circuits.

5.4. Comparison with Other FPGA Implementations

The proposed implementation in this paper was compared with some existing FPGAs and the results are shown in Table 6.

For FPGA designs, resource consumption varies across architectures. The excellent throughput performance achieved in hardware may be mainly due to the heavy use of hardware resources. Compared to other work, our acceleration in inference phase achieved 95.5% of RES recognition and had a more reasonable relationship between power consumption, resource consumption, and throughput, thus enabling real-time processing.

Table 6. Performance comparison of our work with other accelerators.

	Our Work	[10]	[11]	[29]	[30]
Platform	Xilinx XCKU040	Xilinx AC701	Xilinx ZCU102	Xilinx ZCU102	Artix7 TSBG484
Frequency (MHz)	120	200	153.9	300	100
Throughput (GOPS)	7.34	23.06	-	102	22
Power (W)	5.022	3.407	4.7	11.8	7.53
Energy efficiency (GOPS/W)	1.46	6.77	3.49	8.64	2.92

6. Conclusions

In this paper, we constructed a 1D-CNN-LSTM model, which can be deployed on an FPGA-based resource multiplexing computing acceleration platform while ensuring high RES recognition accuracy. The experimental results show that the model is highly adaptable for the recognition of RES pulses with randomly varying lengths. For the computational acceleration platform, we designed the pulsed array to realize the parallel acceleration of both 1D convolution and vector matrix multiplication operations, which reduced the processing latency of the RES identification system and improved the utilization of FPGA computational resources at the same time. Further experiments and correlation analysis illustrate the contribution of each of our improvements and demonstrate their effectiveness. Deploying the system on a Xilinx XCKU040 FPGA development board achieved a data throughput rate of up to 7.34 GOPS with a power consumption of 5.022 W at a recognition rate of 97.53% for the RES modulation method. In conclusion, our experimental results demonstrate that the FPGA-based CNN-LSTM RES recognition system effectively meets the requirements of low-power RES scenarios while maintaining computational accuracy. With these features, our solution holds significant potential for RES recognition.

In the future, we will continue to improve our models in terms of recognition algorithms to improve the accuracy of RES recognition. Additionally, we will keep optimizing our FPGA-based hardware acceleration platform for more NN models and accelerated computation for low-power implementation.

Author Contributions: Conceptualization, B.W. and X.W.; methodology, B.W. and X.W.; software, X.W.; validation, B.W., X.W. and Y.G.; formal analysis, P.L. and Y.G.; investigation, P.L. and Y.G.; resources, P.L., Y.G., J.S. and N.A.-D.; data curation, B.W. and X.W.; writing—original draft preparation, X.W.; writing—review and editing, B.W., J.S. and N.A.-D.; visualization, B.W., and X.W.; supervision, P.L. and Y.G.; project administration, B.W. and Y.G. All authors have read and agreed to the published version of the manuscript.

Funding: This research was funded by the Science and Technology on Electronic Information Control Laboratory: JS20230100020.

Institutional Review Board Statement: Not applicable.

Informed Consent Statement: Not applicable.

Data Availability Statement: Not applicable.

Acknowledgments: The authors would like to show their gratitude to the editors and the reviewers for their insightful comments.

Conflicts of Interest: The authors declare no conflicts of interest.

References

1. Tian, X.; Liu, Y.; Pan, X.; Chen, W. Intra-pulse Intentional Modulation Recognition of Radar Signals at Low SNR. In Proceedings of the 2018 IEEE 2nd International Conference on Circuits, System and Simulation (ICCSS), Guangzhou, China, 14–16 July 2018; pp. 66–70. [CrossRef]
2. Li, H.; Jing, W.; Bai, Y. Radar emitter recognition based on deep learning architecture. In Proceedings of the 2016 CIE International Conference on Radar (RADAR), Guangzhou, China, 10–13 October 2016; pp. 1–5. [CrossRef]

3. Sun, W.; Wang, L.; Sun, S. Radar Emitter Individual Identification Based on Convolutional Neural Network Learning. *Math. Probl. Eng.* **2021**, *2021*, 5341940. [CrossRef]
4. Pan, Y.; Yang, S.; Peng, H.; Li, T.; Wang, W. Specific Emitter Identification Based on Deep Residual Networks. *IEEE Access* **2019**, *7*, 54425–54434. [CrossRef]
5. Huang, J.; Li, X.; Wu, B.; Wu, X.; Li, P. Few-Shot Radar Emitter Signal Recognition Based on Attention-Balanced Prototypical Network. *Remote Sens.* **2022**, *14*, 6101. [CrossRef]
6. Liu, Q.; Han, L.; Tan, R.; Fan, H.; Li, W.; Zhu, H.; Du, B.; Liu, S. Hybrid Attention Based Residual Network for Pansharpening. *Remote Sens.* **2021**, *13*, 1962. [CrossRef]
7. Zhang, S.; Pan, J.; Han, Z.; Guo, L. Recognition of Noisy Radar Emitter Signals Using a One-Dimensional Deep Residual Shrink-age Network. *Sensors* **2021**, *21*, 7973. [CrossRef] [PubMed]
8. Chen Y, Zhu L, Yu L, et al. Individual identification of communication radiation sources based on Inception and LSTM network. *J. Phys. Conf. Ser.* **2020**, *1682*, 012052. [CrossRef]
9. Wei, G.; Hou, Y.; Zhao, Z.; Cui, Q.; Deng, G.; Tao, X. Demo: FPGA-Cloud Architecture For CNN. In Proceedings of the 2018 24th Asia-Pacific Conference on Communications (APCC), Ningbo, China, 12–14 November 2018; pp. 7–8. [CrossRef]
10. Yan, T.; Zhang, N.; Li, J.; Liu, W.; Chen, H. Automatic Deployment of Convolutional Neural Networks on FPGA for Spaceborne Remote Sensing Application. *Remote Sens.* **2022**, *14*, 3130. [CrossRef]
11. Zhang, J.; Huang, Y.; Yang, H.; Martinez, M.; Hickman, G.; Krolik, J.; Li, H. Efficient FPGA Implementation of a Convolutional Neural Network for Radar Signal Processing. In Proceedings of the 2021 IEEE 3rd International Conference on Artificial Intelligence Circuits and Systems (AICAS), Washington, DC, USA, 6–9 June 2021; pp. 1–4. [CrossRef]
12. He, D.; He, J.; Liu, J.; Yang, J.; Yan, Q.; Yang, Y. An FPGA-Based LSTM Acceleration Engine for Deep Learning Frameworks. *Electronics* **2021**, *10*, 681. [CrossRef]
13. Kala, S.; Jose, B.R.; Mathew, J.; Nalesh, S. High-Performance CNN Accelerator on FPGA Using Unified Winograd-GEMM Architecture. *IEEE Trans. Very Large Scale Integr. (VLSI) Syst.* **2019**, *27*, 2816–2828. [CrossRef]
14. Bai, L.; Zhao, Y.; Huang, X. A CNN Accelerator on FPGA Using Depthwise Separable Convolution. *IEEE Trans. Circuits Syst. II Express Briefs* **2018**, *65*, 1415–1419. [CrossRef]
15. Wang, Z.; Xu, K.; Wu, S.; Liu, L.; Liu, L.; Wang, D. Sparse-YOLO: Hardware/Software Co-Design of an FPGA Accelerator for YOLOv2. *IEEE Access* **2020**, *8*, 116569–116585. [CrossRef]
16. Szegedy, C.; Wei Liu; Yangqing Jia; Sermanet, P.; Reed, S.; Anguelov, D.; Erhan, D.; Vanhoucke, V.; Rabinovich, A. Going deeper with convolutions. In Proceedings of the 2015 IEEE Conference on Computer Vision and Pattern Recognition (CVPR), Boston, MA, USA, 7–12 June 2015; pp. 1–9. [CrossRef]
17. Yuan, S.; Wu, B.; Li, P. Intra-Pulse Modulation Classification of Radar Emitter Signals Based on a 1-D Selective Kernel Convolu-tional Neural Network. *Remote Sens.* **2021**, *13*, 2799. [CrossRef]
18. Zhou, Y.; Wang, C.; Zhou, R.; Wang, X.; Wang, H.; Yu, Y. A Specific Emitter Identification Method Based on RF-DNA and XGBoost. In Proceedings of the 2022 7th International Conference on Intelligent Computing and Signal Processing (ICSP), Xi'an, China, 15–17 April 2022; pp. 1530–1533. [CrossRef]
19. Wu, B.; Yuan, S.; Li, P.; Jing, Z.; Huang, S.; Zhao, Y. Radar Emitter Signal Recognition Based on One-Dimensional Convolutional Neural Network with Attention Mechanism. *Sensors* **2020**, *20*, 6350. [CrossRef] [PubMed]
20. Wang, X.; Huang, G.; Zhou, Z.; Gao, J. Radar emitter recognition based on the short time fourier transform and convolu-tional neural networks. In Proceedings of the 2017 10th International Congress on Image and Signal Processing, BioMedical Engineering and Informatics (CISP-BMEI), Shanghai, China, 14–16 October 2017; pp. 1–5. [CrossRef]
21. Sainath, T.N.; Vinyals, O.; Senior, A.; Sak, H. Convolutional, Long Short-Term Memory, fully connected Deep Neural Networks. In Proceedings of the 2015 IEEE International Conference on Acoustics, Speech and Signal Processing (ICASSP), South Brisbane, QLD, Australia, 19–24 April 2015; pp. 4580–4584. [CrossRef]
22. Kolen, J.F.; Kremer, S.C. Gradient Flow in Recurrent Nets: The Difficulty of Learning LongTerm Dependencies. In *A Field Guide to Dynamical Recurrent Networks*; IEEE: Piscataway, NJ, USA, 2001; pp. 237–243. [CrossRef]
23. Zeng, S.; Guo, K.; Fang, S.; Kang, J.; Xie, D.; Shan, Y.; Wang, Y.; Yang, H. An Efficient Reconfigurable Framework for General Purpose CNN-RNN Models on FPGAs. In Proceedings of the 2018 IEEE 23rd International Conference on Digital Signal Processing (DSP), Shanghai, China, 19–21 November 2018; pp. 1–5. [CrossRef]
24. Yuan, S.; Li, P.; Wu, B. Towards Single-Component and Dual-Component Radar Emitter Signal Intra-Pulse Modulation Classifica-tion Based on Convolutional Neural Network and Transformer. *Remote Sens.* **2022**, *14*, 3690. [CrossRef]
25. Kung, H.-T. Why systolic architectures? *IEEE Comput.* **1982**, *15*, 37–46. [CrossRef]
26. Chang, K.-W.; Chang, T.-S. Efficient Accelerator for Dilated and Transposed Convolution with Decomposition. In Proceedings of the 2020 IEEE International Symposium on Circuits and Systems (ISCAS), Seville, Spain, 12–14 October 2020; pp. 1–5. [CrossRef]
27. Wu, X.; Ma, Y.; Wang, M.; Wang, Z. A Flexible and Efficient FPGA Accelerator for Various Large-Scale and Lightweight CNNs. *IEEE Trans. Circuits Syst. I Regul. Pap.* **2022**, *69*, 1185–1198. [CrossRef]
28. Wu, S.; Xu, Z.; Wang, F.; Yang, D.; Guo, G. An Improved Back-Projection Algorithm for GNSS-R BSAR Imaging Based on CPU and GPU Platform. *Remote Sens.* **2021**, *13*, 2107. [CrossRef]

29. Zhang, S.; Cao, J.; Zhang, Q.; Zhang, Q.; Zhang, Y.; Wang, Y. An FPGA-Based Reconfigurable CNN Accelerator for YOLO. In Proceedings of the 2020 IEEE 3rd International Conference on Electronics Technology (ICET), Chengdu, China, 8–12 May 2020; pp. 74–78. [CrossRef]
30. Zhang, Q.; Cao, J.; Zhang, Y.; Zhang, S.; Zhang, Q.; Yu, D. FPGA Implementation of Quantized Convolutional Neural Networks. In Proceedings of the 2019 IEEE 19th International Conference on Communication Technology (ICCT), Xi'an, China, 16–19 October 2019; pp. 1605–1610. [CrossRef]

Article

An Application of Analytic Wavelet Transform and Convolutional Neural Network for Radar Intrapulse Modulation Recognition

Marta Walenczykowska *, Adam Kawalec * and Ksawery Krenc

Faculty of Mechatronics, Armament, and Aerospace, Military University of Technology, 00-908 Warsaw, Poland
* Correspondence: marta.walenczykowska@wat.edu.pl (M.W.) adam.kawalec@wat.edu.pl (A.K.)

Abstract: This article analyses the possibility of using the Analytic Wavelet Transform (AWT) and the Convolutional Neural Network (CNN) for the purpose of recognizing the intrapulse modulation of radar signals. Firstly, the possibilities of using AWT by the algorithms of automatic signal recognition are discussed. Then, the research focuses on the influence of the parameters of the generalized Morse wavelet on the classification accuracy. The paper's novelty is also related to the use of the generalized Morse wavelet (GMW) as a superfamily of analytical wavelets with a Convolutional Neural Network (CNN) as classifier applied for intrapulse recognition purposes. GWT is used to obtain time–frequency images (TFI), and SqueezeNet was chosen as the CNN classifier. The article takes into account selected types of intrapulse modulation, namely linear frequency modulation (LFM) and the following types of phase-coded waveform (PCW): Frank, Barker, P1, P2, and Px. The authors also consider the possibility of using other time–frequency transformations such as Short-Time Fourier Transform(STFT) or Wigner–Ville Distribution (WVD). Finally, authors present the results of the simulation tests carried out in the Matlab environment, taking into account the signal-to-noise ratio (SNR) in the range from −6 to 0 dB.

Keywords: radar signal recognition; artificial neural network (ANN); continuous wavelet transform (CWT); analytic wavelet transform (AWT); analytic Morse wavelet; intrapulse modulation recognition; feature extraction; phase-coded waveforms

Citation: Walenczykowska, M.; Kawalec, A.; Krenc, K. An Application of Analytic Wavelet Transform and Convolutional Neural Network for Radar Intrapulse Modulation Recognition. *Sensors* **2023**, *23*, 1986. https://doi.org/10.3390/s23041986

Academic Editors: Janusz Dudczyk, Piotr Samczyński and Ram M. Narayanan

Received: 14 December 2022
Revised: 30 January 2023
Accepted: 6 February 2023
Published: 10 February 2023

1. Introduction

Nowadays, Electronic Warfare (EW) is an important element of battlefields. Information about the location of hostile emission sources allows for effective mission planning and ensuring the safety of one's own resources. At the same time, an increasing number of emissions and extensive research on the implementation of new types of radar waveform [1,2] reflect the complexity of the source identification problem and require the use of flexible solutions allowing for adaptation to changing conditions. For this reason, Artificial Intelligence (AI) algorithms, and in particular Machine Learning (ML) methods, which can be trained with new data appearing during the operations, seem to be an inseparable element of a modern EW system. However, in order for the application of ML to be effective, appropriate analyses and research works related to signal processing and the feature extraction process should be carried out. Recently, the use of TFI obtained with STFT, WVD or continuous wavelet transform (CWT) is often considered [3–6]. The proposal for a radar signal recognition method based on TFI and high-order spectra analysis is presented in [7].

A very important aspect that should be considered when using TFI for signal recognition algorithms is the analysis of the influence of the type, shape, and length of the window, in the cases of STFT or WVD, or the type and parameters of wavelet in the case of CWT, specifically AWT. Transformation parameters, when selected incorrectly, may significantly affect the classification capabilities. For this reason, in this article, the authors presented their simulation studies related to the influence of generalized Morse wavelet (GMW) parameters on the classification accuracy of the proposed method. This allows to obtain comparable or even higher accuracy than when using WVD, which is currently considered very effective in this respect [6]. At the same time, the use of GWT is characterized by much lower computational complexity, resulting in a reduction in time in both the classification process and the network learning. It is worth noticing that all the simulation tests have been carried out using smoothed pseudo-WVD (SPWVD) with a Kaiser window of various shape parameters (α).

The number of algorithms based on TFIs and deep learning proposed in the literature, e.g., [8–10] has been dynamically increasing in recent years. However, there is a visible lack of analysis in the field of parameters values selection for proposed time–frequency transforms. The type of applied atom (such as wavelet) or smoothing window used in energy distributions determines the possibility of adjusting to certain characteristics of the signal or the required level of interference reduction. For different parameter values, the same differences are observed in recognition ability for selected waveform types, so it is considered by authors using a set of networks trained with different wavelet types and parameter values and/or adding other methods used for signal recognition purposes and then applying data fusion methods. The problem of intrapulse modulation recognition with a fusion network is presented in [4]. Data fusion techniques are successfully used in areas such as tracking systems [11] or multisensory systems for unnamed aerial vehicles (UAV) detection [12].

An integral part of the EW system is signal intelligence (SIGINT). In the case of communications intelligence (COMINT) systems, the key information is the type and level of the modulation system used by the enemy. Intercepting this information not only enables to identify the source but also facilitates demodulation, decoding, and decryption of the transmitted signal. Typically, the ability to recognize common modulation types such as Phase Shift Keying (PSK), Frequency Shift Keying (FSK), Amplitude Shift Keying (ASK) or Quadrature Amplitude Modulation (QAM) is considered [13,14]. There are also studies taking into account the possibility of detection of OFDM transmission, among other communication signals [15,16]. In the case of electronic intelligence (ELINT), the main task of the system is to detect, classify, and determine the location of the emission sources other than communication. Therefore, these are particularly radars. The structure and parameters of the radar signals depend primarily on their intended use, no matter if they are early warning radars, short-, medium-, and long-range missile systems radars, as well as aircraft radars, jamming systems, or any other. Typical radar waveforms considered in most studies are continuous wave (CW) or pulses with LFM, stepped frequency modulation (SFM), and phase coding (PC) or traditional pulses with no intrapulse modulation applied. However, more complex types of waveform such as those with nonlinear frequency modulation or mixed signals are increasingly being considered [2,16,17].

In ELINT systems, many emission source parameters, such as radio frequency (RF), direction of arrival (DOA), time of arrival (TOA), pulse width (PW), pulse repetition frequency (PRF), intrapulse modulation, etc., are determined. These parameters are stored in the database and constitute the basis of the process of recognizing the emissions detected by the EW system. In order to determine the basic parameters of radar signals, it is necessary to perform signal processing and the selection of features allowing for effective classification. This issue is a dynamically developing area of research [18–22], while measuring the parameters of specific radar signals in real time still remains a challenge.

The radar emission identification process is usually carried out using a knowledge-based approach. In [18], methods for determining specific radar signal parameters (signatures) are discussed. In [20], the authors analyze the role of estimation accuracy of the arrival time of each step of the pulse. As it turns out, it may affect the determination correctness of such parameters as PW and PRF. Moreover, the application of the wavelet transform (WT) as well as Haar wavelet as tools for sorting radar signals has also been proposed. Compared with the traditional pulse repetition interval algorithm based on a statistics histogram, the method based on WT is characterized by the high accuracy of the arrival time. In [19], three characteristic parameters, namely CWT eigenvalue, frequency domain moment kurtosis coefficient, and frequency domain moment skewness coefficient, have been used in order to recognize the radar signal. The following signals have been considered: polynomial phase signal, pseudocode phase modulation and sinusoidal frequency modulation, product composite of pseudocode phase modulation and LFM, and convolution composite of pseudocode phase modulation and LFM. For a signal-to-noise ratio (SNR) higher then 0 dB, the probability of signal recognition is claimed to be greater then 98%.

The aim of this article is to present an algorithm for automatic intrapulse modulation recognition with use of an AWT and CNN. The proposed solution is based on the use of generalized Morse wavelet (GMW) as a superfamily of analytic wavelets. The properties of the GMW are described in Section 2.3. A significant advantage of using CWT for signal recognition purposes is the possibility of using CWT coefficients at individual stages of signal analysis, e.g., to estimate selected parameters of a radar signal. The CWT is successfully used in algorithms for sorting radar pulses and determining parameters such as pulse repetition time (PRT), pulse width (PW) or time of arrival (TOA) [20].

The research presented in the paper brings a new contribution to the area of radar signal recognition techniques with the use of TFI. In particular, the authors consider:

- Using AWT (with GMW applied) for radar intrapulse recognition purposes instead of popular WVD, SPWVD or STFT;
- Performance metrics comparing methods with AWT and SPWVD used as TFI;
- The influence of GMW parameters values on classification accuracy;
- Applying SqueezeNet as a CNN classifier.

A significant part of the conducted simulation works has been focused on the study of the influence of generalized Morse wavelet parameters on the properties of the analyzed method. The simulation results are presented in Section 3. The usefulness of AWT is mainly due to the ability to observe the instantaneous amplitude, phase, and frequency of signals simultaneously. The appropriate selection of the wavelet parameters allows, in turn, to obtain the required resolution in frequency and to emphasize specific features of the analyzed signals. The simulations, necessary to test the proposed method and confirm its potential effectiveness, were carried out in the Matlab environment.

The use of TFI enables the simultaneous observation of signal features corresponding to the nature of changes in amplitude, frequency or phase over time. This allows for the complete or partial replacement of traditionally used features, calculated usually with the use of FFT, higher-order statistics, instantaneous amplitude, frequency and phase parameters, cepstral analysis, phase constellation, frequency histogram, etc. Moreover, the algorithms of time–frequency signal decompositions with lower computational complexity and lower memory requirements are presented in the literature [23,24]. The computational complexity of time–frequency distributions is presented in [25]. CNNs, on the other hand, are successfully used in image recognition. This is the main motivation for authors to use one of them as the classifier. According to the authors' knowledge, at present, there are CNN implementations with a significantly reduced structure. An example of such a CNN is SqueezeNet, the implementation of which is available in the Matlab environment. The primary advantages of SqueezeNet, according to [26], are, e.g.,:

- More efficient distributed training;
- Feasible FPGA and embedded deployment;
- Less overhead when exporting new models to clients.

The SqueezeNet structure implementation in Matlab is 18 layers deep. The network has an image input size of 227 by 227, so TFIs rescaling must be applied. A comparison of classification accuracy using TFI with other CNN-type structures is presented in [8].

2. Materials and Methods

The number of studies related to the possibility of using ML for signal recognition for electronic warfare and, e.g., cognitive radio, has significantly increased in recent years. When analyzing the literature on signal analysis for ELINT systems, one can see the division of the proposed algorithms into two areas of application. The first one focuses on the analysis of signals, with their duration covering a certain number of symbols or a sequence of radar pulses, and determining their characteristic parameters such as PRI, PW, TOA, etc. [18,20–22,27,28]. The second concerns the problem of recognizing intrapulse modulation, which is another important parameter characterizing the [5,29,30] emission source. In both cases, the usefulness of using AI algorithms is emphasized.

There are propositions of algorithms using CWT and ML for recognizing modulations of typical signals used in communications such as ASK, PSK, FSK or QAM [13,14,31–34]. Moreover, in the field of ELINT, more and more papers are published related to the application of AI for the purpose of recognizing selected types of radar waveforms [3,6,9,10,28,29,35].

Of particular interest is the application of CNNs, where TFIs obtained using STFT, WVD or CWT are fed to the network input. In this context, an important problem that requires special attention is the appropriate selection of the type of transformation and its parameters, so that the input data in the form of TFIs contain features that ensure the required classification accuracy of selected signals. The following subsections present issues related to the usefulness of using CWT to generate TFI and CNN as an classifier of selected pulsed radar signals, with particular emphasis on the influence of the parameters of the generalized Morse wavelet on the classification accuracy.

2.1. Radar Waveforms

Applying the pulse compression technology has resulted in a significant increase in the number of signals used in radar technology. Currently, there is a noticeable increase in the number of research works devoted to specific forms of radar signals [1,2], i.e., more and more complex types of radar waveforms ensuring the best possible resolution in time (range) and Doppler frequency (velocity), as well as a high compression ratio. This involves, among other things, the analysis of the ambiguity function, which is also a very useful way of analyzing signal properties in the time–frequency space.

The general division of radar waveforms most often taken into account in the research area for automatic recognition of radar signals is presented in Figure 1. Traditional pulsed radar waveforms include rectangular pulses (RP) with continuous wave (CW), LFM, and phase-coded waveforms (PCW). Other types of often considered radar waveforms are NLFM and SFM.

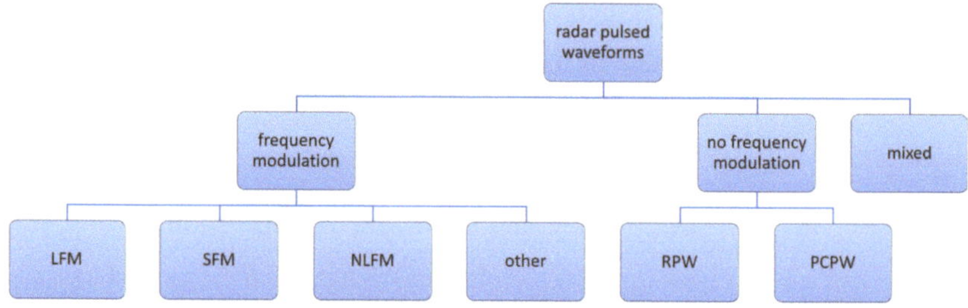

Figure 1. General division of radar waveforms.

To an NLFM group of signals, authors also assigned quadratic frequency modulation (QFM), sinusoidal frequency modulation (SiFM), etc. There are also works [15] considering the application of OFDM signals for radar purposes. In [2], there is an algorithm for identifying so-called exotic modulations, which include signals modeled as a combination of LFM and Biphase modulation (BPM). In [30], there is also recognition of the same hybrid waveforms as LFM-BPSK and FSK-BPSK. An analysis of combined signals is also presented in [1], in which authors took into account a group called "mixed", which includes hybrid modulation types (Figure 1). All listed waveforms can be considered in continuous or pulsed form.

For the purpose of this article, the following types of signals were taken into account for simulation: LFM, RP, and PC waveform. In the studies carried out so far, presented in [31], a significant problem was distinguishing between the different types of signals with phase modulation (PM). In most of the presented research results, the classification accuracy in this group of signals is much worse for both COMINT (M-ary PSK and M-QAM signals) [36] and ELINT (PCW) [37]. In order to emphasize the effectiveness of the proposed method, the following types of codes were included in the phase-coded waveform (PCW) group: Barker, Frank, P1, P2, and Px. According to [38], the complex envelope of the PC pulse, with duration T and M bits in the pulse, is given by equation:

$$x(t) = \frac{1}{\sqrt{T}} \sum_{m=1}^{M} x_m rect\left[\frac{t - (m-1)t_b}{t_b}\right] \tag{1}$$

where M is called "chip", $t_b = \frac{T}{M}$ is chip number, $x_m = exp(j\phi_m)$, the set of M phases $[\phi_1, \phi_2, \ldots, \phi_M]$ is the phase code associated with $x(t)$, and $rect()$ is a rectangular function. The first family of phase codes taken into account in our research is Barker. There were $N = 13$ lengths of code chosen for simulations, witch ensures the lowest level of ambiguity function side-lobes ($\frac{1}{N}$) in this code family. There are no longer Barker codes found [39]. The choice of polyphase codes used for simulations (Frank, P1, P2, and Px) is motivated by the fact that difficulties in separating them were expected. The P1, P2, and Px codes are derived from the Frank code [38,39], and they are considered by authors as one family.

The elements of the polyphase codes can be complex depending on the value ϕ_m. For Frank polyphase codes, witch have a length that is perfect square ($M = L^2$, where L is integer), the definition of its elements $s_m (1 \leq m \leq M)$ according to [37,38] is

$$s_{(n-1)L+k} = exp(j\phi_{n,k}) \tag{2}$$

where $1 \leq n \leq L, 1 \leq k \leq L$ and:

$$\phi_{n,k} = 2\pi(n-1)(k-1)/L \tag{3}$$

According to Equation (2), for $L = 4$, we have phase values:
$$\begin{bmatrix} 0 & 0 & 0 & 0 & 0 & \frac{\pi}{2} & \pi & \frac{3\pi}{2} & 0 & \pi & 0 & \pi & 0 & \frac{3\pi}{2} & \pi & \frac{\pi}{2} \end{bmatrix}$$

The Frank, P1, P2, and Px codes are applicable only for perfect square length ($M = L^2$). The phase of P1, P2, and Px is given by [37,38]:

$$P1: \quad \phi_{n,k} = \frac{2\pi}{L}[(L+1)/2 - n][(n-1)L + (k-1)] \tag{4}$$

$$P2: \quad \phi_{n,k} = \frac{2\pi}{L}[(L+1)/2 - k][(L+1)2 - n] \tag{5}$$

only for even L, and:

$$Px: \quad \phi_{n,k} = \begin{cases} \frac{2\pi}{L}[(L+1)/2 - k][(L+1)2 - n], & L \quad even \\ \frac{2\pi}{L}[L/2 - k][(L+1)2 - n], & L \quad odd \end{cases} \tag{6}$$

It should be noted that in the case of the P_x code for even L, $\phi_{n,k}$ takes the values as for the $P2$ code. For this reason, the simulation parameters should be carefully selected to avoid problems with distinguishing the above codes. For the same reason, the classification was carried out only taking into account the P_x code for odd L. It was show in [37], where for radar waveform recognition purposes Choi–Williams Distribution (CWD) was considered, that P1 and P2 codes get confused with each other. A similar problem was observed in studies where individual levels of PSK were classified for the communication system recognition algorithm [31]. Similarly, it is expected that, in noisy conditions, the differences in this group of phase codes can be blurred.

2.2. The Time–Frequency Transforms in Signal Analysis

Time–frequency (or decomposition) transforms are a very convenient way to present features of different types of waveforms simultaneously in time and frequency domains. The increase in interest in CNNs, particularly observable in the recent years, has made TFIs more and more frequently used schemes in solving the problem of signal classification [3,6]. However, not enough attention is always paid to the influence of the chosen transform parameters on the possibility of enhancing the characteristic features of the signals. This, in turn, may lead to the loss of valuable information and, as a consequence, affect the classification process efficiency. Similarly to the classic approach, the stage of feature extraction determines the effectiveness of the recognition algorithm.

Some of the most popular transforms such as STFT, WVD, and CWT are presented in [40,41]. We can divide them in two classes of solutions: atomic decompositions and energy distributions. The first class includes STFT and CWT, while the second one includes WVD, pseudo-WVD, and SPWVD.

STFT according to [42] is defined as

$$STFT_x(t,v) = \int_{-\infty}^{\infty} x(u)h^*(u-t)e^{-j2\pi vu}du \tag{7}$$

where $h(t)$ is a short time analysis window localized around $t = 0$ and $v = 0$. The multiplication by the relatively short window $h^*(u-t)$ effectively suppresses the signal outside a neighborhood of time point $u = t$, so the STFT is a "local" spectrum of the signal $x(u)$ around t. The time resolution of the STFT is proportional to the effective duration of the analysis window h. In turn, the resolution of the STFT in the frequency is proportional to the effective bandwidth of the analysis window h. According to the above, using the STFT cannot achieve good resolution in time and frequency simultaneously. A good time resolution requires a short window $h(t)$, and a good frequency resolution requires a narrow-band filter, and so a long window $h(t)$.

The CWT of a signal $x(t)$ is defined as [42,43]

$$CWT(t,a) = \int_{-\infty}^{\infty} x(u)\psi_{t,a}^*(u)du = \frac{1}{\sqrt{|a|}} \int_{-\infty}^{\infty} x(u)\psi^*\left(\frac{u-t}{a}\right)du \tag{8}$$

where the $\psi(u)$ is called mother wavelet and a is the scaling constant. $\psi_{t,a}^*$ is called baby wavelet and is the translated and scaled version of $\psi(u)$.

CWT projects a signal $x(t)$ on a family of zero-mean functions, called the wavelets, which are translated and dilated versions of the elementary function, called the mother wavelet. When the scale factor is being modified, the duration and the bandwidth of the wavelet are both changed, but its shape remains the same. In contrast to the STFT, which uses a single analysis window, the CWT uses short windows at high frequencies and long windows at low frequencies. This partially overcomes the resolution limitation of the STFT. STFT and CWT are linear transforms of the signal. Another way of signal analysis consists in distributing the energy of the signal along the two variables of time and frequency. This leads to energy time–frequency distributions, which are naturally quadratic transforms of time and frequency. For STFT, we have a spectral energy density of the locally windowed signal $x(u)h * (u-t)$, defined as [42,43]

$$S_x(t,v) = |STFT_x(t,v)|^2 \tag{9}$$

According to Equation (5) *spectrogram* is a real-valued and non-negative distribution and satisfies the global energy distribution property:

$$\int_{-\infty}^{\infty} \int_{-\infty}^{\infty} S_x(t,v)\,dt\,dv = E_x \tag{10}$$

In the *CWT* case, the scalogram can be defined [44]:

$$T_x(t,a) = |CWT(t,a)|^2 \tag{11}$$

CWT also preserves energy:

$$\int_{-\infty}^{\infty} \int_{-\infty}^{\infty} T_x(t,a)\,dt\,\frac{da}{a^2} = E_x \tag{12}$$

An example of scalograms for selected waveforms are presented on Figures 2–4.

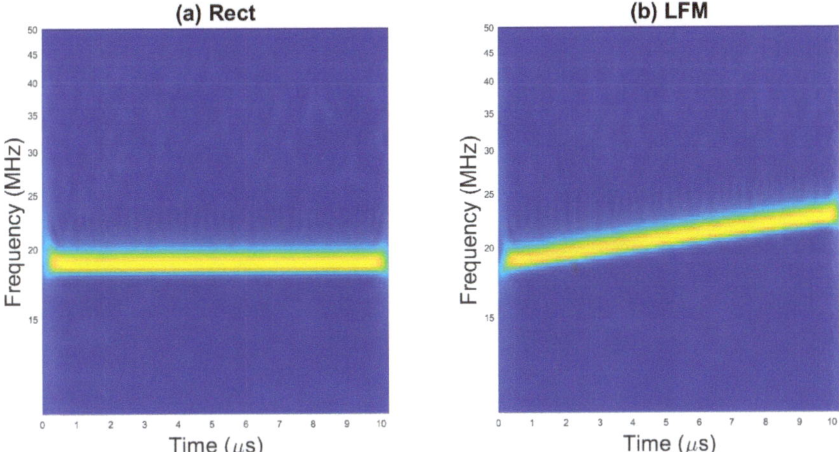

Figure 2. Scalogram for (**a**) rectangular pulse and (**b**) LFM.

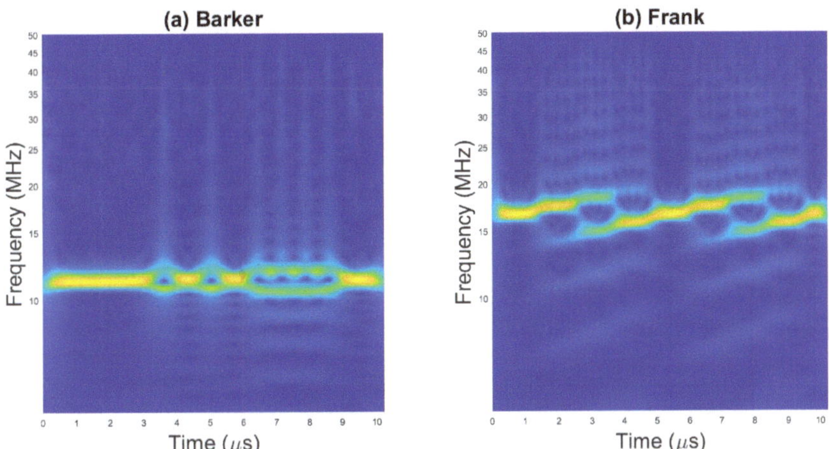

Figure 3. Scalogram for (**a**) pulses with Barker code and (**b**) pulses with Frank code.

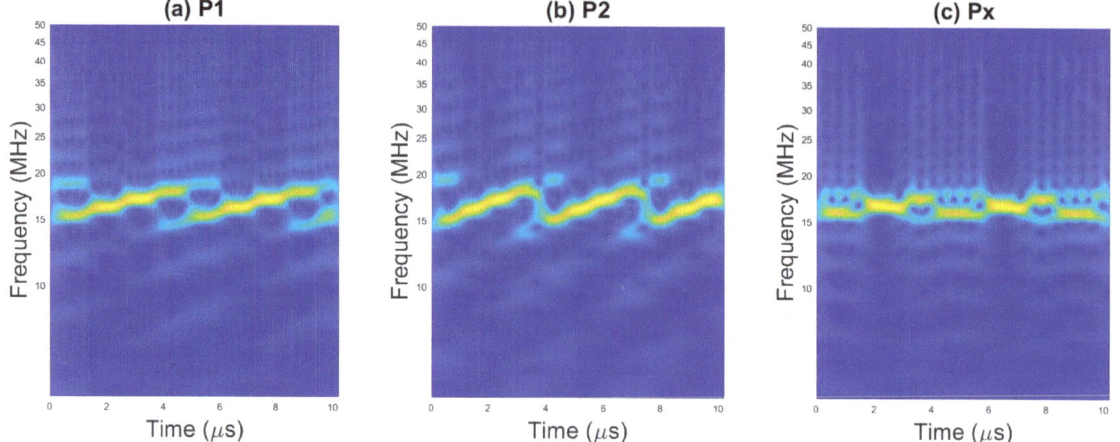

Figure 4. Scalogram for pulses with (**a**) P1 code, (**b**) P2 code, and (**c**) P_x code.

The second class of time–frequency representations is energy distributions. Some of the most commonly used energy distributions for signal recognition are WVD and CWD. The WVD is defined as [42]

$$WV(t,v) = \int_{-\infty}^{\infty} x(t + \frac{\tau}{2})x^*(t - \frac{\tau}{2})e^{-jv\tau}d\tau \tag{13}$$

and is a nonlinear decomposition, so the spectrum of two combined signals is not the sum of their spectra but includes so-called cross-spectrum:

$$WV_{x+y}(t,v) = WV_x(t,v) + WV_y(t,v) + 2\Re WV_{x,y}(t,v) \tag{14}$$

where:

$$WV_{x,y}(t,v) = \int_{-\infty}^{\infty} x(t + \frac{\tau}{2})y^*(t - \frac{\tau}{2})e^{-jv\tau}d\tau \tag{15}$$

The WVD interference terms will be nonzero regardless of the time–frequency distance between the two signal terms [42]. This could be troublesome and make it difficult to visually interpret the WVD image. A common way to reduce interferences is to use a new distribution, called *pseudo-WV*, defined as:

$$PWV(t,v) = \int_{-\infty}^{\infty} h(\tau)x(t + \frac{\tau}{2})x^*(t - \frac{\tau}{2})e^{-jv\tau}d\tau \tag{16}$$

where $h(t)$ is regular window. The windowing operation is equivalent to frequency smoothing. Another modification of WVD is the smoothed pseudo-Wigner–Ville Distribution (SPWVD), defined as

$$SPWV(t,v) = \int_{-\infty}^{\infty} h(\tau) \int_{-\infty}^{\infty} g(s - t)x(s + \frac{\tau}{2})x^*(s - \frac{\tau}{2})ds \ e^{-jv\tau}d\tau \tag{17}$$

According to [42], the compromise of the spectrogram between time and frequency resolutions is in case of SPWVD being replaced by a compromise between the joint time–frequency resolution and the level of the interference terms. Stronger smoothing in time and/or frequency runs into poorer resolution in time and/or frequency.

The examples of TFI images for selected radar waveforms obtained with WVD are presented in Figures 5–7.

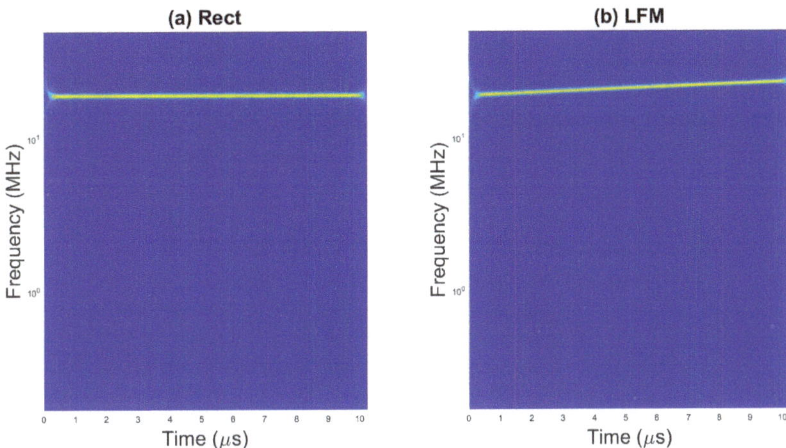

Figure 5. WVD for (**a**) rectangular pulse; (**b**) LFM.

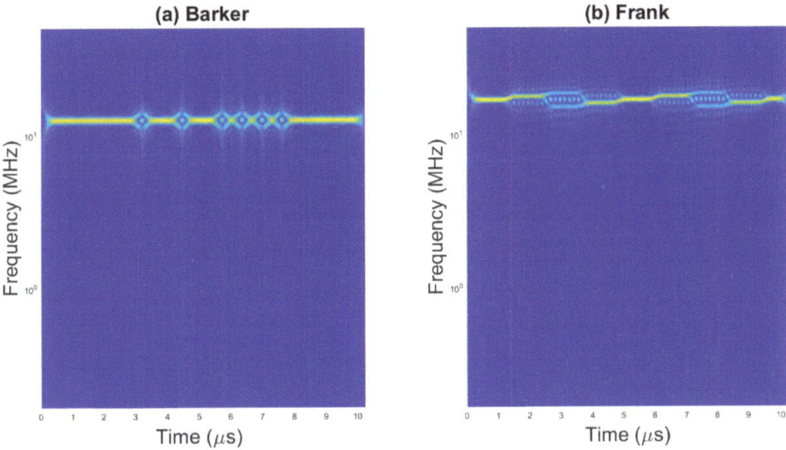

Figure 6. WVD for (**a**) pulse with Barker code; (**b**) pulses with Frank code.

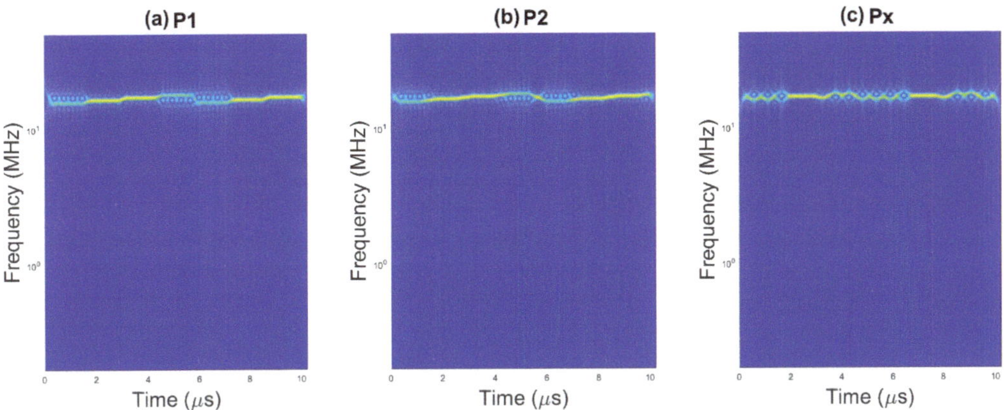

Figure 7. WVD for pulses with (**a**) P1 code, (**b**) P2 code, and (**c**) P_x code.

2.3. The Analytic Wavelet Transform (AWT) and Generalized Morse Wavelet (GMW)

The continuous wavelet transform (CWT) enables extraction of transient information associated with amplitude/frequency changes and phase shifts, which are characteristics of modulated signals. According to [45,46], it is generally the most useful to describe oscillatory signals and time-localized events in noisy environment, where $1/s$ normalization is more appropriate. Taking the above into account, the definition of the CWT of a signal $x(t)$ takes the form:

$$CWT(\tau, a) = \frac{1}{a} \int_{-\infty}^{\infty} x(t)\psi^* \left(\frac{t - \tau}{a} \right) dt \tag{18}$$

where the $\psi(t)$ is called mother wavelet and a is the scaling constant. ψ_a^* is called baby wavelet and is the translated and scaled version of $\psi(t)$. In this case, rescaling time in the input signal as $x(\frac{t}{\rho})$ rescales both the time and the scale of CWT but without changing its magnitude. For analytic wavelets considered in [45–47], we have $\psi(\omega) = 0$ for $\omega < 0$, which means that wavelets vanish for negative frequencies. The AWT is represented in the frequency domain as

$$CWT(\tau, a) = \frac{1}{2\pi} \int_{0}^{\infty} \Psi^*(a\omega)X(\omega)e^{j\omega\tau}d\omega \tag{19}$$

For the analytic wavelet $\Psi(\omega)$, maximum amplitude occurs in the frequency domain at $\omega = \omega_\psi$, which is called the *peak frequency*. In the case of using an analytic wavelet, such as GMW, there is AWT term used. According to [45], if the value of the wavelet in the peak frequency is set to $\Psi(\omega_\psi) = 2$, then for signal $x(t) = a_0 cos(\omega t)$, we obtain result $|W_\psi(t, \omega_\psi/\omega_0)| = |a_0|$.

During the simulation studies, the generalized Morse wavelet was taken into account. According to [45,47,48], the Morse wavelet is defined as follows:

$$\psi_{\beta,\gamma}(\omega) = U(\omega)a_{\beta,\gamma}\omega^\beta e^{-\omega^\gamma} \tag{20}$$

where $a_{\beta,\gamma}$ is a normalization constant, $U(\omega)$ is the unit step function, and β and γ are parameters controlling the wavelet form. The normalization constant is defined in [47] as

$$a_{\beta,\gamma} \equiv 2(\frac{e\gamma}{\beta})^{\frac{\beta}{\gamma}} \tag{21}$$

It follows from the analysis presented in [48] that, by varying the β and γ parameters, the generalized Morse wavelets can take a wide variety of forms. For example, the $\gamma = 1$ family corresponds to the Cauchy (or Paul) wavelet, the $\gamma = 2$ correspond to analytic Derivative of Gaussian wavelets, and $\gamma = 3$ corresponds to Airy wavelets family [47].

On Figures 8 and 9 are presented the TFIs of radar signal with Frank code ($N = 16$), obtained with AWT and Morse wavelet for different values of γ and β and two values for SNR. There is a significant impact of the parameters of the Morse wavelet on the form of CWT and on the way in which TFI is affected by the increased level of noise. On this basis, it can be assumed that a higher accuracy of the classification can be obtained for selected parameters γ and β. The simulation results described in the following subsections confirm the above statement.

In Matlab Wavelet Toolbox parameters for parameterized analytic Morse wavelet are defined as γ and the time–bandwidth product $P^2 = \beta\gamma$, and they correspond to the analysis in [45,47,48].

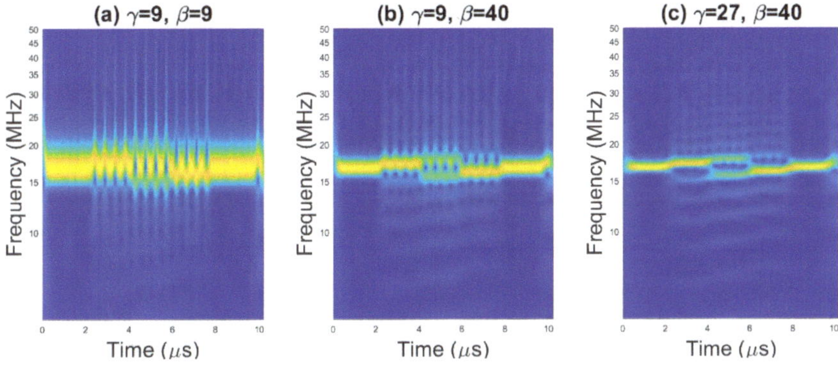

Figure 8. Scalogram for the pulse with Frank code for SNR = 20 dB.

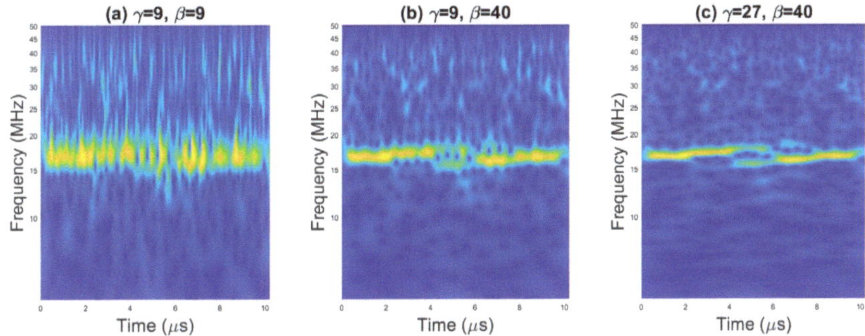

Figure 9. Scalogram for the pulse with Frank code for SNR = 0 dB.

3. Results

In order to carry out the research works, the following equipment (hardware and software) was used:

- Dell Precision 3561, i7-1180H, 32GB RAM, NVidia T600, Win11, manufacturer: Dell, Warsaw, Poland;
- MATLAB Version: 9.12.0.2039608 (R2022a) Update 5 with Toolboxes, manufacturer: MathWorks, Inc., Natick, MA, USA.

The simulations were performed for recognition between continuous wave (CW), LFM, and phase-coded waveforms. For signals with phase coding, the same number of chips in one pulse was selected for the *Frank*, *P1*, *P2* code. Due to the fact that in the *Px* code for even values of L the phase $\phi_{n,k}$ takes the same values as in *P2* code, the L was set to the odd one. The simulation parameters of phase-coded waveforms are listed in Table 1.

Table 1. Phase-coded waveform parameters.

Code Name	Chip Number
Barker	13
Frank	16
P1	16
P2	16
Px	9

The conducted research was based on AWT and the use of the Morse wavelet. The authors focused on the analysis of the impact of the wavelet parameters on the obtained classification accuracy. The general simulation parameters are presented in Table 2.

Table 2. General simulation parameters.

Parameter Name	Parameter Value
Carrier frequency F_c	16.7 MHz to 20 MHz
Sample frequency F_s	100 MHz
Bandwidth (for LFM Waveforms)	from 5 to 6.25 MHz
SNR	from -6 to 0 [dB]
Rician Channel Parameters	PathDelays [0 1.8 3.4]/Fs, AveragePathGains [0–2–10]
Number of pulses in signal	1–2
Number of samples (in observation window)	1024
Number of each type of signal	3000
Number of cycles per phase code (in one Chip)	[7, 8] for Barker, *Px*, [4, 5] for Frank, *P1*, *P2*
Wavelet type	Morse

The simulation tests were carried out taking into account selected values of the γ parameter, and for each of them, the β parameter was changed in the loop. Two parameters per set were considered. The first one was with the analogy with Morse wavelet parameters presented in [45,47,48] and listed in Table 3. Additionally, $\beta = 40$ was applied. The view of the wavelets corresponding to the parameters γ and β from Table 3 is shown in the Figure A1 in Appendix A.

Table 3. The γ and β values taken into account in simulation.

γ	$P^2 = \gamma\beta$ (β)
3	9 (3), 27 (9), 81 (27), 120 (40)
9	27 (3), 81 (9), 243 (27), 360 (40)
27	81 (3), 243 (9), 729 (27), 1080 (40)

The second wavelet parameters set is presented in Table 4. The number of oscillations and duration of wavelets was increased (according to change in γ and β) to find out if they could better match the recognized waveform. However, there was no significant increase in classification accuracy noticed.

Table 4. Additional γ and β values taken into account in simulation.

γ	$P^2 = \gamma\beta$ (β)
9	81 (9), 99 (11), 117 (13), 135 (15), 153 (17), 171 (19), 189 (21), 207 (23), 225 (25), 243 (27), 261 (29), 279 (31), 297 (33), 315 (35), 333 (37), 351 (39)
27	728 (27), 783 (29), 831 (31), 891 (33), 945 (35), 999 (37), 1053 (39)
50	1500 (30), 1550 (31), 1600 (32), 1650 (33), 1700 (34), 1750 (35), 1800 (36), 1850 (37), 1900 (38), 1900 (39), 1950 (40)

Only for $\gamma = 50$ and $\beta = 37$, the same small increase in classification accuracy was observed. Lower classification correctness also occurred for the same types of PCW: P1 and P2 codes. Generally, classification accuracy remained at the level of about 85–95% (Table 5).

Table 5. Total accuracy for parameters corresponding to Table 3.

Parameter Value	$\beta = 3$	$\beta = 9$	$\beta = 27$	$\beta = 40$
$\gamma = 3$	83.7%	90.1%	93.6%	96.4%
$\gamma = 9$	86.6%	94.8%	**98.2%**	97.3%
$\gamma = 27$	86.1%	91.7%	93.8%	91.9%

In spite of the lack of significant improvement, it is worth noticing that frequency resolution is much better for higher wavelet parameters values. This can be useful for frequency-modulated waveform recognition and methods for optimal scale selection for emitter parameters estimation e.g., pulse duration (PD), sweep frequency, pulse repetition interval (PRI), time of arrival (TOA), and others.

The confusion matrix for parameters with the highest total accuracy corresponding to Table 3 is presented in Figure 10. The overall classification accuracy for $\gamma = 9$, $\beta = 27$ was 98.2%.

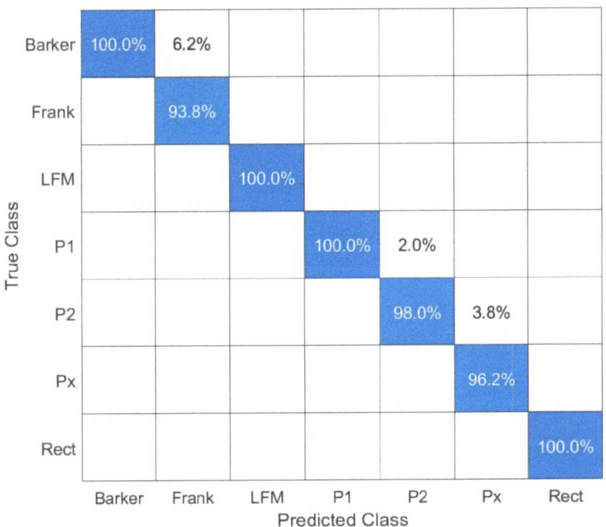

Figure 10. Confusion matrix for AWT with Morse wavelet parameters: $\gamma = 9$, $\beta = 27$.

Additionally, in Figure 11, receiver operating characteristic (ROC) curves for the method with AWT applied (with Morse wavelet parameters: $\gamma = 9$, $\beta = 27$) are presented.

The shape of the one-versus-all ROC curve for Barker confirms that their classifier for high-value true positives gives a significant level of false positives, which corresponds to the confusion matrix presented in Figure 10. In the case of Frank, P1, and P2 signals, the one-versus-all ROC curves indicate slightly better performance.

For comparative purposes, the classification process was carried out with SPWVD as TFI and with the Kaiser window with the same parameter values for the smoothing window in frequency and time: window length L = 101 and shape factor N with the selected values. The classification accuracy obtained for each value of N is presented in Table 6.

Table 6. Total accuracy for parameters for selected values of shape factor N.

Shape Factor N	0.5	1	1.5	2	3	4	5
Accuracy [%]	95.9%	96.3%	94.4%	95.6%	94.4%	94.9%	88.9%

A similar correctness of the classification was obtained for shape factor N = 1, and the confusion matrix is shown in Figure 12. The overall classification accuracy for shape factor N was 96.3%. Lower classification accuracy occurred for P1 and P2 codes, as well as some problems with distinguishing the waveform with the Barker and Frank code. This is a similar situation as for the method with AWT applied.

Figure 13 presents ROC curves for the method with SPWVD applied. Similarly, as for the method with AWT applied, there are observable levels of false positives for Barker.

In the case of Frank, P1, and P2 signals, the one-versus-all ROC curves indicate slightly worse classification possibilities, almost the same as for the method with AWT applied.

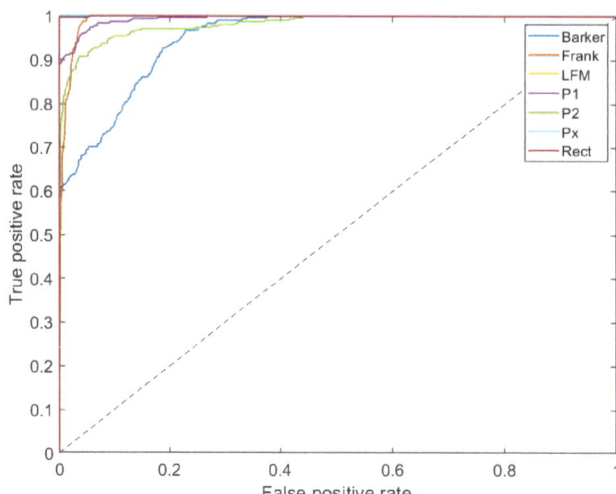

Figure 11. One-versus-all ROC curves for each class for the method with AWT with Morse wavelet parameters: $\gamma = 9$, $\beta = 27$.

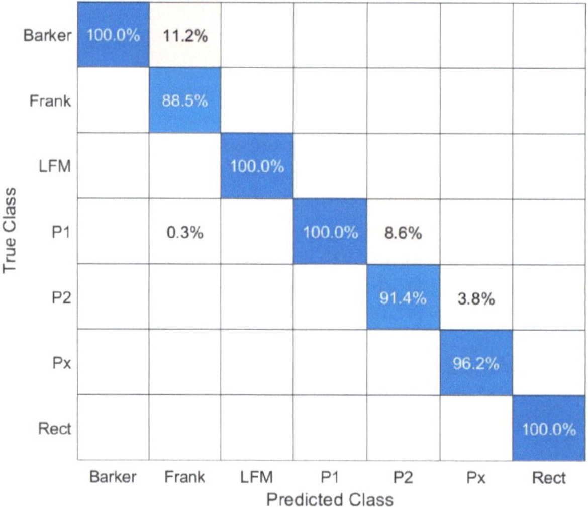

Figure 12. Confusion matrix for WVD with Kaiser window parameters $L = 101$, $\alpha = 1$.

Tables 7 and 8 list the metrics obtained for both methods, with AWT and SPWVD applied for TFI calculation. Consequently, it is noticeable that the classifier has the ideal ability to distinguish LFM and typical rectangular radar signals (F1-score is equal 1).

Table 7. Classification report of the model with AWT applied (Precision, Recall, and F1-Score).

Signal Type	Barker	Frank	LFM	P1	P2	Px	Rect
Precision	0.9412	1	1	0.9800	0.9622	1	1
Recall	1	0.9373	1	1	0.9796	0.9615	1
F1-Score	0.9697	0.9677	1	0.9899	0.9708	0.9804	1

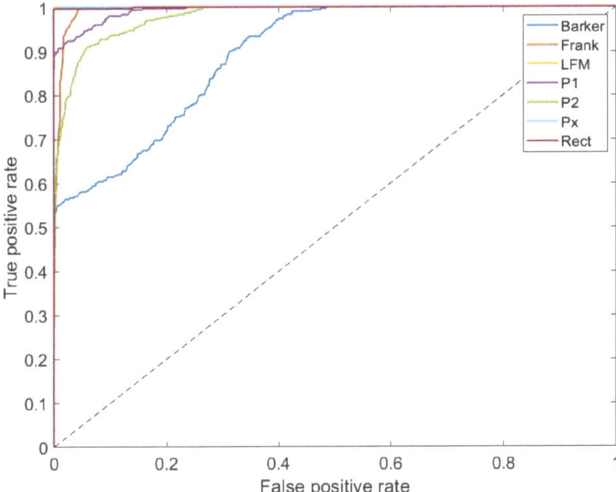

Figure 13. One-versus-all ROC curves for each class for the method with SPWVD with Kaiser window and $\alpha = 1$.

Table 8. Classification report of the model with SPWVD applied (Precision, Recall, and F1-Score).

Signal Type	Barker	Frank	LFM	P1	P2	Px	Rect
Precision	0.8992	1	1	0.9186	0.9596	1	1
Recall	1	0.8850	1	1	0.9143	0.9615	1
F1-Score	0.9469	0.9390	1	0.9575	0.9364	0.9804	1

4. Conclusions

The actual direction of the research works presented in this paper was drawn in one of our previous works [49]. The method of feature extraction based of AWT seems to be very versatile. The results presented in [31] show that using CWT and artificial neural networks (ANN) performs as an effective way to recognize communication signals such as ASK, PSK, FSK, and QAM. The usefulness of the wavelet transform in signal analysis and the more and more commonly used TFI prompted the authors to verify the possibility of using AWT (with parameterized GMW) in combination with a reduced CNN structure (SqueezeNet). Thus, the conducted research works were focused on the influence of the GMW parameters on classification accuracy. The obtained results confirm the satisfactory effectiveness of the proposed algorithm for SNR in the range from -6 to 0 dB. The highest obtained classification accuracy is 98.2% for $\gamma = 9$ and $\beta = 27$.

For comparative purposes, the classification process with TFIs based on SPWVD was conducted. It was observed that both algorithms allow for approximate classification accuracy. The advantage of CWT over SPWVD resides in the ability to use a filter bank to improve the efficiency of calculations. On the other hand, for SPWVD, the use of smoothing windows improves the readability of TFI but requires additional time, which was noticed during the simulation tests.

Future work will include the application of a wider range of selected signals with specific parameters such as pulse width, pulse envelope, pulse repetition time, intrapulse modulation, phase coding types, etc. It seems to be reasonable to consider applying adaptive algorithms for wavelet (in CWT/AWT) or smoothing window parameters selection. Another important problem is selecting the type of classifier. Neural network training is a time-consuming process. In the EW systems domain, there is a continuous need for fast methods. Therefore, faster adaptive methods, such as the one presented in [50], even if with relatively lower classification accuracy, are preferable over those that are slower with higher classification rates. Moreover, an interesting alternative worthy of consideration is information fusion methods, referred to in [11], which, apart from the neural-network based solutions, deliver analytical methods. This may be particularly attractive due to the mentioned speed requirement assessed on the optimizing methods.

Summarizing the presented considerations, the method based on AWT, applied for TFI calculation with the reduced CNN structure (e.g., SqueezeNet), would be more appropriate for hardware implementations then those with STFT or WVD (SPWVD) applied for TFI calculation and with other types of CNN (ResNet, ALexNet, etc.) as classifiers.

Furthermore, by selecting the appropriate wavelet parameters' values, it is possible to achieve comparable or even better performance test results than those presented in the references.

Author Contributions: Conceptualization, M.W. and K.K.; methodology, M.W. and K.K.; software, M.W.; validation, M.W.; formal analysis, M.W. and K.K.; investigation, M.W.; resources, M.W.; data curation, M.W.; writing—original draft preparation, M.W. and K.K.; writing—review and editing, M.W., K.K. and A.K.; visualization, M.W.; supervision, A.K.; project administration, A.K.; funding acquisition, M.W. All authors have read and agreed to the published version of the manuscript.

Funding: The methods and results presented in this paper have been obtained during research works conducted within the university research project entitled "Selected problems of detection, recognition and classification of signals as well as analysis of imaging and radar data". This work was financed by the Military University of Technology (Warsaw, PL), in 2022, under research project UGB-778.

Informed Consent Statement: Not applicable.

Data Availability Statement: Not applicable.

Conflicts of Interest: The authors declare no conflict of interest.

Appendix A

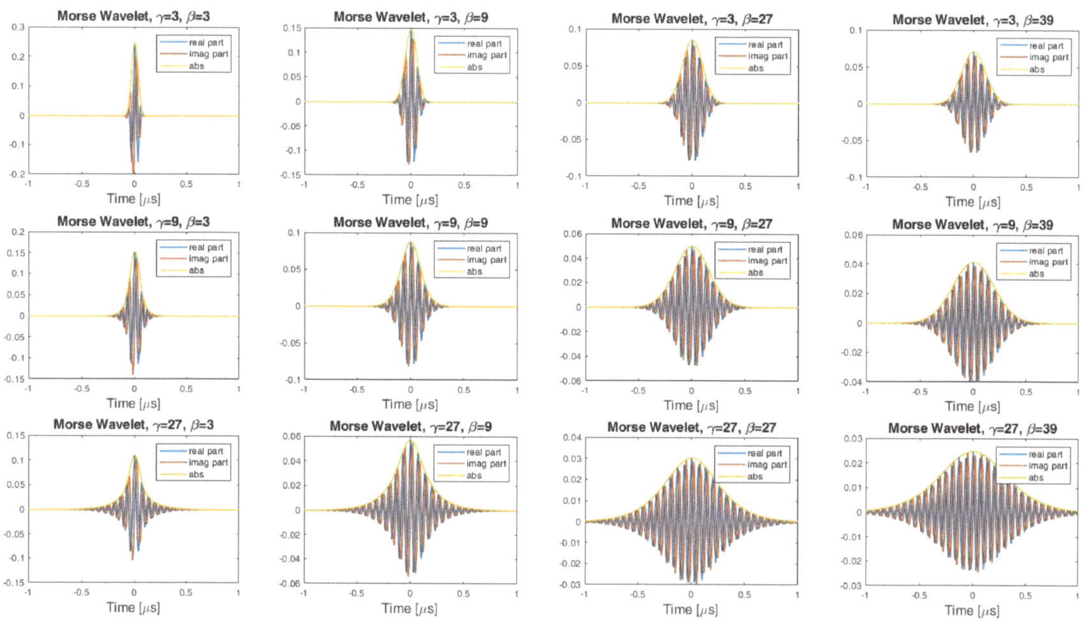

Figure A1. Morse wavelets corresponding to parameters from Table 3 for selected wavelet center frequency.

References

1. Li, H.; Zhao, J. Analysis of a combined waveform of linear frequency modulation and phase coded modulation. In Proceedings of the 2016 11th International Symposium on Antennas, Propagation and EM Theory (ISAPE), Guilin, China, 18–21 October 2016; pp. 539–541. [CrossRef]
2. Niranjan, R.; Rao, C. R.; Singh, A. Real-Time Identification of Exotic Modulated Radar Signals for Electronic Intelligence Systems. In Proceedings of the Emerging Trends in Industry 4.0 (ETI 4.0), Raigarh, India, 19–21 May 2021; pp. 1–4. [CrossRef]
3. Wang, C.; Wang, J.; Zhang, X. Automatic radar waveform recognition based on time-frequency analysis and convolutional neural network. In Proceedings of the IEEE International Conference on Acoustics, Speech and Signal Processing (ICASSP), New Orleans, LA, USA, 5–9 March 2017; pp. 2437–2441. [CrossRef]
4. Gao, L.; Zhang, X.; Gao, J.; You, S. Fusion Image Based Radar Signal Feature Extraction and Modulation Recognition. *IEEE Access* **2019**, *7*, 13135–13148. [CrossRef]
5. Wei, S.; Qu, Q.; Su, H.; Shi, J.; Zeng, X.; Hao, X. Intra-pulse modulation radar signal recognition based on Squeeze-and-Excitation networks. *Signal Image Video Process.* **2020**, *14*, 1133–1141. [CrossRef]
6. Si, W.; Wan, C.; Deng, Z. Intra-Pulse Modulation Recognition of Dual-Component Radar Signals Based on Deep Convolutional Neural Network. *IEEE Commun. Lett.* **2021**, *25*, 3305–3309. [CrossRef]
7. Ren, M.; Tian, Y. Radar signal cognition based time-frequency transform and high order spectra analysis. In Proceedings of the IEEE International Conference on Signal Processing, Communications and Computing (ICSPCC), Xiamen, China, 22–25 October 2017; pp. 1–4. [CrossRef]
8. Li, J.; Zhang, H.; Ou, J.; Wang, W. A Radar Signal Recognition Approach via IIF-Net Deep Learning Models. *Comput. Intell. Neurosci.* **2020**, *2020*, 8858588. [CrossRef]
9. Xia, Y.; Ma, Z.; Huang, Z. Over-the-Air Radar Emitter Signal Classification Based on SDR. In Proceedings of the 6th International Conference on Intelligent Computing and Signal Processing (ICSP), Xi'an, China, 9–11 April 2021; pp. 403–408. [CrossRef]
10. Xu, T.; Darwazeh, I. Deep Learning for Over-the-Air Non-Orthogonal Signal Classification. In Proceedings of the IEEE 91st Vehicular Technology Conference (VTC2020-Spring), Virtual Event, 25–28 May 2020; pp. 1–5.
11. Krenc, K.; Gradolewski, D.; Dziak, D.; Kawalec, A. Application of Radar Solutions for the Purpose of Bird Tracking Systems Based on Video Observation. *Sensors* **2022**, *22*, 3660. [CrossRef] [PubMed]
12. Dudczyk, J.; Czyba, R.; Skrzypczyk, K. Multi-Sensory Data Fusion in Terms of UAV Detection in 3D Space. *Sensors* **2022**, *22*, 4323. [CrossRef]
13. Hassan, K.; Dayoub, I.; Hamouda, W.; Berbineau, M. Automatic modulation recognition using wavelet transform and neural network. In Proceedings of the 9th International Conference on Intelligent Transport Systems Telecommunications (ITST), Lille, France, 20–22 October 2009; pp. 234–238. [CrossRef]
14. Hassan, K.; Dayoub, I.; Hamouda, W.; Berbineau M. Automatic Modulation Recognition Using Wavelet Transform and Neural Networks in Wireless Systems. *EURASIP J. Adv. Signal Process.* **2010**, *2010*, 532898. [CrossRef]
15. Dash, D.; Sa, K. D.; Jayaraman, V. Time Frequency Analysis of OFDM-LFM Waveforms for Multistatic Airborne Radar. In Proceedings of the 2018 Second International Conference on Inventive Communication and Computational Technologies (ICICCT), Coimbatore, India, 20–21 April 2018; pp. 865–870. [CrossRef]
16. Wang, Y.; Shi, Z.; Ma, X.; Liu, L. A Joint Sonar-Communication System Based on Multicarrier Waveforms. *IEEE Signal Process. Lett.* **2022**, *29*, 777–781. [CrossRef]
17. Leśnik, C. Nonlinear Frequency Modulated Signal Design. *Acta Phys. Pol. A* **2009**, *116*, 351. 10.12693/APhysPolA.116.351. [CrossRef]
18. Matuszewski, J. The radar signature in recognition system database. In Proceedings of the 19th International Conference on Microwaves, Radar and Wireless Communications, Warsaw, Poland, 21–23 May 2012; pp. 617–622. [CrossRef]
19. Shi, Q.; Zhang, J. Radar Emitter Signal Identification Based on Intra-pulse Features. In Proceedings of the IEEE 6th Information Technology and Mechatronics Engineering Conference (ITOEC), Chongqing, China, 15–17 September 2022; pp. 256–260. [CrossRef]
20. Zheng, Z.; Qi, C.; Duan, X. Sorting algorithm for pulse radar based on wavelet transform. In Proceedings of the IEEE 2nd Information Technology, Networking, Electronic and Automation Control Conference (ITNEC), Chengdu, China, 15–17 December 2017; pp. 1166–1169. [CrossRef]
21. Chan, Y.T.; Plews, J.W.; Ho, K.C. Symbol Rate Estimation by the Wavelet Transform. In Proceedings of the IEEE International Symposium on Circuits and Systems (ISCAS), Hong Kong, China, 12 June 1997; Volume 1, pp. 177–180. [CrossRef]
22. Gao, Y.; Li, M.; Huang, Z.; Lu J. A symbol rate estimation algorithm based on Morlet wavelet transform and autocorrelation. In Proceedings of the IEEE Youth Conference on Information, Computing and Telecommunication, Beijing, China, 20–21 September 2009; pp. 239–242. [CrossRef]
23. O'Toole, J.M.; Boashash, B. Fast and memory-efficient algorithms for computing quadratic time-frequency distributions. *Appl. Comput. Harmon. Anal.* **2013**, *35*, 350–358. [CrossRef]
24. Munoz, A; Ertle, R.; Unser, M. Continuous wavelet transform with arbitrary scales and O(N) complexity. *Signal Process.* **2002**, *82*, 749–757. [CrossRef]

25. Vishwanath, M.; Owens, R.M.; Irwin, M.J. The computational complexity of time-frequency distributions. In Proceedings of the IEEE Sixth SP Workshop on Statistical Signal and Array Processing, Victoria, BC, Canada, 4–6 October 1992; pp. 444–447 [CrossRef]
26. Iandola, F.N.; Moskewicz, M.W.; Ashraf, K.; Han, S.; Dally, W.J.; Keutzer, K. SqueezeNet: AlexNet-level accuracy with 50x fewer parameters and <1 MB model size. *arXiv* **2016**, arXiv:1602.07360.
27. Nedyalko, P.; Jordanov, I.; Roe, J. Radar Emitter Signals Recognition and Classification with Feedforward Networks. *Procedia Comput. Sci.* **2013**, *22*, 1192–1200. [CrossRef]
28. Granger, E.; Rubin, M.A.; Grossberg, S.; Lavoie, P. A What-and-Where fusion neural network for recognition and tracking of multiple radar emitters. *Neural Netw.* **2001**, *14*, 325–344. [CrossRef]
29. Walenczykowska, M.; Kawalec, A. Application of Continuous Wavelet Transform and Artificial Naural Network for Automatic Radar Signal Recognition. *Sensors* **2022**, *22*, 7434. [CrossRef]
30. Zhang, X.; Zhang, J.; Luo, T.; Huang, T.; Tang, Z.; Chen, Y.; Li, J.; Luo, D. Radar Signal Intrapulse Modulation Recognition Based on a Denoising-Guided Disentangled Network. *Remote Sens.* **2022**, *14*, 1252. [CrossRef]
31. Walenczykowska, M., Kawalec, A. Type of modulation identification using Wavelet Transform and Neural Network. *Bull. Pol. Acad. Sci. Tech. Sci.* **2016**, *64*, 257–261. [CrossRef]
32. Ho, K.C.; Prokopiw, W.; Chan, Y. Modulation identification of digital signals by the wavelet transform. *IEE Proc. Radar Sonar Navig.* **2000**, *147*, 169–176. [CrossRef]
33. Lv, Y.; Liu, Y.; Liu, F.; Gao, J.; Liu, K.; Xie, G. Automatic Modulation Recognition of Digital Signals Using CWT Based on Optimal Scales. In Proceedings of the IEEE International Conference on Computer and Information Technology, Washington, DC, USA, 11–13 September 2014; pp. 430–434. [CrossRef]
34. Sobolewski, S.; Adams, W.L.; Sankar, R. Universal Nonhierarchical Automatic Modulation Recognition Techniques For Distinguishing Bandpass Modulated Waveforms Based On Signal Statistics, Cumulant, Cyclostationary, Multifractal And Fourier-Wavelet Transforms Features. In Proceedings of the IEEE Military Communications Conference, Baltimore, MD, USA, 6–8 October 2014; pp. 748–753. [CrossRef]
35. Wang, X.Z. Electronic radar signal recognition based on wavelet transform and convolution neural network. *Alex. Eng. J.* **2022**, *61*, 3559–3569. [CrossRef]
36. Ghodeswar, S.; Poonacha, P.G. An SNR estimation based adaptive hierarchical modulation classification method to recognize M-ary QAM and M-ary PSK signals. In Proceedings of the 3rd International Conference on Signal Processing, Communication and Networking (ICSCN), Chennai, India, 26–28 March 2015; pp. 1–6. [CrossRef]
37. Lunden, J.; Koivunen, V. Automatic Radar Waveform Recognition. *IEEE J. Sel. Top. Signal Process.* **2007**, *1*, 124–136. [CrossRef]
38. Levanon, N.; Mozeson, E. Phase-Coded Pulse. In *Radar Signals*; Wiley-Interscience; IEEE Press: Piscataway, NJ, USA, 2004. [CrossRef]
39. Farnane, K.; Minaoui, K.; Rouijel, A.; Aboutajdine, D. Analysis of the ambiguity function for phase-coded waveforms. In Proceedings of the 2015 IEEE/ACS 12th International Conference of Computer Systems and Applications (AICCSA), Marrakech, Morocco, 17–20 November 2015; pp. 1–4. [CrossRef]
40. Boggiatto, P.; Cordero, E.; de Gosson, M.; Feichtinger, H.; Nicola, F.; Oliaro, A.; Tabacco, A. *Landscapes of Time-Frequency Analysis*; Applied and Numerical Harmonic Analysis; Springer International AG: New York, NY, USA, 2019.
41. Majkowski, A.; Kołodziej, M.; Rak, R. Joint Time-Frequency And Wavelet Analysis-An Introduction. *Metrol. Meas. Syst.* **2014**, *21*, 741–758. [CrossRef]
42. Auger, F.; Flandrin, P.; Goncalves, P.; Lemoine, O. Time—Frequency toolbox for use with MATHLAB. Available online: https://tftb.nongnu.org/tutorial.pdf (accessed on 12 December 2022).
43. Addison, P.S. *The Illustrated Wavelet Handbook, Introuctory Theory and Applications in Science, Engeneering, Medicine and Finance*; Taylor and Francis Group, LLC: New York, NY, USA, 2002.
44. Rioul, O.; Vetterli, M. Wavelets and signal processing. *IEEE Signal Process. Mag.* **1991**, *8*, 14–38. [CrossRef]
45. Lilly, J.M.; Olhede, S.C. On the Analytic Wavelet Transform. *IEEE Trans. Inf. Theory* **2010**, *56*, 4135–4156. [CrossRef]
46. Lilly, J. Element analysis: A wavelet-based method for analysing time-localized events in noisy time series. *Proc. R. Soc. A Math. Phys. Eng. Sci.* **2017**, *473*, 20160776. [CrossRef]
47. Lilly, J.M.; Olhede, S.C. Higher-Order Properties of Analytic Wavelets. *IEEE Trans. Signal Process.* **2009**, *57*, 146–160. [CrossRef]
48. Lilly, J.M.; Olhede, S.C. Generalized Morse Wavelets as a Superfamily of Analytic Wavelets. *IEEE Trans. Signal Process.* **2012**, *60*, 6036–6041. [CrossRef]
49. Walenczykowska, M.; Kawalec, A. Radar signal recognition using Wavelet Transform and Machine Learning, In Proceedings of the 23rd International Radar Symposium (IRS), Gdansk, Poland, 12–14 September 2022; pp. 492–495. [CrossRef]
50. Rybak, Ł.; Dudczyk, J. Variant of Data Particle Geometrical Divide for Imbalanced Data Sets Classification by the Example of Occupancy Detection. *Appl. Sci.* **2021**, *11*, 4970. [CrossRef]

Article

Low Complexity Radar Gesture Recognition Using Synthetic Training Data

Yanhua Zhao [1,2,*], Vladica Sark [1], Milos Krstic [1,3] and Eckhard Grass [1,2]

1   IHP—Leibniz-Institut für Innovative Mikroelektronik, 15236 Frankfurt, Germany
2   Institute of Computer Science, Humboldt University of Berlin, Rudower Chaussee 25, 12489 Berlin, Germany
3   Design and Test Methodology, University of Potsdam, August-Bebel-Straße 89, 14482 Potsdam, Germany
*   Correspondence: zhao@ihp-microelectronics.com

Abstract: Developments in radio detection and ranging (radar) technology have made hand gesture recognition feasible. In heat map-based gesture recognition, feature images have a large size and require complex neural networks to extract information. Machine learning methods typically require large amounts of data and collecting hand gestures with radar is time- and energy-consuming. Therefore, a low computational complexity algorithm for hand gesture recognition based on a frequency-modulated continuous-wave (FMCW) radar and a synthetic hand gesture feature generator are proposed. In the low computational complexity algorithm, two-dimensional Fast Fourier Transform is implemented on the radar raw data to generate a range-Doppler matrix. After that, background modelling is applied to separate the dynamic object and the static background. Then a bin with the highest magnitude in the range-Doppler matrix is selected to locate the target and obtain its range and velocity. The bins at this location along the dimension of the antenna can be utilised to calculate the angle of the target using Fourier beam steering. In the synthetic generator, the Blender software is used to generate different hand gestures and trajectories and then the range, velocity and angle of targets are extracted directly from the trajectory. The experimental results demonstrate that the average recognition accuracy of the model on the test set can reach 89.13% when the synthetic data are used as the training set and the real data are used as the test set. This indicates that the generation of synthetic data can make a meaningful contribution in the pre-training phase.

Keywords: FMCW radar; gesture sensing; machine learning; mmWave; synthetic features

Citation: Zhao, Y.; Sark, V.; Krstic, M.; Grass, E. Low Complexity Radar Gesture Recognition Using Synthetic Training Data. *Sensors* **2023**, *23*, 308. https://doi.org/10.3390/s23010308

Academic Editors: Janusz Dudczyk and Piotr Samczyński

Received: 24 October 2022
Revised: 19 December 2022
Accepted: 25 December 2022
Published: 28 December 2022

1. Introduction

Low-cost, miniaturised radars have become increasingly popular in recent years. This has led to a large number of radar-based applications. For example, automotive radar can play an important role in collision avoidance systems. In addition to this, the potential of the radar in the field of medical applications is also being investigated. Applications such as weather radar and ground-penetrating radar reveal a need for such applications and research on radar technology and algorithms is highly desirable.

Traditional human–computer interaction mediums such as buttons, mice and keyboards are not always convenient in certain situations, such as operations in clean rooms. Contactless human–computer interaction requires less touching and is more hygienic. It can also further enhance the user experience. Hand gestures are an important medium for contactless human–computer interaction [1].

The outstanding privacy-protecting character of radar makes it preferable over cameras and its ability to be unaffected by light conditions is again preferable to LIDAR. The frequency-modulated continuous-wave (FMCW) radar is able to detect the distance, velocity and angle of several objects at the same time; hence, it is employed for hand gesture recognition in our work. FMCW radars can suffer from mutual interference. If there are other radars as sources of interferences, the methods in [2] for finding the range, velocity and angle of the target can be referred to. In this paper, only one radar is employed and interference from other radars is not considered.

Hand gesture recognition based on FMCW radar can be grouped into two categories. The first is based on raw data, meaning that the raw data are fed directly into a neural network to classify the hand gestures. The second is based on features, such as range, velocity and angle. These three features are obtained by certain data pre-processing methods and then combined with machine learning methods to achieve hand gesture classification. In this work, feature-based hand gesture recognition will be considered since the size of the radar raw data is large and will increase the training complexity of the neural network.

1.1. Related Work of Heat Map-Based Recognition

The features of a hand gesture can be represented in the form of a heat map. Many researchers have made various contributions to the field of heat map-based hand gesture recognition. Heat map-based gesture recognition is a relatively common approach and in [3–11] all authors use heat map-based recognition.

In [3], the authors conducted an experiment on gesture sensing using an FMCW radar with a centre frequency of 25 GHz, without considering the recognition of gestures. Later in [4], the authors collected 1714 gesture samples using FMCW radar, which contains ten types of gesture. Then they extracted time-Doppler heat maps from the radar raw data and trained a deep convolutional neural network model. An average recognition rate of 89.1% was achieved. Deep convolutional networks can be very challenging to implement at the hardware level. This is because they have a large number of weights and although it is possible to remove some of the unimportant weights, by pruning and other methods, it does require a lot of computational resources.

In [5], Lien et al. developed a small, low-power radar with a center frequency of 60 GHz. In contrast to [4], a time-Doppler heat map and a time-range heat map were extracted. With a classical random forest classifier, the average recognition accuracy of the four micro-hand gestures was up to 92.10%.

From the raw data of the FMCW radar not only time-distance and time-Doppler heat maps can be extracted, but also a time–angle heat maps. These three types of heat maps are employed as the basis for hand gesture recognition in [9]. The authors adopted background modelling to separate the static background and the moving target so that the feature heat maps were cleaned efficiently. The average recognition rate of six gestures was over 98.93%. The authors employed a pre-trained model, but a large number of weights of the model made implementation on an FPGA problematic.

Similarly, the authors used the multi-stream convolutional neural network (MS CNN) model in [11] to learn features of the dataset for on-air writing recognition. Although the accuracy achieved was very impressive, the drawback, as before, was that the MS CNN would require a lot of hardware resources.

Given the above references, we can state that heat map-based hand gesture recognition can achieve encouraging results, but its disadvantages should not be ignored. Firstly, the process of constructing a heat map is relatively complex and time-consuming. A hand gesture is made up of several frames of data, each of which will form one or more heat maps depending on the combination of features selected. A single hand gesture sample can produce a large number of heat maps, which leads to complex processing. In addition, the hand gesture features in the form of heat maps need to be further fed into a deep convolutional network to extract features, which requires even more computational efforts.

1.2. Motivation for Synthetic Data Generation and Related Work

Machine learning-based hand gesture recognition faces data scarcity issues and only a few open radar datasets are available. Many scientists spend a lot of effort and time collecting data. In Google's Soli project [5], its team collected 5000 samples with 5 participants. For [10], 7200 gesture samples were collected from 20 people. The authors of [6] collected

2750 samples, which involved 11 participants. A total of 1500 and 1200 gesture samples were collected in [9] and [8], respectively.

In [12], the authors utilised a sparse point cloud extraction method and a Message Passing Neural Network (MPNN)-based graphical convolution method for real-time gesture recognition. Despite the reduced computational complexity, all datasets were collected by other teams manually, which involved a lot of time and effort.

Collecting gesture samples with radar is very challenging. It is needed to perform specific hand gestures repetitively, which is time-consuming and labour-intensive. An efficient gesture feature synthesiser would be highly beneficial.

Human motion simulation is a good start for synthetic gestures. Some researchers have already combined human motion simulation with radar sensors. The authors have developed a human walking model in [13]. Afterwards, in [14], this model was employed to construct micro-Doppler spectrograms for gait analysis.

Reference [15] proposed a radar data synthesis process for four hand gestures. The authors used the 3D computer graphics software Blender [16] to build a simple human hand animation that captured the motion trajectories of the hand gestures. The motion trajectories were then utilised to synthesise radar data. The drawback of this work is that only micro-Doppler spectrograms were considered and the model was not tested with real data.

In [17], the authors proposed a radar data synthesis flow for macro-gestures. Seven gestures were simulated as a training set, which was used to train the Multi-Layer Perceptron model and the real data were employed to test the model with an average recognition accuracy of 84.2%. The drawback is that the ranges and angles of the gestures were not taken into account.

A human target model for the flexible simulation of various modalities of gesture was constructed in [18]. It covered the main parts of the body. In a similar way to the work in [17], only the Doppler spectrum was simulated. A CNN model was applied to classify eight macro-gestures with an accuracy of 80.4%.

The authors in [19] converted video footage of human activity into realistic, synthetic Doppler radar data by means of a cross-domain conversion approach to achieve the goal of synthesising radar training data for human activity. Other features such as range and angle were not synthesised.

In view of this, the gesture synthetic training data generator proposed by the authors in [20] can generate range–time heat maps, velocity–time heat maps and angle–time heat maps. Six gestures were synthesised and real data were also employed to test their validity. The authors used the VGG19 [21] pre-trained model to extract the features from heat maps. After that, the XGBoost [22] and Random Forest [23] classifiers were employed to recognise the hand gestures. The achieved average accuracy was 84.93% and 87.53%, respectively.

1.3. Contributions

To reduce the complexity of processing and tackle data scarcity, the main contributions of this paper are as follows.

- A simplified gesture recognition algorithm is proposed. The features of the gestures are represented as one-dimensional vectors instead of images.
- A simplified synthetic hand gesture feature generator is presented. As the hand gesture features extracted from the real data are simplified to one dimension, the synthesis processes from [15,17–20] are no longer needed. In our simplified synthetic hand gesture feature generator, the generation of radar raw data is skipped.
- The impact of range, velocity and angle features, extracted from a real data set, on the accuracy of gesture recognition is analysed. The experimental results reveal that all three features have a positive effect on gesture recognition. For the evaluation scenarios with a single feature, the average recognition rate based on the velocity feature alone achieves the highest recognition rate on the test set, with a support vector machine (SVM) classifier, which is 87.59%. For the different feature combination evaluation

scenarios, the average recognition rate based on the three features yields the best result with an average recognition rate of 98.48%.

- The impacts of the synthetic data set on the recognition accuracy of the real gesture data set are investigated. The SVM classifier trained with the synthetic data has an average recognition accuracy of 89.13% on the real data.

The remainder of the paper is organised as follows: Section 2 describes the FMCW radar system. Section 3 introduces low computational complexity algorithms for extracting features from radar raw data and Section 4 presents a synthetic hand gesture feature generator. Section 5 presents the experiments and results. The conclusions are given in Section 6.

2. FMCW Radar System

The system architecture of the FMCW radar is illustrated in Figure 1.

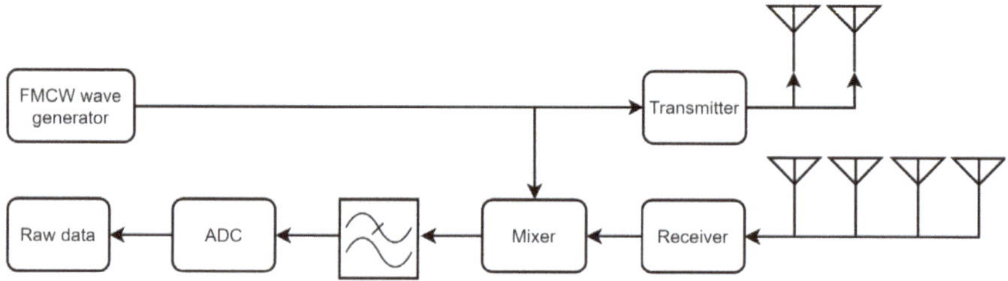

Figure 1. FMCW radar system.

The classical waveforms of FMCW radar are rectangular, upward sawtooth, triangular and staircase voltage waves. Since the hand gesture speed is not as high as an aircraft's, the upward sawtooth waveform is used in our radar. Firstly, the waveform generator generates an upward sawtooth wave as depicted in Figure 2. The bandwidth of the waveform is B. The solid blue line represents the transmitted wave and the dashed black line is the received wave. The frequency slope of the waveform is:

$$s = \frac{B}{T_c}. \tag{1}$$

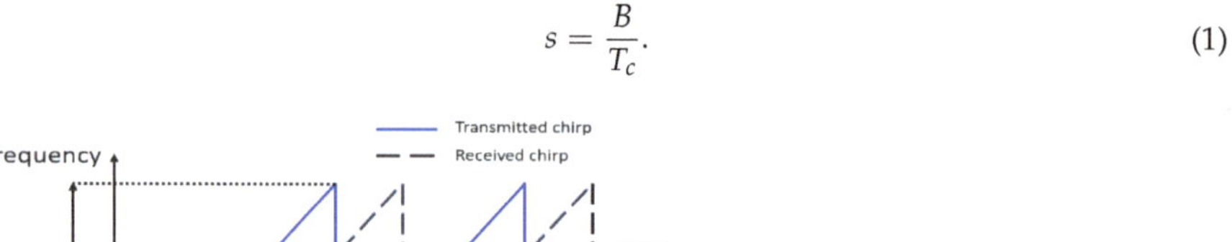
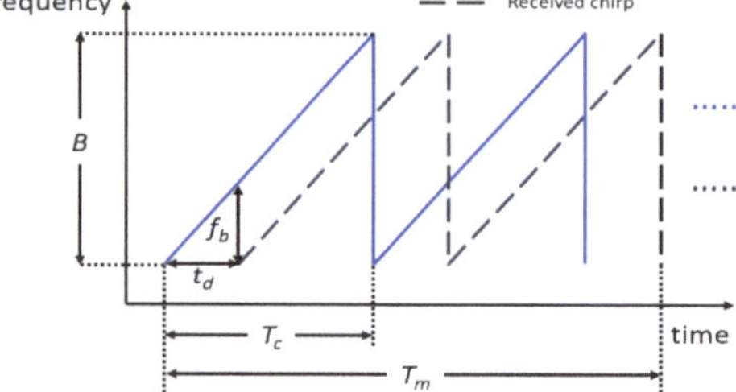

Figure 2. FMCW waveform.

The signal is emitted into space by the transmitter via the transmitting antennas. Our radar has two transmitting and four receiving antennas. The distance between the transmitting antennas is two wavelengths and the distance between the receiving antennas is half a wavelength. A transmitted wave can also be referred to as a chirp. The equation of the transmit chirp is given in (2):

$$T_X(t) = A \exp\left(j(2\pi f_c t + \pi s t^2)\right). \tag{2}$$

where the amplitude of the chirp is denoted as A and the starting frequency is represented by f_c. T_c is the duration of the chirp.

When a wave encounters any object in space, it will be reflected back to the receiver. The object could be our target or any other irrelevant object. The distance between the target and the radar is assumed as r_0, then the time delay t_d between the received and transmitted waves can be represented as:

$$t_d = \frac{2r_0}{c}.\tag{3}$$

where c is the velocity of light.

As the waveform loses energy as it travels through the air and is bounced by objects, there is an amplitude attenuation μ and phase shift in the received waveform. The received wave is defined as follows:

$$R_x(t) = \mu A \exp\left(j\left(2\pi f_c\left(t - \frac{2r_0}{c}\right) + \pi s\left(t - \frac{2r_0}{c}\right)^2\right)\right).\tag{4}$$

In the next step, the received and transmitted waves will be mixed in a mixer and passed through a low-pass filter to remove the high-frequency components and preserve the low-frequency signal. The remaining signal at this stage is known as the intermediate frequency signal or beat signal. As shown in Figure 2, the slope and time delay of the waveform are s and t_d, respectively, and the frequency of that beat signal can then be given as f_b:

$$f_b = s t_d = \frac{2Br_0}{cT_c}.\tag{5}$$

The beat signal can be defined as follows:

$$\begin{aligned}
B(t) &= \mu A^2 \exp\left(j\left(2\pi f_c\frac{2r_0}{c} + \pi s\frac{4r_0}{c}t - \pi s\left(\frac{2r_0}{c}\right)^2\right)\right)\\
&= \mu A^2 \exp j\left(\pi s\frac{4r_0}{c}t + 2\pi f_c\frac{2r_0}{c} - \pi s\left(\frac{2r_0}{c}\right)^2\right)\\
&= \mu A^2 \exp\left(j\left(2\pi s t_d t + \underbrace{2\pi f_c\frac{2r_0}{c} - \pi s\left(\frac{2r_0}{c}\right)^2}_{\phi(t)}\right)\right)\\
&= \mu A^2 \exp\left(j\left(2\pi f_b t + \phi(t)\right)\right)\\
&= \mu A^2 \exp\left(j\left(\frac{4\pi Br_0}{cT_c}t + \phi(t)\right)\right).
\end{aligned}\tag{6}$$

In the final step, the beat signal is digitalised by an analogue-to-digital converter (ADC).

3. Feature Extraction from Radar Raw Data

In this section, the process of extracting gesture features from the radar raw data is presented.

3.1. Range and Velocity Extraction

As illustrated in Figure 3, the structure of a frame of radar raw data has three dimensions, namely range, chirps and antennas. There are eight virtual antenna channels in one frame of the data. Each antenna channel has M chirps and each chirp has N range bins.

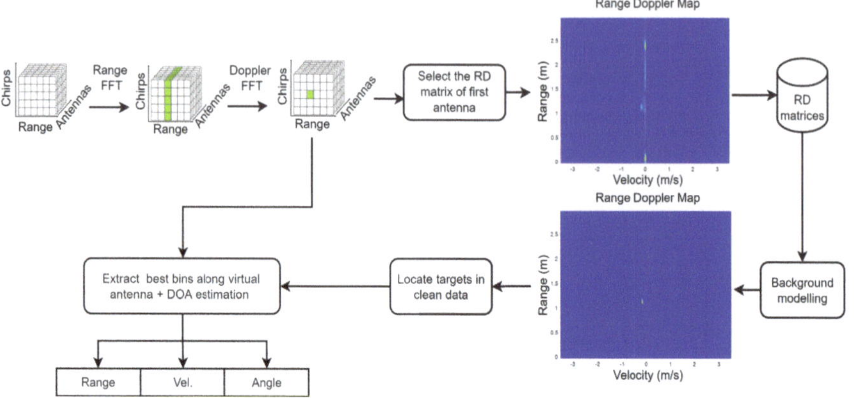

Figure 3. Processing chain for feature extraction from radar raw data.

To derive range and velocity information, a Fast Fourier Transform (FFT) is applied to the range and chirp dimensions. As depicted in Figure 3, after the range FFT, the range information of the target is highlighted. Then after the Doppler FFT, the range-Doppler (RD) matrix is formulated and the velocity information of the target is enhanced.

FMCW radar detects not only dynamic objects but also static ones. The radar data contain static clutter and static backgrounds, which can be disruptive to gesture recognition. Therefore, it is mandatory to take measures to combat irrelevant noise and static background.

Background modelling based on the Greedy bilateral smoothing (GreBsmo) [24] is employed to remove static clutter and static background objects, while the dynamic objects are retained. The data X can be decomposed to the background L, clean data S and noise G.

$$X = L + S + G. \tag{7}$$

After obtaining the RD matrices, the RD matrix for the first antenna channel is selected and saved in the data container. After executing the data for a whole hand gesture, the background modelling is carried out. The two heat maps in Figure 3 indicate the contribution of background modelling. After background modelling, the RD heat map becomes clean and only the target object remains. Thus the range and velocity of the target can be located more accurately in the RD matrix. It is assumed that the moving target in the data set has only one hand and is the main component. The index of that best bin can be derived by finding the maximum value in the RD matrix. Once the index is found, it is possible to calculate the range and velocity of the target.

3.2. Angle Extraction

The target bin is defined as the index of the RD matrix where the target is located. The best bin is searched along the antenna index. The bins being extracted are a 1×8 vector, denoted by $\mathbf{e}$.

By performing Fourier beam steering (FB) [25,26] on $\mathbf{e}$, angle information is derived. The virtual beam steering matrix is represented as:

$$\mathbf{V}(\Theta, q) = \exp\left(j2\pi\left(-\frac{Q-1}{2} + q\right)\frac{\Delta d}{\lambda}\sin\left(\frac{\pi}{180}\Theta\right)\right). \tag{8}$$

where Q is the total number of virtual antennas, $q \in \{1, 2, 3, \ldots, 8\}$. λ stands for the wavelength. Δd is the spacing distance between the receiving antennas and its value is $\frac{1}{2}\lambda$. The scanning scope of the angle Θ is $[-90, 90]$ and the step size is 1.

$$I = \mathbf{V} \cdot \mathbf{e}^T. \tag{9}$$

It is assumed that there is only one target and therefore the angle corresponds to the maximum value in I, as depicted in Figure 4.

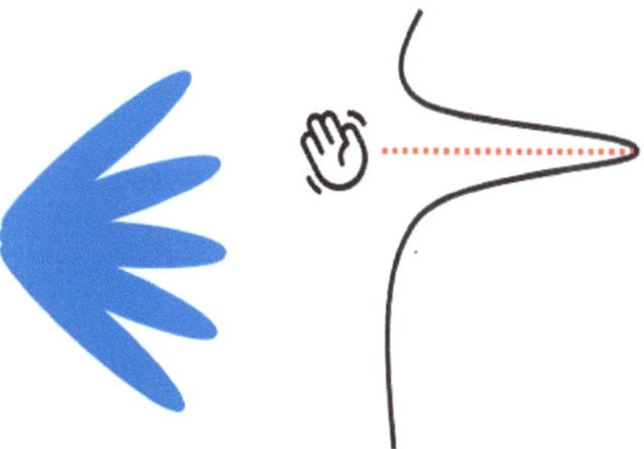

Figure 4. Fourier beam steering.

The processing sequence of our approach makes the angle estimation more stable and accurate. By selecting the best bins in the RD matrices, the angle information of irrelevant objects will be excluded. This is exactly the opposite of the processing order of unmanned aerial vehicle (UAV) swarms detection using the radar in [27,28]. In contrast to UAV swarm detection, only one target is taken into account in gesture recognition and a linear array of antennas is used in the radar.

In contrast to heat map-based gesture recognition, a gesture sample has 32 frames only, leading to a 1×96 feature vector. The features of a sample are shown in Figure 5. The overall process of extracting features from the radar raw data is shown in Algorithm 1.

Algorithm 1: Feature extraction from radar raw data.

Data: Radar raw data: Data
Result: Range, velocity and angle feature vectors
initialisation;
for $frame\_counter = 1, 2, \ldots, 32$ **do**
    Perform range FFT to the Data;
    Perform Doppler FFT to the Data;
    Data_container$(:,:,frame_counter)$ = Data $(:,:,1)$;
end
Perform background modelling ;
for $frame\_counter = 1, 2, \ldots, 32$ **do**
    Locate the peak in the clean RD matrix, compute range and velocity;
    Extract bins along antenna dimension;
    Perform Fourier beam steering;
    Locate the peak, compute the angle;
end

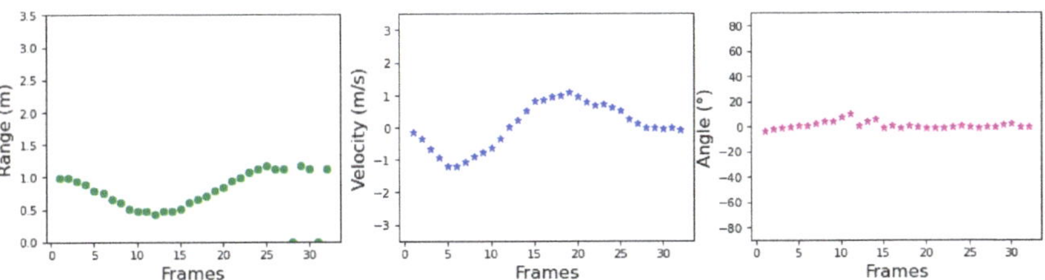

Figure 5. An example of gesture "push pull".

4. Synthetic Feature Generator

To match the gesture features with the real data, a synthetic generator of hand gestures is adapted. In this section, the workflow of the gesture feature synthesiser is proposed as illustrated in Figure 6a.

4.1. Generator Architecture

Our work uses Blender to animate hands. Blender provides the armature API and Python scripting. Python scripts make it easier to produce a large number of hand gestures. In Blender, the skeletal structure and joints of the hand are constructed, as depicted in Figure 6b.

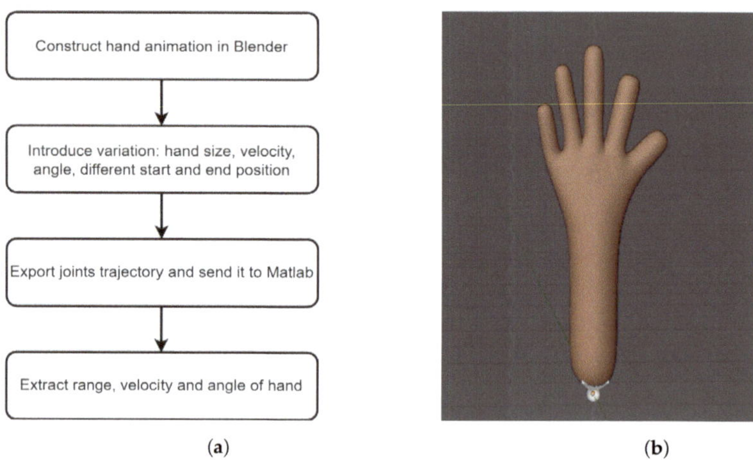

(**a**)   (**b**)

Figure 6. (**a**) The workflow of the gesture feature synthesiser; (**b**) an example of hand animation.

Human hands vary in size and have different habits of movement. Therefore, the simulator can reproduce different joint types and the hand can be simulated with varying velocities, angles and start–stop positions. The trajectories are fed into Matlab to calculate the features of the hand movement, namely range, velocity and angle.

4.2. Feature Extraction

The process of extracting features from a trajectory is illustrated in Algorithm 2. Figure 7 displays the radar and hand in the 3D space. The orange dot represents the radar with the location represented as (x_r, y_r, z_r). The position of one joint is denoted as (x_i, y_i, z_i). The distance between these two points in space is calculated by (10):

$$d = \sqrt{(x_i - x_r)^2 + (y_i - y_r)^2 + (z_i - z_r)^2}. \tag{10}$$

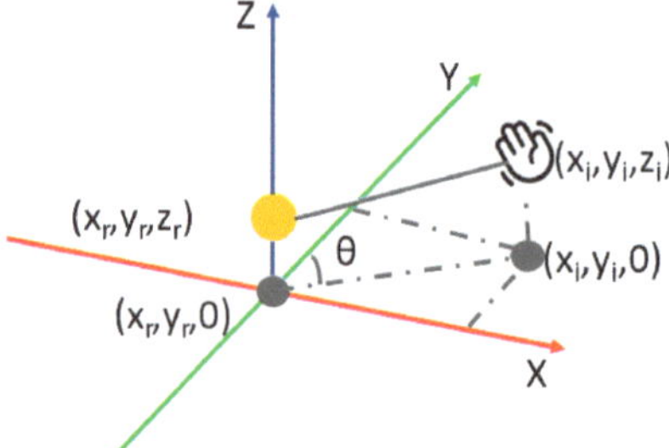

Figure 7. Target and radar in the coordinate system.

The distance from each joint of the simulated hand to the radar is calculated according to (10) and averaged as the distance between the hand and the radar for one frame.

For velocity, the initial value of velocity is set to 0. The formula for the calculation is given in (11). d_k is the distance between the radar and the hand in the current frame. d_{k-1} is the distance in the previous frame while T_m represents the duration of a frame. The difference between the distances of adjacent frames divided by the frame duration gives the hand velocity in the current frame.

$$v = \frac{d_k - d_{k-1}}{T_m}, k \in \{2, 3, \ldots, 32\}. \tag{11}$$

Algorithm 2: Synthetic feature extraction.

Data: Trajectories: Target_pos
Result: Range, velocity and angle feature vectors
initialisation;
for $frame\_counter = 1, 2, \ldots, 32$ **do**
  if $frame\_counter == 1$ **then**
    v(1) = 0;
    d($frame\_counter$) = Euclidean distance(Target_pos($frame\_counter$,:), Rad_pos);
    $\theta(frame\_counter) = \frac{180}{\pi} \arctan\left(\frac{\text{Target\_pos}(frame\_counter,1)}{\text{Target\_pos}(frame\_counter,2)}\right)$;
  else
    d($frame\_counter$) = Euclidean distance(Target_pos($frame\_counter$,:), Rad_pos);
    $v(frame\_counter) = \frac{d(frame\_counter) - d(frame\_counter-1)}{T_m}$;
    $\theta(frame\_counter) = \frac{180}{\pi} \arctan\left(\frac{\text{Target\_pos}(frame\_counter,1)}{\text{Target\_pos}(frame\_counter,2)}\right)$;
  end
end

Our radar can only detect the azimuth of the object. In Figure 7, the radar and the hand are projected onto the same plane. The grey dot is the projection of the hand. The projection of the radar is then at the origin of the coordinate system and the azimuth angle θ between radar and a joint is derived by (12). The angle between the hand and the radar is averaged over all joints of the hand.

$$\theta = \frac{180}{\pi} \arctan\left(\frac{x_i}{y_i}\right). \tag{12}$$

To make the synthesised hand gesture features more realistic, random noise is added to the extracted features.

4.3. Recognition Pipeline

The features of the hand gestures are fed directly into the support vector machine (SVM) [29] after they have been extracted based on the approach described previously. The support vector machine algorithm is particularly efficient in terms of memory and it performs better if there is a significant margin of separation between hand gestures.

5. Experiment and Evaluation

5.1. Radar Settings

The radar used for the experiments is a Texas Instruments (TI) AWR1642 [30], which operates at a starting frequency of 77 GHz and a maximum bandwidth of 4 GHz. It has two transmitting and four receiving antennas. The specific parameters used for the experiments are listed in Table 1. This radar also needs a raw data acquisition board. For this reason, we use TI DCA1000EVM [31].

5.2. Dataset

Two datasets are collected in this study, a real dataset based on AWR1642 and a synthetic dataset synthesised by the gesture feature simulator. There are six hand gestures in the dataset as shown in Figure 8: "grab", "to left", "to right", "move close", "move away" and "push pull". These are gestures that are commonly used in daily life. The real data set contains 250 samples of each gesture gathered from two participants in an indoor environment. Our low computational complexity approach takes an average of 0.0157 s to extract features from the raw data of each sample. The synthetic data set has 2700 samples per gesture.

Table 1. Radar parameters.

Parameter	Value
Starting frequency	77 GHz
Transmitting antennas	2
Receiving antennas	4
Number of range bins	64
Number of chirps per frame	255
Bandwidth	3.8 GHz
Chirp duration	38 µs
Frequency slope	100 MHz/µs
Frame duration	71 ms
Number of frames per gesture	32

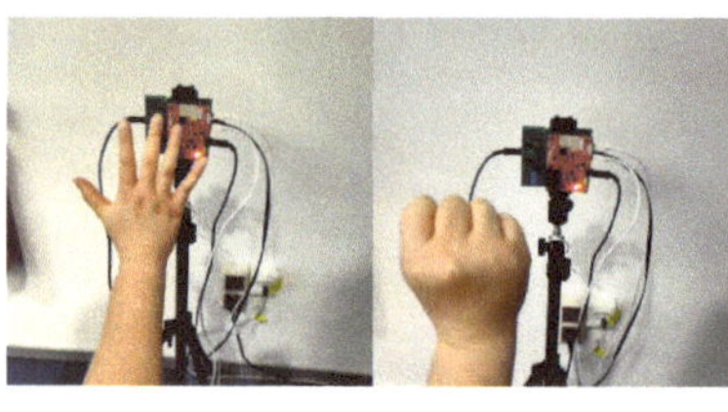

(**a**) Grab.

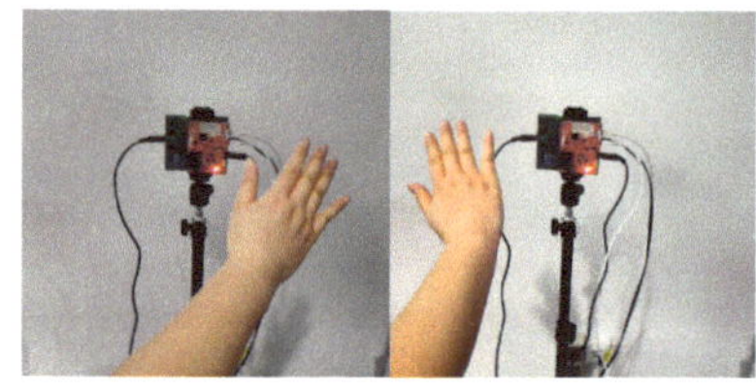

(**b**) To left.

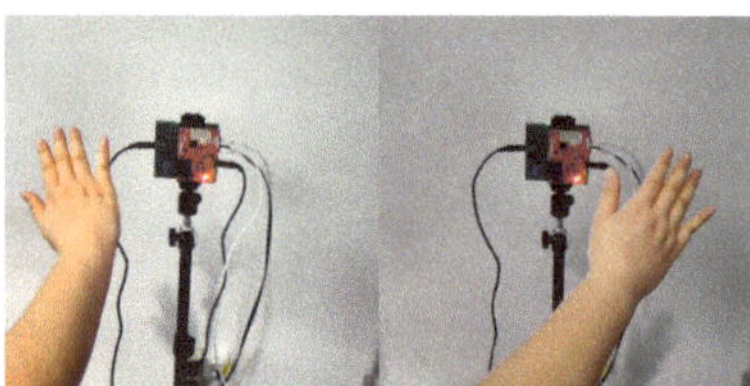

(**c**) To right.

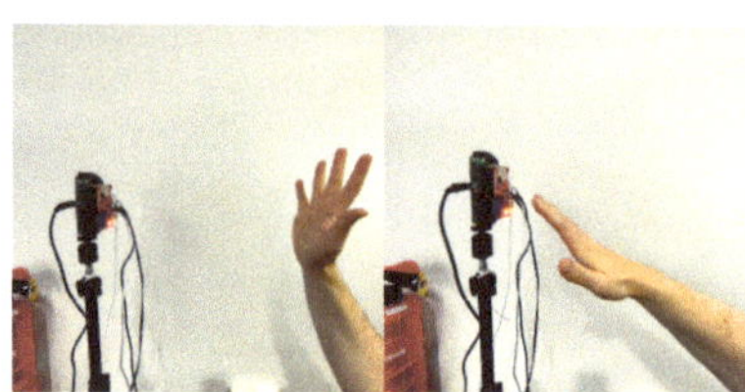

(**d**) Move close.

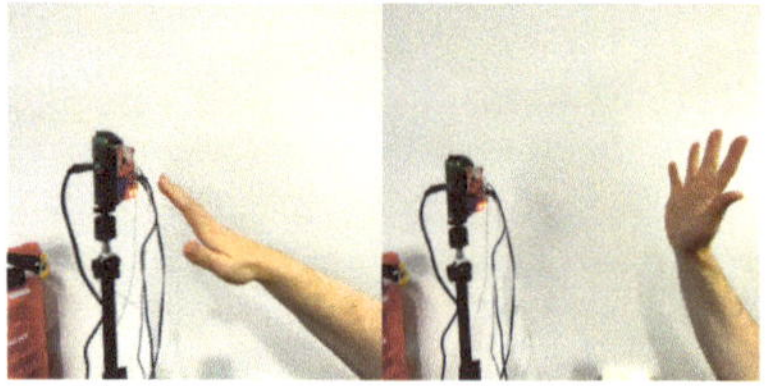

(**e**) Move away.

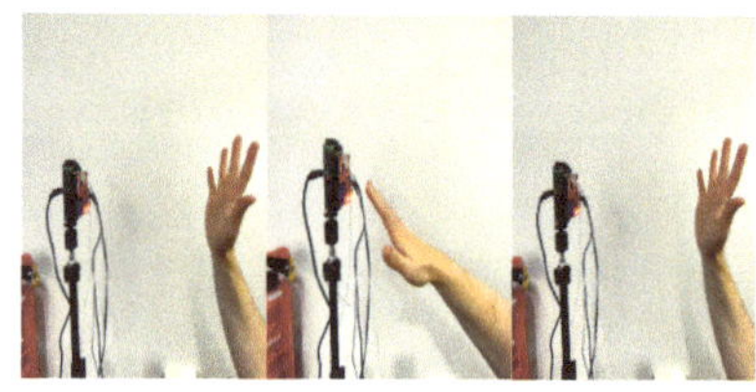

(**f**) Push pull.

Figure 8. Hand gesture type.

5.3. Feature Comparison

Next, features from the real dataset and synthetic features are analysed and compared. The features of the hand gestures are illustrated in a 3D scatter plot. For readability, only five samples of each gesture are displayed. A gradient colour has been used for the feature scatter plot, from dark to light. The starting frame of the feature for each sample is the darkest colour and the last frame is the lightest colour.

Figure 9a,b indicate the features of the real "grab" and the synthetic "grab". This gesture does not change much in terms of range, velocity and angle and that is because the grabbing, with the fingers slowly closing together, does not change much in motion.

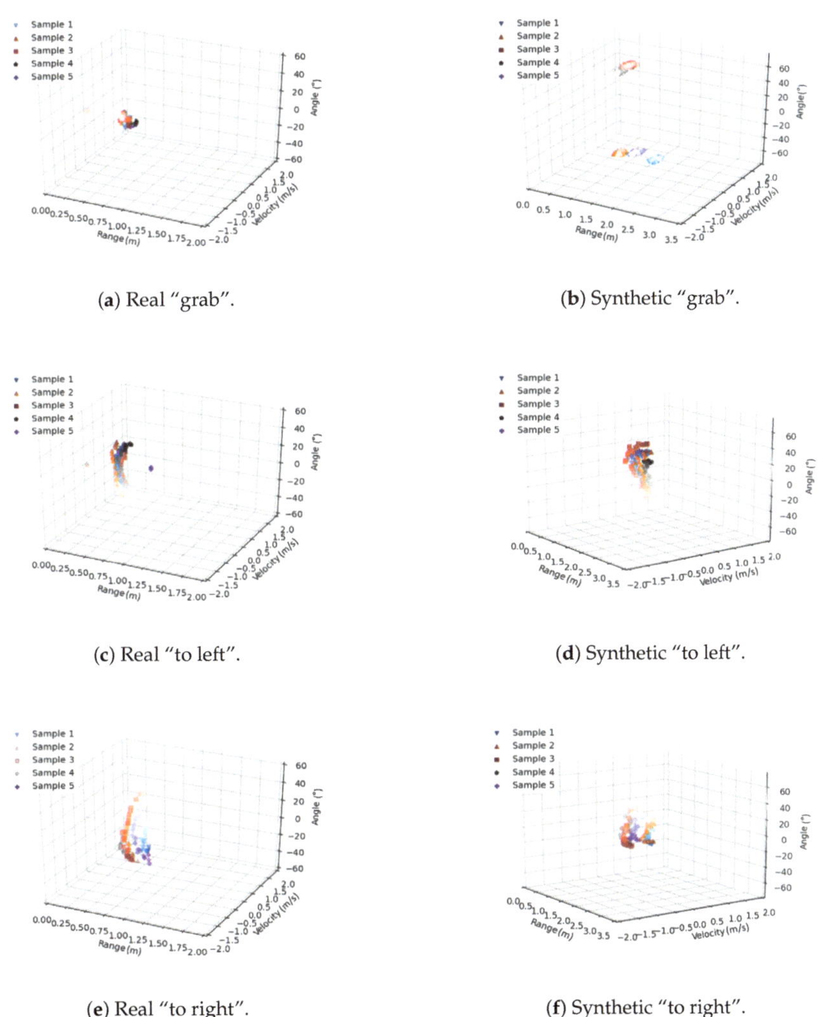

(**a**) Real "grab".

(**b**) Synthetic "grab".

(**c**) Real "to left".

(**d**) Synthetic "to left".

(**e**) Real "to right".

(**f**) Synthetic "to right".

Figure 9. Comparison of "grab", "to left" and "to right".

"To left" gesture is compared in Figure 9c,d. The real "to left" and the synthetic "to left" both have a drastic change in the dimension of the angle. A trend from larger to smaller angles can be seen on both figures. The hand moves from right to left and there will surely be a changing of angle relative to the radar. In contrast, there is not a lot of variation but a small decreasing and increasing trend in the range. This comes from the fact that the midpoint of the gesture is closer to the radar than the start and end points. Some small variation is expected because human movements are not perfectly aligned with the radar. The same is true for the velocity because only movements towards or away from the radar will influence the measured velocity.

The change in angle features for "to right" is the opposite of "to left" in Figure 9e,f. The real "to right" angle changes from a negative to a positive value. The other features of the gesture are identical to the "to left" gesture.

Figure 10a,b present the 3D features of the real "move close" and the synthetic "move close". The characteristic element of this gesture is the change of the range from larger to smaller values. This is because the hand is gradually moving closer to the radar. The velocity changes from zero to a negative value and then back to zero when the movement is finished. There is almost no change in angle. Some variation can be observed because the target is not perfectly positioned to move directly towards the radar.

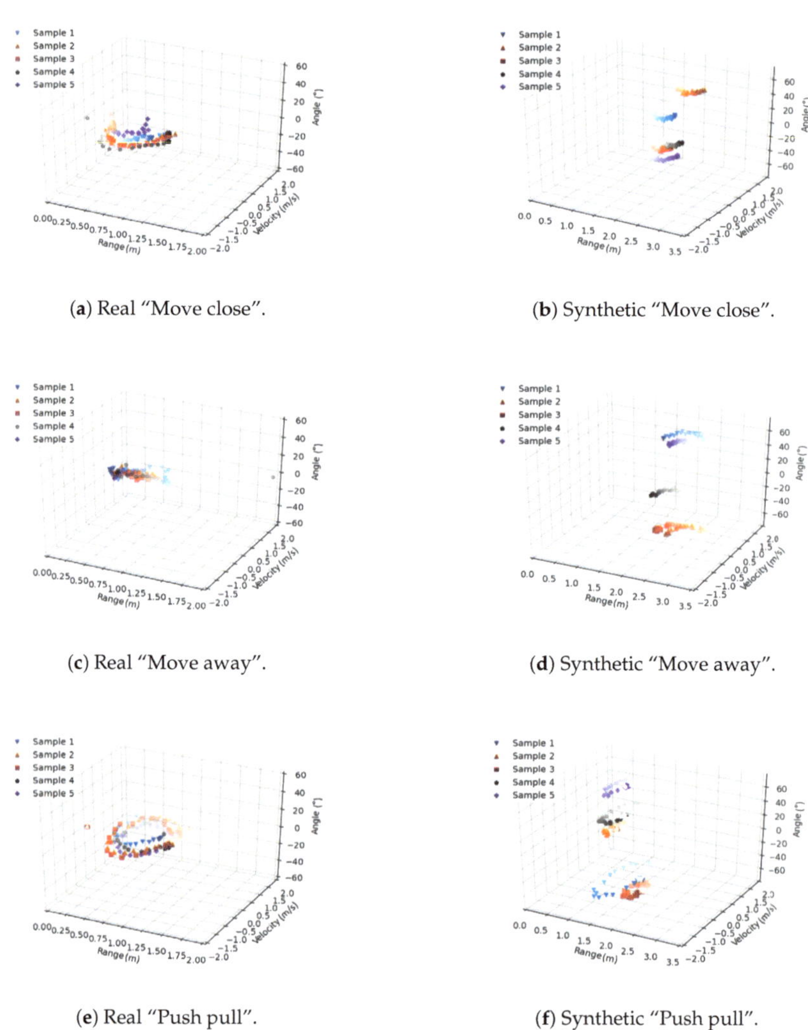

(**a**) Real "Move close".

(**b**) Synthetic "Move close".

(**c**) Real "Move away".

(**d**) Synthetic "Move away".

(**e**) Real "Push pull".

(**f**) Synthetic "Push pull".

Figure 10. Comparison of "move close", "move away" and "push pull".

For the "move away" gesture, the range between the hand and radar gradually increases as the hand moves in Figure 10c,d. In addition, the trend of the velocity is the opposite of that of "move close". The angle does not have much variation and the little variation observed behaves in the opposite manner of "move close".

"Push pull" is a combination of "move close" and "move away". The hand first approaches the radar and then moves away. The distance decreases and then increases.

The corresponding velocity changes towards positive and negative values. The angle does not change obviously. The feature of "push pull" in Figure 10e,f is in the form of a closed circle. There is a high similarity between the real and synthetic features.

5.4. Feature Distribution

To show the distribution of features for one gesture from the entire data set, the distribution of features for the gesture "push pull" is illustrated in Figure 11.

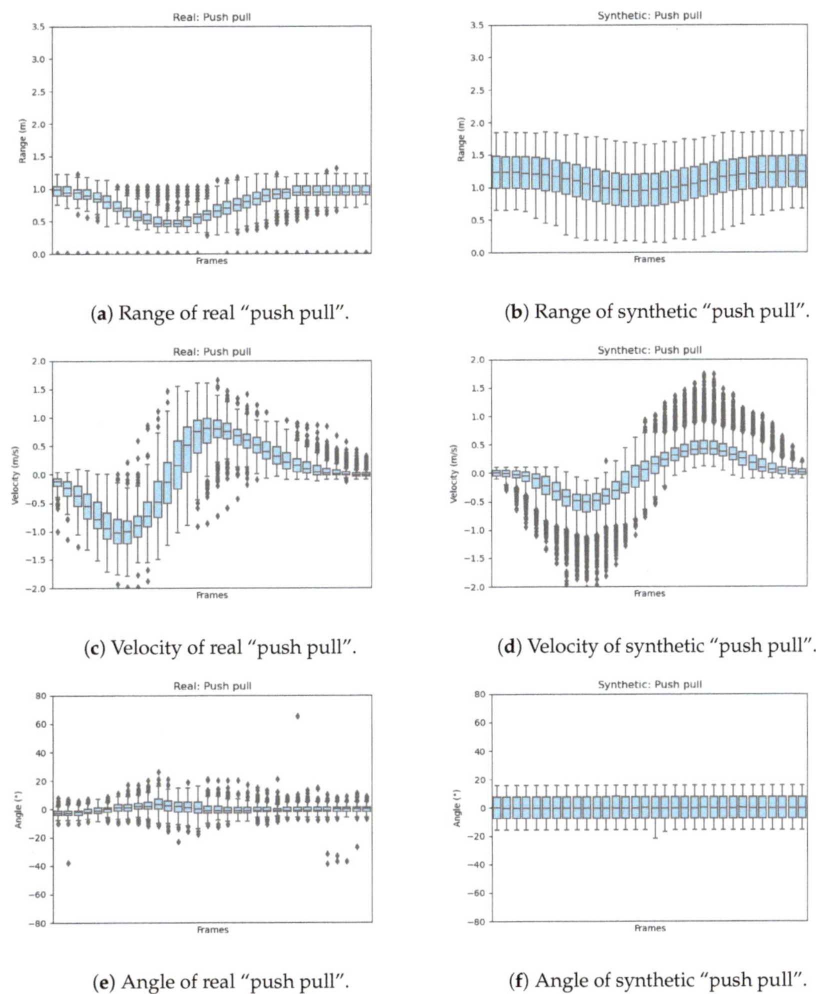

(**a**) Range of real "push pull".　　　　　(**b**) Range of synthetic "push pull".

(**c**) Velocity of real "push pull".　　　　　(**d**) Velocity of synthetic "push pull".

(**e**) Angle of real "push pull".　　　　　(**f**) Angle of synthetic "push pull".

Figure 11. Box-plot for feature distribution of real and synthetic "push pull".

This "push pull" distribution is based on 250 samples in the real data set and 2700 samples in the synthetic data set. It can be concluded from Figure 11a,b that the distribution zone of the range features for most of the "push pull" samples first decreases and then increases over time. The distribution of synthetic range features is relatively wider; this is due to the richness of the hand start and end position variations during the simulation of the trajectory. The synthetic data are purposefully created with as much variation as possible while still performing the hand gesture. The goal of this is to represent as many different ways to perform this gesture as possible. The speed of the real gestures seems to be higher than most of the synthetic data but the velocity of the real gesture is still included in the synthetic data. The velocity pattern for this gesture can be seen in both data sets. For the angle, the synthetic data cover a wider range of values; this will lead to a better classification performance if the gesture is performed from different angles. The overall pattern is similar to the real data. It can be seen that the real gesture was usually measured directly from the front. The synthetic data also cover the gesture if it is performed at an angle.

5.5. Feature Impact Analysis

From the radar raw data, three features have been extracted, namely range, velocity and angle. In this subsection, the effect of feature combinations on the accuracy of gesture recognition will be analysed. To analyse the impact of one type of feature, a data set containing a single feature is fed into the SVM for training and testing. A total of 50% of the real data set was randomly selected as the training set and the other 50% as the test set.

The experiment is repeated ten times and the recognition rates for the test set are summarised in Figure 12 and Table 2. As can be seen in the figure, velocity plays a significant role. For velocity alone, the average recognition accuracy in the test set was as high as 87.59%.

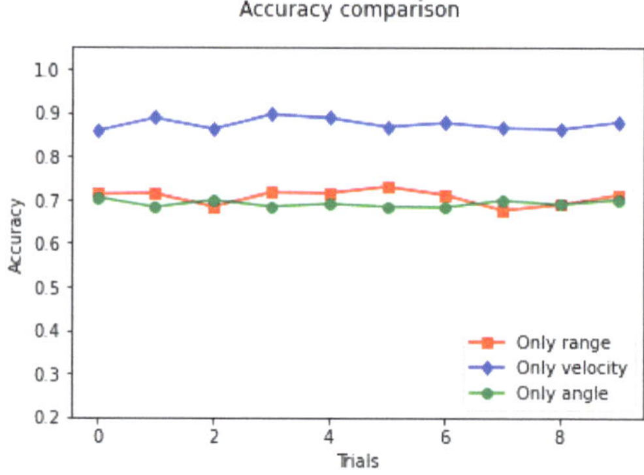

Figure 12. Accuracies based on single feature.

Table 2. Comparison of accuracy based on a single feature.

Feature Type	Average Accuracy	Standard Deviation
Only range	70.69%	1.59%
Only velocity	87.59%	1.23%
Only Angle	69.28%	0.78%

Furthermore, the combination of features is evaluated. The results based on the combination of different features are illustrated in Figure 13 and Table 3. Recognition rates based on multiple features are higher compared to a single feature. The average accuracy based on the three features is as high as 98.48%. The combination of velocity and angle can achieve accuracies of up to 98.15% on the test set.

From the outcomes, it can be derived that the velocity is more recognisable among the three features. However, based on velocity alone, the recognition rate of gestures is below 90%. When range and angle are also taken into account, the recognition accuracy on the test set is improved considerably.

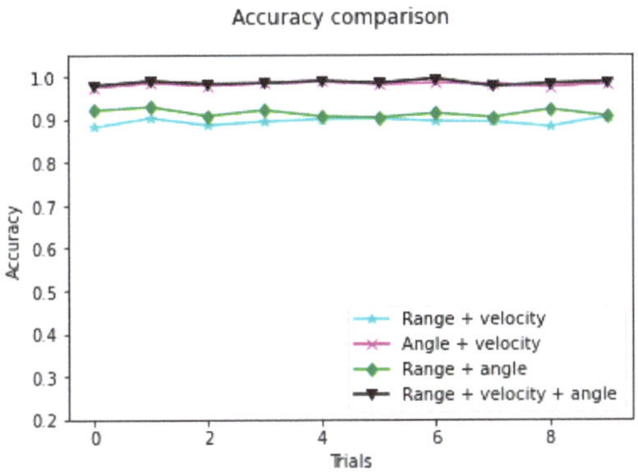

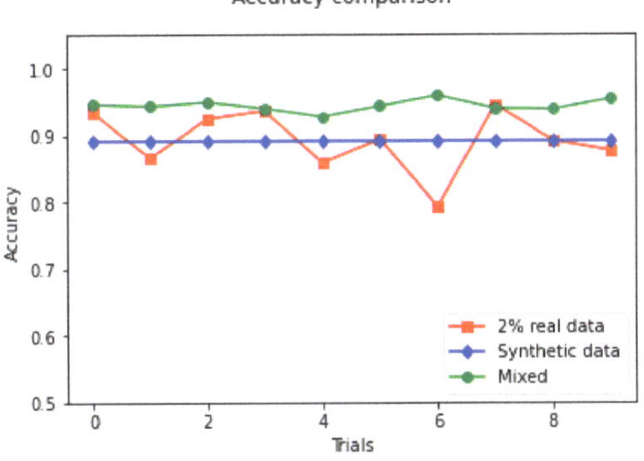

Figure 13. Accuracies based on multi features.

Table 3. Comparison of accuracy based on feature combination.

Feature Combination	Average Accuracy	Standard Deviation
Range + velocity	89.43%	0.76%
Velocity + angle	98.15%	0.42%
Range + angle	91.37%	0.87%
Range + velocity + angle	98.48%	0.52%

5.6. Synthetic Feature Impact Analysis

Three evaluation scenarios are defined. Scenario I is a random selection of 2% of the real data as the training set and the rest of the real data are used as the test set. Scenario II uses the synthetic data set as the training set and the entire real data set as the test set. Scenario III takes 2% of the real data and the entire synthetic data set as the training set and the remaining 98% of the real data set as the test set.

The recognition accuracy of the SVM on the test set is given in Figure 14 and Table 4. The experiment is repeated ten times.

Figure 14. Accuracy of test set.

Table 4. Comparison of accuracy.

Scenarios	Average Accuracy	Standard Deviation
Scenario I: 2% real data	89.21%	4.43%
Scenario II: Synthetic data	89.13%	0.00%
Scenario III: Mixed	94.43%	0.82%

The red line represents the accuracy of scenario I on the test set. It can be seen that the high fluctuation is due to a low amount of real data and the quality of the selected data varies. If the 2% randomly selected real samples from the real data set cover a large variation and have a high quality, then it is likely to yield good results on the test set. If the randomly selected samples have low quality or little variation, there is a chance that worse results can be obtained on the test set. The results for scenario II are indicated by the blue line. The blue line is stable because the test set and training set have not changed. The green line represents the result of scenario III. These results are the best among the scenarios. This shows that the synthetic data are able to enhance data sets of real data to achieve a performance that is higher than the individual data sets.

The average accuracy of ten attempts for the three scenarios is given in Table 4. When the SVM classifier uses only the synthetic data set as the training set, the model has an average accuracy of 89.13% on the test set. When the synthetic data set is combined with a small amount of real data, better performance can be obtained, with an average recognition rate of 94.43%.

One more advantage of synthetic data is that they cover a wider range of possible gestures. The training and test sets are very similar, so a small amount of real data can lead to high performance in the test set. A test set that differs in these features but still includes valid gestures may result in poorer performance for a model trained on real data. Synthetic data has a wider range of feature values, so it is able to generalise better and cover different test sets in a better way.

5.7. Comparison with Other Works on Synthetic Feature Generators

In Table 5, a comparison with other works is summarised. Previous work such as [17] combined animation modelling in Blender with a simulated FMCW radar sensor to simulate velocity–time heatmap features for seven types of macro-gesture. In [18], the human model was constructed and Doppler spectrograms for eight macro-gestures were simulated. The authors synthesised six activities in [19] by transforming the data from the camera into Doppler heat maps. The number of samples in their test set is 720.

Table 5. Comparison with other works.

Ref.	Sample Size	Real Samples	Testset Average Accuracy
[17]	-	1050	84.2%
[18]	-	3354	80.4%
[19]	-	720	81.4%
[20]	8–12 KB	1500	87.53%
This work	1 KB	1500	89.13%

Reference [20] simulated range–time, velocity–time and angle–time heatmap features for six types of hand gestures by combining gestures constructed in Blender and a simulated FMCW radar. For both, the classifiers were trained with synthetic data sets and the models were tested with real data sets. However, this does not suit our features extracted from the radar raw data.

The features of our sample are saved in a csv [32] file. In Table 5, it can be observed that the features extracted from a sample of [20] are 8 to 12 times larger than the features extracted by our new approach. Compared to [17–20], the average recognition accuracy on the real data set is higher, which indicates the strength of this work. In addition to this, our synthetic feature generator does not contain radar signal simulation, which reduces the computational effort significantly. More importantly, the dataset extracted by our method does not need to be fed further into the machine learning algorithm for feature extraction. This will help to save hardware resources significantly.

6. Conclusions and Future Work

In this work, a low computational complexity hand gesture feature extraction methodology based on FMCW radar and a synthetic hand gesture feature generator are proposed.

The impact of range, velocity and angle on the accuracy of gesture recognition is analysed. The combination of the three features leads to the highest accuracy.

Compared to other works, our synthetic gesture feature process avoids the generation of radar raw data and is more straightforward. In contrast to heat map-based recognition, additional feature extraction can be skipped in the training phase.

The synthetic data set is used to train a SVM classifier to recognise six different hand gestures in the real data set, which are acquired with an AWR1642 FMCW radar. The algorithm achieves an average recognition accuracy of 89.13% on the real data set. By combining a small amount of the real data set with the synthetic data set, an average recognition accuracy of 94.43% is obtained on the real data set. Thus, we demonstrate the effectiveness of our raw data pre-processing approach and feature synthetisation process.

In the future, more gesture types, different frequency bands and cross-platform gesture recognition should be investigated. We would like to verify the performance of our selected models in different aspects in future work, for example, by adding more datasets with different characteristics.

Author Contributions: Data curation, Y.Z.; Methodology, Y.Z.; Supervision, V.S., M.K. and E.G.; Writing—original draft, Y.Z.; Writing—review & editing, V.S., M.K. and E.G. All authors have read and agreed to the published version of the manuscript.

Funding: This work is partially funded by the Federal Ministry of Education and Research of Germany (BMBF) within the iCampus Cottbus project (grant number: 16ME0424) and the "Open6GHub" project (grant number: 16KISK009).

Institutional Review Board Statement: Not applicable.

Informed Consent Statement: Not applicable.

Data Availability Statement: Not applicable.

Conflicts of Interest: The authors declare no conflict of interest.

References

1. Ahmed, S.; Kallu, K.D.; Ahmed, S.; Cho, S.H. Hand Gestures Recognition Using RADAR Sensors for Human-Computer-Interaction: A Review. *Remote Sens.* **2021**, *13*, 527. [CrossRef]
2. Xu, Z. Bi-Level l_1 Optimization-Based Interference Reduction for Millimeter Wave Radars. *IEEE Trans. Intell. Transp. Syst.* **2022**. [CrossRef]
3. Molchanov, P.; Gupta, S.; Kim, K.; Pulli, K. Short-range FMCW monopulse RADAR for hand-gesture sensing. In Proceedings of the 2015 IEEE RADAR Conference, Johannesburg, South Africa, 27–30 October 2015.
4. Molchanov, P.; Gupta, S.; Kim, K.; Pulli, K. Multi-sensor system for driver's hand-gesture recognition. In Proceedings of the 2015 11th IEEE International Conference and Workshops on Automatic Face and Gesture Recognition (FG), Ljubljana, Slovenia, 4–8 May 2015.
5. Lien, J.; Gillian, N.; Karagozler, M.E.; Amihood, P.; Schwesig, C.; Olson, E.; Raja, H.; Poupyrev, I. Soli: Ubiquitous gesture sensing with millimeter wave RADAR. *ACM Trans. Graph.* **2016**, *35*, 142. [CrossRef]
6. Wang, S.; Song, J.; Lien, J.; Poupyrev, I.; Hilliges, O. *Interacting with Soli: Exploring Fine-Grained Dynamic Gesture Recognition in the Radio-Frequency Spectrum*; ACM: New York, NY, USA, 2016; pp. 851– 860.
7. Wang, Y.; Shu, Y.; Jia, X.; Zhou, M.; Xie, L.; Guo, L. Multifeature Fusion-Based Hand Gesture Sensing and Recognition System. *IEEE Geosci. Remote Sens. Lett.* **2021**. [CrossRef]
8. Wang, Y.; Ren, A.; Zhou, M.; Wang, W.; Yang, X. A Novel Detection and Recognition Method for Continuous Hand Gesture Using FMCW RADAR. *IEEE Access* **2020**, *8*, 167264–167275. [CrossRef]
9. Zhao, Y.; Sark, V.; Krstic, M.; Grass, E. Novel Approach for Gesture Recognition Using mmWave FMCW RADAR. In Proceedings of the 2022 IEEE 95th Vehicular Technology Conference: (VTC2022-Spring), Helsinki, Finland, 19–22 June 2022; pp. 1–6. [CrossRef]
10. Sun, Y.; Fei, T.; Li, X.; Warnecke, A.; Warsitz, E.; Pohl, N. Realtime RADAR-based gesture detection and recognition built in an edge computing platform. *IEEE Sens. J.* **2020**, *20*, 10706–10716. [CrossRef]
11. Ahmed, S.; Kim, W.; Park, J.; Cho, S.H. Radar-Based Air-Writing Gesture Recognition Using a Novel Multistream CNN Approach. *IEEE Internet Things J.* **2022**, *9*, 23869–23880. [CrossRef]

12. Salami, D.; Hasibi, R.; Palipana, S.; Popovski, P.; Michoel, T.; Sigg, S. Tesla-Rapture: A Lightweight Gesture Recognition System from mmWave Radar Sparse Point Clouds. *IEEE Trans. Mob. Comput.* **2022**. [CrossRef]
13. Boulic, R.; Thalmann, N.M.; Thalmann, D. A global human walking model with real-time kinematic personification. *Vis. Comput.* **1990**, *6*, 344–358. [CrossRef]
14. Trommel, R.P.; Harmanny, R.I.A.; Cifola, L.; Driessen, J.N. Multitarget human gait classification using deep convolutional neural networks on micro-doppler spectrograms. In Proceedings of the 2016 European RADAR Conference (EuRAD), London, UK, 5–7 October 2016; pp. 81–84.
15. Ishak, K.; Appenrodt, N.; Dickmann, J.; Waldschmidt, C. Human motion training data generation for RADAR based deep learning applications. In Proceedings of the 2018 IEEE MTT-S International Conference on Microwaves for Intelligent Mobility (ICMIM), Munich, Germany, 16–17 April 2018; pp. 1–4.
16. Blender. Free and Open Source 3D Creation Suite. Available online: https://www.blender.org/ (accessed on 27 July 2022).
17. Ninos, A.; Hasch, J.; Alvarez, M.; Zwick, T. Synthetic RADAR Dataset Generator for Macro-Gesture Recognition. *IEEE Access* **2021**. [CrossRef]
18. Kern, N.; Aguilar, J.; Grebner, T.; Meinecke, B.; Waldschmidt, C. Learning on Multistatic Simulation Data for Radar-Based Automotive Gesture Recognition. *IEEE Trans. Microw. Theory Tech.* **2022**, *70*, 5039–5050. [CrossRef]
19. Ahuja, K.; Jiang, Y.; Goel, M.; Harrison, C. Vid2Doppler: Synthesising Doppler radar data from videos for training privacy-preserving activity recognition. In Proceedings of the 2021 CHI Conference on Human Factors in Computing Systems, Yokohama, Japan, 8–13 May 2021; pp. 1–10.
20. Zhao, Y.; Sark, V.; Krstic, M.; Grass, E. Synthetic Training Data Generator for Hand Gesture Recognition Based on FMCW RADAR. In Proceedings of the 2022 23rd International Radar Symposium (IRS), Gdansk, Poland, 12–14 September 2022; pp. 463–468. [CrossRef]
21. Simonyan, K.; Zisserman, A. Very Deep Convolutional Networks for Large-Scale Image Recognition. *arXiv* **2014**, arXiv:1409.1556.
22. Chen, T.; Guestrin, C. XGBoost: A scalable tree boosting system. In Proceedings of the 22nd ACM SIGKDD International Conference on Knowledge Discovery and Data Mining, San Francisco, CA, USA, 13–17 August 2016; pp. 785–794.
23. Breiman, L. Random Forests. *Mach. Learn.* **2001**, *45*, 5–32.:1010933404324. [CrossRef]
24. Zhou, T.; Tao, D. Greedy Bilateral Sketch, Completion; Smoothing. In Proceedings of the Sixteenth International Conference on Artificial Intelligence and Statistics, Scottsdale, AZ, USA, 29 April–1 May 2013; pp. 650–658.
25. Sit, Y.L. MIMO OFDM Radar-Communication System with Mutual Interference Cancellation. Ph.D. Thesis, Karlsruher Institut für Technologie (KIT), Karlsruhe, Germany, 2017. [CrossRef]
26. Zhao, Y.; Sark, V.; Krstic, M.; Grass, E. Low Computational Complexity Algorithm for Hand Gesture Recognition using mmWave RADAR. In Proceedings of the 2022 International Symposium on Wireless Communication Systems (ISWCS), Hangzhou, China, 19–22 October 2022; pp. 1–6. [CrossRef]
27. Zheng, J.; Chen, R.; Yang, T.; Liu, X.; Liu, H.; Su, T.; Wan, L. An Efficient Strategy for Accurate Detection and Localization of UAV Swarms. *IEEE Internet Things J.* **2021**, *8*, 15372–15381. [CrossRef]
28. Zheng, J.; Yang, T.; Liu, H.; Su, T.; Wan, L. Accurate Detection and Localization of Unmanned Aerial Vehicle Swarms-Enabled Mobile Edge Computing System. *IEEE Trans. Ind. Inform.* **2021**, *17*, 5059–5067. [CrossRef]
29. Chang, C.; Lin, C. LIBSVM: A library for support vector machines. *ACM Trans. Intell. Syst. Technol.* **2011**, *2*, 27. [CrossRef]
30. AWR1642. Single-Chip 76-GHz to 81-GHz Automotive RADAR Sensor Evaluation Module. Available online: http://www.ti.com/tool/AWR1642BOOST (accessed on 8 July 2022).
31. DCA1000EVM. Real-Time Data-Capture Adapter for RADAR Sensing Evaluation Module. Available online: https://www.ti.com/tool/DCA1000EVM (accessed on 8 July 2022).
32. *RFC 4180: Common Format and MIME Type for Comma-Separated Values (CSV) Files*; SolidMatrix Technologies, Inc.: Staten Island, NY, USA, 2005. [CrossRef]

Article

Recognition of Targets in SAR Images Based on a WVV Feature Using a Subset of Scattering Centers

Sumi Lee [1] and Sang-Wan Kim [2,*]

1   Department of Geoinformation Engineering, Sejong University, 209 Neungdong-ro, Gwangjin-gu, Seoul 05006, Korea
2   Department of Energy Resources and Geosystems Engineering, Sejong University, 209 Neungdong-ro, Gwangjin-gu, Seoul 05006, Korea
*   Correspondence: swkim@sejong.edu

Abstract: This paper proposes a robust method for feature-based matching with potential for application to synthetic aperture radar (SAR) automatic target recognition (ATR). The scarcity of measured SAR data available for training classification algorithms leads to the replacement of such data with synthetic data. As attributed scattering centers (ASCs) extracted from the SAR image reflect the electromagnetic phenomenon of the SAR target, this is effective for classifying targets when purely synthetic SAR images are used as the template. In the classification stage, following preparation of the extracted template ASC dataset, some of the template ASCs were subsampled by the amplitude and the neighbor matching algorithm to focus on the related points of the test ASCs. Then, the subset of ASCs were reconstructed to the world view vector feature set, considering the point similarity and structure similarity simultaneously. Finally, the matching scores between the two sets were calculated using weighted bipartite graph matching and then combined with several weights for overall similarity. Experiments on synthetic and measured paired labeled experiment datasets, which are publicly available, were conducted to verify the effectiveness and robustness of the proposed method. The proposed method can be used in practical SAR ATR systems trained using simulated images.

Keywords: synthetic aperture radar; scattering center; world view vector; recognition; SAMPLE

Citation: Lee, S.; Kim, S.-W. Recognition of Targets in SAR Images Based on a WVV Feature Using a Subset of Scattering Centers. *Sensors* **2022**, *22*, 8528. https://doi.org/10.3390/s22218528

Academic Editors: Janusz Dudczyk and Piotr Samczyński

Received: 26 August 2022
Accepted: 27 October 2022
Published: 5 November 2022

Publisher's Note: MDPI stays neutral with regard to jurisdictional claims in published maps and institutional affiliations.

1. Introduction

Over the last few years, automatic target recognition using synthetic aperture radar (SAR ATR) has increasingly become important as a crucial means of surveillance [1–4]. SAR images can be obtained in most weather types, day and night, and at a high resolution [5–8]. Based on these characteristics, SAR ATR algorithms have been evaluated on several targets, including ground-based vehicles, aircrafts, and vessels, which are challenging for military operations [9]. However, the sensitivity of SAR to sensor parameters, target configurations, and environmental conditions make it challenging to implement SAR ATR [10–12].

SAR ATR algorithms can be divided into three basic steps: detection, discrimination, and classification [13,14]. The first two steps are intended to extract potential target areas and remove false alarms [15]. The purpose of target classification is to automatically classify each input target image obtained by target detection and discrimination [16]. A large number of efforts have been made to achieve robust SAR ATR. However, the classification performances under extended operating conditions (EOCs) are insufficient for practical applications. In real-world cases, most targets are likely to be camouflaged or blocked by surrounding obstacles [11,17]. To improve the performance under EOCs within a SAR ATR system, this paper focused on the classification stage.

Target classification methods are mainly divided into model-based and feature-based methods [5]. Feature-based methods involve pattern recognition and rely solely on features to represent the target, with many SAR target templates stored in advance. Once the features of the input target are extracted, the test target is classified as the category of the matched template.

One of the promising features, attributed scattering center (ASC), provides a physically relevant description of the target [4,18–21]. According to electromagnetic theory, the high frequency scattering response of the target can be approximated by a sum of responses from multiple ASCs [22]. The classification method based on ASCs reflects the specific scattering structure of the target. In Refs. [16,23], an ASC-based model was proposed for target classification, and they achieved higher accuracy than when not using the ASC model. It has also been proven that the local descriptions contained in ASCs are beneficial for reasoning under EOCs [19,24]. However, there are some problems associated with point-pattern matching between test ASCs and template ASCs. Two ASC sets, even in the same class, can have different numbers of points and subtly different positions according to the SAR sensor operating conditions [25]. Therefore, one-to-one correspondence is impractical, and similarity measurements of the two point sets are highly complex.

To solve these problems, several researchers have attempted to design optimal ASC-based methods. In Ref. [26], the researchers introduced the world view vector (WVV) to reconstruct the ASC features, then the weighted bipartite graph model (WBGM) was used. As the WVV provides a robust description of the point location and spatial relationship between the two sets, each weight on the line was allocated by computing the WVV-based similarity of the two ASC sets. Therefore, the direct one-to-one matching problem was solved. In Ref. [24], a similarity evaluation method for two ASC sets was introduced by exploiting the structural information contained in the ASC set. Once ASC matching was conducted using the Hungarian algorithm, point similarity and structure similarity were fused to evaluate the overall similarity of the two ASC sets based on the Dempster–Shafer theory of evidence. Meanwhile, a target reconstruction based on ASC was employed to avoid precise one-to-one correspondence and complex similarity measures [19,27]. Using the neighbor matching algorithm, only the template ASCs related to the test ASCs were selected to reconstruct the template image, whereas all the test ASCs were used to reconstruct the test image. Then, image correlation was employed to effectively measure the similarity between the two ASC sets.

As a dataset for experiments, moving and stationary target acquisition and recognition (MSTAR), which consists of a collection of one-foot-resolution SAR images, has been widely used in the past two decades [5,9,22,28–30]. Unfortunately, the amount of measured data needed to build a SAR ATR system is insufficient because of practical limitations in operating sensors and targets [31]. Furthermore, measured data typically represent limited environmental conditions, with very few articulation, configuration, and clutter changes, along with few sensor collection geometry variations [8,32]. In this regard, computer-generated synthetic data could serve as a valuable tool in the development of SAR ATR systems, leading to training and testing using only synthetic data and measured data, respectively [31,33]. Synthetic data are created using electromagnetic prediction software involving a computer-aided design (CAD) model of the MSTAR. By replacing measured data with synthetic data, many EOCs can be considered close to representing the real world, which is necessary to enhance classification performance.

However, an apparent distribution gap between synthetic and measured distributions exists due to differences in the CAD model and the real model, or the type of simulator used. Despite effort to decrease the gap [33], some parts of the target that did not appear in the measured data tend to be visible in the corresponding synthetic data (Figure 1). This could be regarded as a target occlusion situation if we only considered the target appearance of the image, which commonly happens in the real world [17,34]. Target occlusion inevitably occurs due to radar sensor operation and the external environment, such as artificial or natural objects, which make it difficult to classify the target using the traditional feature-based method. Comprehensively, a similarity measurement method where the unique features of the target can be captured is highly recommended for robustness to target occlusion in practical SAR ATR.

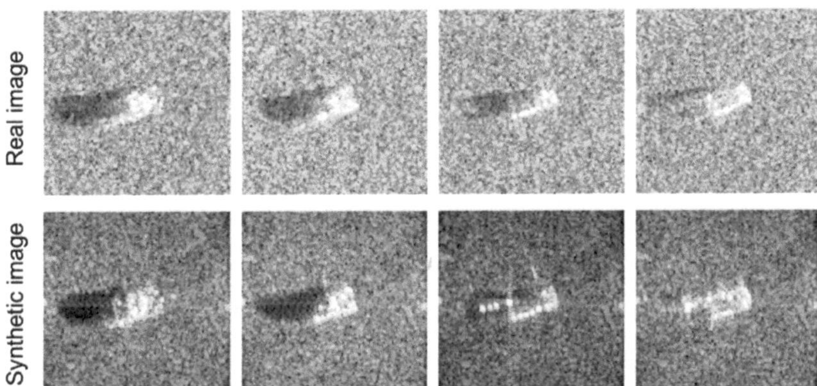

Figure 1. Examples of real and synthetic SAR images in SAMPLE, demonstrating substantial target differences located in the middle of the images.

The previously proposed WVV-based method has a robust ATR capability using all extracted ASCs [35]. The WVV is insensitive to translation, rescaling, random perturbation, and random addition and deletion of points. However, the method is sensitive to partial differences that may arise due to simulation limitations. Further, local feature matching under imbalance from two sets is still a challenging and important point for classification problems [36,37]. It is necessary to design a classification method that is less sensitive to the partial difference between real and synthetic images. Therefore, we propose an improved WVV-based ATR method using a subset of ASCs instead of using all ASCs to focus on local similarity.

2. Target Classification with a WVV-Based Feature Set

To classify the target in SAR images, we designed the classification algorithm in terms of local features matching. The ASCs were used as the unique features of the target in the SAR image. Figure 2 illustrates a flowchart of the proposed method. The flowchart is divided into five main steps: extraction of the scattering centers, sub-sampling of scatter centers based on amplitude, neighbor matching, WVV-based feature reconstruction, and similarity measurement. The ASCs of the template dataset extracted from each template image were prepared offline in advance. The value of T in Figure 2 is the number of template samples. In the classification algorithm, the extraction of scattering center from a test image was first employed. To analyze the local similarity and the imbalance problem in two ASC sets, some of the template ASCs were then selected by amplitude-based subsampling, where the number of test ASCs was exploited to adjusting the number of template ASCs. Afterwards, neighbor matching was applied to partially concentrate on the related points between the test ASCs and template ASCs. The subset of ASCs was used to reconstruct the world view vector feature set, thus considering the point similarity and structure similarity simultaneously. Later, the matching scores between the two ASC sets were calculated using weighted bipartite graph matching and then combined with several weights for overall similarity. Regarding the WVV-based similarity as the weight of bipartite graph matching, we found the optimal matching between the two sets. Finally, the overall similarity was determined to recognize the target by combining the matching score with several weights related to the matched/unmatched number of ASCs and selected/unselected number of ASCs, which could not be exclusively applied to the WVV-based similarity only. By repeating the process T times, the test image was consequently classified.

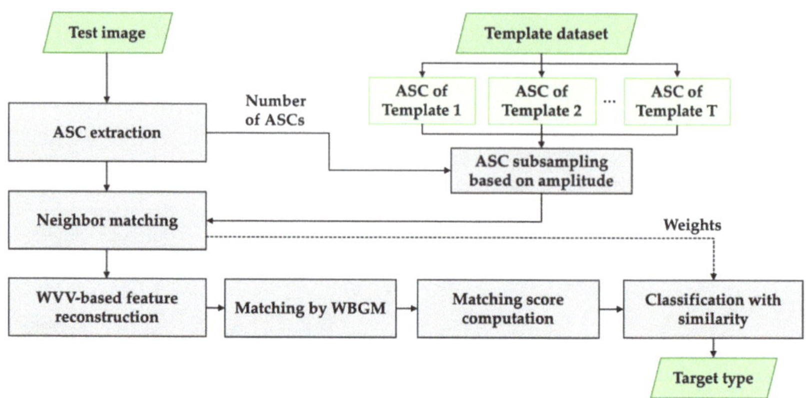

Figure 2. Flowchart of the proposed method.

2.1. Extraction of Scattering Centers

Several algorithms are available for extracting scatter centers in the image domain [38,39]. CLEAN is one of the most used algorithms for extracting scatter centers from SAR images [16,17,19,24,26,27,40–43]. The CLEAN algorithm employs a filter derived from the point spread function (PSF), which is given by:

$$\text{PSF}(x, y) = \left(e^{j\frac{4\pi f_c}{c}(x+\theta_c y)} \frac{4f_c B\Omega}{c^2} \right) \sin c\left(\frac{2B}{c}x \right) \sin c\left(\frac{2f_c\Omega}{c}y \right) \tag{1}$$

where (x, y) denotes the position of the scattering center; c is the speed of light; f_c is the center frequency; θ_c is the center azimuth; B is the frequency bandwidth of the radar; and Ω is the azimuth aperture. To extract the scattering centers properly, the filter used in the CLEAN algorithm has to incorporate the same smoothing window $w(x, y)$ used during image formation, resulting in:

$$h(x, y) = \text{psf}(x, y)w(x, y) \tag{2}$$

This work used a -35 dB Taylor Window that was employed by the SAMPLE dataset [31].

The CLEAN algorithm searches for the highest-amplitude pixel in the SAR image and records its amplitude A_i and its image coordinates (x_i, y_i). Then, the filter $h(x, y)$ shifts to the center of the pixel location and is multiplied by A_i. Assuming a point spread function (PSF) with a corresponding amplitude, the response of the imaging system is removed from the complex image [44]. In general, the iterative process is repeated with the residual image until a predetermined number of ASCs are extracted, or the amplitude of the extracted scattering point is less than the threshold value.

In this study, we extracted scattering centers located in a target region if its amplitude was greater than or equal to a threshold, which was intended to limit the number of ASCs. The target region and shadow of a SAR image can be separated by some conventional segmentation algorithms, including the K-means, the Otsu's method, and the iterated conditional modes (ICM). In this experiment, the ICM in Ref. [45] was used, but the ICM processing speed was improved by using the initial segmented image by the K-means, instead of the SAR image, as the input data of the ICM. Figure 3 shows the results of our scattering-center extraction task using CLEAN. The reconstructed SAR image, using a total of N scattering centers, shown in Figure 3b, was very similar to the real SAR image. The segmentation results and the extracted scattering centers located only inside the target region are shown in Figure 3c. At the end of the extraction process, the scattering centers were stored in an N × 3 matrix, $\{A_i, x_i, y_i \mid i = 1, 2, \ldots, N\}$. For the extracted scattering centers, the coordinates of the range (in other words, the slant range) were converted to the ground range to facilitate the matching analysis of scattering centers between SAR images taken from different depression angles, as follows:

$$x_i^g = x_i / cos\left(\theta_{depression}\right) \tag{3}$$

where $\theta_{depression}$ denotes the depression angle. Hereafter, x_i is referred to as ground range coordinated, x_i^g.

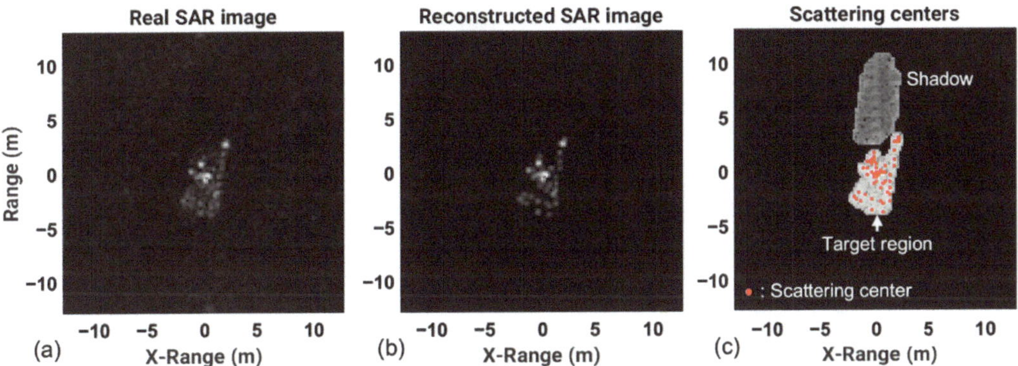

Figure 3. Examples of scattering centers extracted using CLEAN: (**a**) SAR image, (**b**) reconstructed SAR image, and (**c**) extracted scattering centers with the segmentation results.

In this paper, we used the ASC location (x_i, y_i) and normalized amplitude ($|A_i^{Norm}|$) for the SAR ATR. The locations describe the spatial distribution of ASCs, and the normalized amplitudes reflect the relative intensities of different ASCs. Therefore, the direct relevance to the target physical structures benefits the ATR performance. The N-scattering center feature set G was defined as follows [26]:

$$G = \left\{ \left(x_i, y_i, \left|A_i^{Norm}\right|\right) \mid i = 1, 2, \ldots, N \right\} \tag{4}$$

2.2. Amplitude-Based Selection

The apparent distribution difference between the amplitude distributions of synthetic and real SAR images is a challenge in training synthetic data. The statistical differences between the synthetic and measured data of SAMPLE were investigated by Ref. [33] to plot histograms of the image means and variances. The synthetic data tended to have a lower mean and variance than the measured data. The mean image difference was approximately 0.1 when we extracted scattering centers from real and synthetic images. There was a difference in the number of SCs extracted between the synthetic and real images. This could be due to the clear differences in the overall structure of the targets, as well as fine differences in the target details. However, it could also be due to the use of unequal minimum amplitudes to alleviate the difference in amplitude values. Before neighbor matching, a subset of the scatter points in the template image was selected in the order of their amplitude values.

The number of scatter points in subset M' was equal to the number of scatter points in the test image. However, the amplitude of the scattering points extracted from SAR images varied depending on subtle changes in the target's pose and imaging geometry (e.g., depression angle), and it is desirable to extract the points with a small buffer according to A_{ratio} as follows:

$$M' = min(M, N \times A_{ratio}) \tag{5}$$

where N and M are the numbers of scatter points in the test and template images, respectively. More ASCs in the template (synthetic) image can be expected (i.e., $A_{ratio} > 1$), considering the possible partial and random occlusion of the test image and the full set of scattering centers of the synthetic image (at least no occlusion). In addition, a maximum value should be set to control the imbalance of extracted ASC numbers between the real and synthetic images. In our experiments, the best results were achieved when the A_{ratio} was around 1.3, which was used in the following analysis.

2.3. Neighbor Matching

As shown in Figure 1, scattering centers with strong amplitudes in synthetic images are often invisible in real SAR images. In addition, the scattering centers of a part of the target may not appear in the real image because of the occlusion of the target. Therefore, considering the above reasons (difference from synthetic images and target occlusion), we focused on local similarity rather than identification using all scattering centers. For local similarity, the WVV descriptor should be reconstructed using scattering centers in the target overlap area (i.e., adjacent scattering centers). In this study, after selecting only the adjacent points between two sets of scattering centers through the neighbor-matching algorithm, we evaluated the similarity. First, the positions of the test ASCs were taken as the centers to form a binary region combined by all circles. When the template ASC was in the constructed binary region, it was selected; otherwise, it was discarded. The positions of the template ASCs were then reversed. The radius set for neighbor matching was chosen to be {0.3, 0.4, and 0.5 m} based on the resolution of MSTAR images, as well as the experimental observations [19,27].

Figure 4 presents an illustration of neighbor matching when the radius was set to 0.5 m. Neighbor matching was first conducted on the test ASCs (Figure 3a) and then on the template ASCs (Figure 3b). The corresponding matching results are shown in Figure 3a,b. Matched template ASCs, unmatched template ASCs, matched test ASCs, and unmatched test ASCs are represented by different markers. As shown in Figure 3c, the ratio of unmatched to matched points can be used to distinguish different targets. It is clear that the selected number of ASCs in the template was smaller when the type was not similar. Therefore, the matching results can provide discriminability for correct target recognition.

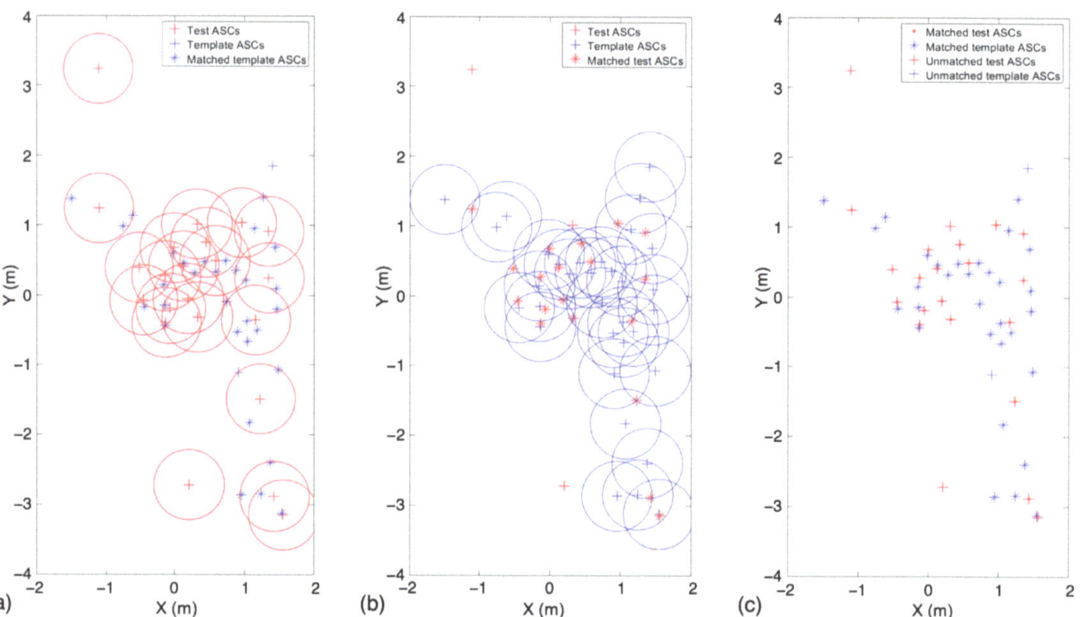

Figure 4. Example of neighbor matching: (**a**,**b**) neighbor matching based on the coordinates of the test ASC and template ASC, respectively. (**c**) Result of neighbor matching.

2.4. WVV-Based Feature Reconstruction

To solve the difficulties of similarity calculation, as mentioned before, the correspondence was established based on the existing feature descriptor, the WVV [26]. Test ASCs and matched template ASCs were used to construct each WVV-based feature set. The detailed procedure of the WVV-based feature reconstruction is illustrated in Algorithm 1 [26]. First, WVV establishes a polar coordinate system, taking the ith scattering center as the origin. The location of i is represented by the polar radii and polar angles of the remaining SCs. Next, the WVV_i is defined by sorting the polar radii according to their polar angles, and the radii are linearly interpolated at 1° intervals. The WVV is mapped into a vector of length 360. Finally, to avoid sensitivity to rescaling, the elements in the interpolated WVV are normalized by the maximum element. After iteration as the number of scattering centers, the scattering center feature set will consequently be the WVV-based feature set.

Algorithm 1: WVV-based scattering center feature reconstruction

---

Input: Scattering center feature set $G = \left\{ \left(x_i, y_i, \left| A_i^{Norm} \right| \right) \middle| i = 1, 2, \ldots, N \right\}$.

For $i = 1{:}N$

1. Establish a polar coordinate system with the origin at ith point.

2. Compute the polar radii r_{ik} and the polar angles θ_{ik} $(k = 1, \ldots, N, \ k \neq i)$ of the rest $N - 1$ points.

3. Sort the polar radii r_{ik} corresponding to their polar angles θ_{ik} and define ith WVV as $WVV_i = \{r_{ik} | k = 1, \ldots, N; k \neq i; \theta_{ik} \leq \theta_{ik+1}\}$.

4. Interpolate linearly the polar radii.

5. Construct the interpolated WVV, WVV_i^{interp} which consists of r_{ij} $(j = 1, \ldots, 360)$.

6. Normalize the r_{ij} by the maximum element, $r_{ij}^{Norm} = r_{ij} / \max\limits_{k=1,\ldots,360} \left\{ r_{ij} \right\}$.

$WVV_i^{interp} = \{ r_{ij}^{Norm} | j = 1, \ldots, 360 \}$.

End

Output: WVV-based feature set, $G' = \left\{ \left(WVV_i^{interp}, \left| A_i^{Norm} \right| \right) \middle| i = 1, 2, \ldots N \right\}$

Figure 5 shows an example of WVV-based feature reconstruction using 2S1 ASCs from the SAMPLE dataset. Figure 5a,b show the 56-ASC set and interpolated 23rd WVV, respectively. In Figure 5b, the blue point at the origin is the 23rd ASC. At this point, the WVV_{23} comprises the polar radii of the remaining points. Subsequently, the WVV_{23} was linearly interpolated at 1° intervals, and the elements were normalized by the maximum element. After iterations, the WVV-based feature set will have 56 interpolated WVV sets and normalized amplitudes. When some ASCs are randomly removed, the WVV-based features, consisting of the remaining ASCs, can maintain their spatial structures. Therefore, the proposed method is not sensitive to random noise removal.

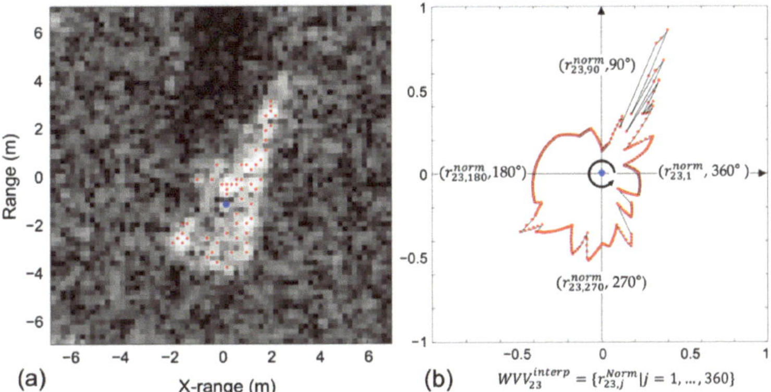

Figure 5. Example of WVV-based feature reconstruction using ASCs: (**a**) 56-ASC set and (**b**) interpolated 23rd WVV.

2.5. Similarity Calculation

In ATR, it is important to design a similarity measurement between the input test image and archive templates. The structural similarity between two sets of scattering centers is obtained through WVV reconstruction of the scattering centers, and several weights related to the number of scattering centers and the number of matches are proposed.

2.5.1. Matching Score

After the WVV-based feature reconstruction, target classification based on point matching was conducted. The WVV provides a robust description of scattering centers. The Euclidean distance between the locations of the scattering centers is generally used when evaluating the point-to-point similarity between two sets of SCs. Meanwhile, spatial structure relationships were characterized by WVV-based features.

Comparing G'_{test}, the WVV-based feature set of the test, and template dataset $\{B'_k | k = 1, \ldots T\}$ with Equation (4) $S(G'_{test}, B'_k)$ below, the most similar template was selected

$$S(G'_{test}, B'_k) = \sum s(g'_l, b'_l), \; l = 1, \ldots, \min(M, N) \tag{6}$$

where $g'_l \in G'_{test}$ and $b'_l \in B'_k$ are the matching pairs and N and M are the number of SCs in the test and k-th template, respectively.

There was no strict one-to-one correspondence between the two sets. We needed to find an optimal assignment to maximize S between the two point sets. Based on weighted bipartite graph matching (WBGM), the similarity of the matching pairs was calculated as

$$s\left(g'_i, b'_j\right) = F\left(g'_i, b'_j\right)\left(360 - \| WVV^{interp}_{g'_i} - WVV^{interp}_{b'_j} \|^2\right) / (360) / (1 + D) \tag{7}$$

$$F\left(g'_i, b'_j\right) = \begin{cases} 1, & D\left(g'_i, b'_j\right) \le R \\ 0, & otherwise \end{cases} \tag{8}$$

where $D\left(g'_i, b'_j\right)$ are the Euclidean distances between the test ASC and template ASC. If D is greater than R, which is the distance (meter) for neighbor matching, 0 is assigned to $s\left(g'_i, b'_j\right)$ to prevent matching between points that are too far away. Accordingly, the number of matched pairs, K_{match} was applied to normalize the similarity measurement.

$$S_{Norm}\left(G'_{test}, B'_k\right) = \frac{S\left(G'_{test}, B'_k\right)}{K_{match}} \tag{9}$$

2.5.2. Weight Design

The previously obtained WVV-based matching score did not consider the difference in the number of scattering centers that occurs when the types of targets are different. Therefore, it was necessary to properly design the weights based on the number of matches and the number of points that were not matched, and reflect them in the overall degree of similarity.

Although the type was different from the test, if the template points around the test point were gathered, the WVV-based similarity was calculated to be high by neighbor matching (Figure 6a). The WVV-based matching score is higher in Figure 6a than that in Figure 6b, but in Figure 6a, the difference between the number of test and template scattering centers is substantial. Because the number of points in the template will be similar for targets of the same type, it is necessary to consider the difference in the number of points between the test and the template.

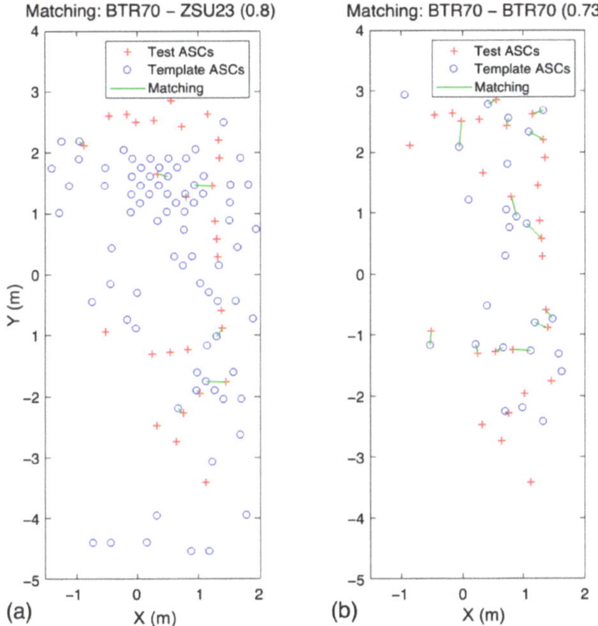

Figure 6. WVV-based similarity matching results between two sets of scattering centers. (**a**) Different types of test and template SCs, (**b**) same target type of test and template SCs.

Unmatched test scattering centers (MA: Missing Alarm) and unmatched template scattering centers (FA, False Alarm) were considered in the matching score. Here, we used the following quadratic weights presented by Ref. [24]:

$$w_a = 1 - \left(\frac{s+q}{N+m} \right)^2 \tag{10}$$

where s and q represent the number of MA and FA, respectively, and $s + q$ can be calculated as $N + m - 2 \times K_{match}$, where m is the number of matched template ASCs by the neighbor matching algorithm. Because of noise and extraction errors, it is common to observe the appearance of a few MAs and FAs. As the proportion of MAs and FAs increases, the weight w_a decreases more quickly.

In addition, when the test and template are of the same type, the number of scattering centers within the radius tends to be large during neighbor matching. Therefore, along with S_{Norm} and w_a, the ratio of the number of selected points to the total number of points was used as a weight. The overall matching score adopted in this study was defined as:

$$MS = S_{Norm}(G'_{test}, B'_k) \times W_a \times n/N \times m/M \tag{11}$$

In Equation (11), n/N is the ratio of the number of scattering centers extracted from the test image to the test ASCs adjacent to the template ASCs, and m/M is the ratio of the number of template ASCs to the number of neighboring ASCs to the number of test ASCs.

The procedure for the proposed target WVV-based matching method is shown in Figure 2. The entire algorithm was programmed in MATLAB® R2021a (9.10), employing Matlab's Parallel Computing Toolbox. First, because it is impossible to define a one-to-one correspondence between the test and template, subsampling based on amplitude and neighbor matching was conducted to reduce the imbalance. The similarity matrix of the WVV sets was obtained using Equation (7) $s\left(g'_i, b'_j \right)$, and then all the similarities were employed as weights for the WBGM. By repeating the above process based on the number of template databases, the type of test target could be determined as a category according to the template with the maximum matching score.

3. Experiments

3.1. Experimental Settings

To evaluate the classification performance of the proposed method, experiments were conducted on SAMPLE datasets under standard operation conditions (SOC) and EOC, including random and partial occlusions. The SAMPLE dataset consisted of real and synthetic SAR images using CAD models of 10-class MSTAR targets, which are listed in Table 1 [31]. Furthermore, the optical images of 10-class targets are shown in Figure 7; they are ground vehicles, carriers, and trucks (you can see more targets in SAR images and types of targets in Refs. [46–49]). The data had a spatial resolution of one foot. Unfortunately, only some parts of the SAMPLE dataset are publicly available, which is appropriate for small-scale operations. They were collected at azimuth angles of 10–80° with depression angles from 14–17°. To validate the proposed method for 10 targets, we used target chips with depression angles of 16° and 17°. In Table 1, the number of SAMPLE datasets for each target class is listed according to the depression angle.

Table 1. Number of SAR images in the test and training sets of the SAMPLE dataset.

Type	2S1	BMP2	BTR70	M1	M2	M35	M60	M548	T72	ZSU23
Test set (real 16°) Train set (syn 16°)	50	55	43	52	52	52	52	51	56	50
Test set (real 17°) Train set (syn 17°)	58	52	49	51	53	53	53	60	52	58

Figure 7. Examples of optical images of SAMPLE dataset targets.

When using CLEAN, scattering centers with a minimum amplitude or higher were extracted. The distribution between the amplitude in the clutter around the target and that of the target may be considered when selecting the threshold value for the extraction of scattering centers. Because a difference in amplitude exists between real and synthetic images, scattering centers with an amplitude of 0.25 or higher for real and 0.14 or more for synthetic images were extracted considering the difference in amplitude between synthetic and real data of SAMPLE [33,50]. The statistical values of the number of extracted scattering centers are shown in Figure 8; evidently, the number of scattering centers extracted by CLEAN was different for each target. BMP2 and BTR70 had an average of 30–50 scattering centers, which was less than that in the other targets. Meanwhile, M548 and M60 had numerous scattering centers. Thus, this imbalance in the number of scattering centers caused matching to be challenging, although valuable information was obtained for target identification.

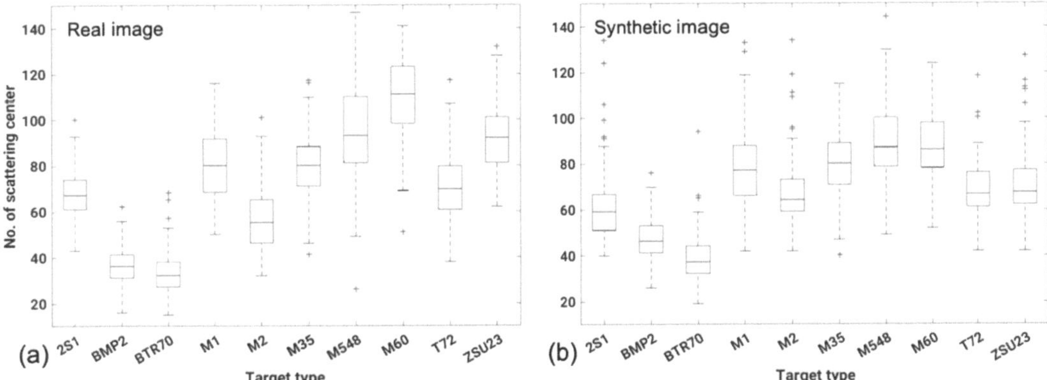

Figure 8. Box plot of the number of scattering centers. (**a**) Real SAR image (16°, 17°) and (**b**) synthetic SAR image (16°, 17°).

For neighbor matching, the radius must be determined. Although the WVV descriptor is not affected by the translation of the scattering point, translation may affect the identification performance because only a part of the template remains based on the test point after neighbor matching. In Refs. [17,24], the researchers applied several different radii (e.g., 0.3, 0.4, and 0.5) for neighbor matching, and then the averages of all the similarities were employed to determine the final similarity between the test and its corresponding template.

In the experiment of this study, when the number of extracted scattering centers was small, the identification rate increased when a high value of 0.5 was used, rather than a low value of 0.3 m. Conversely, when a high value of 0.5 was used, there was no change in the identification rate, even when the number of scattering centers was large. We may need to consider the possibility of a translation transformation between the target and template before applying the neighbor-matching algorithm. Therefore, we used a single value of 0.5 so that the effect of the centering error of the test image could be mitigated.

Figure 9 shows an example of the processing according to the proposed algorithm (Figure 2), in which the test image BTR 70 was well recognized in the ambiguous template images (BTR 70 and ZSU23). First, the scattering points of each template image were selected in order of amplitude so that the number of scattering points of the test image did not exceed 48×1.3 (=62.4), where the parameter 1.3 means A_{ratio}, as mentioned in Section 2.2. Consequently, all 46 scattering points extracted from template BTR70 were selected, and only 62 of the 69 scattering points of template ZSU23 were selected. The point ratio selected by neighbor matching had a lower value for template ZSU23 than BTR70. The test ASCs and matched template ASCs were used to reconstruct the WVV set, and the similarity between the test and template WVV sets was measured. The WVV-based similarity (S_{Norm}) of template ZSU23 was higher than that of template BTR70. Contrastingly, the number of pairs matched by the WBGM of template ZSU23 was smaller than that of template BTR70; therefore, the W_a of ZSU23 was lower. Finally, after WBGM, the matching score was calculated using Equation (9), giving a value of 0.63 for the template image of BTR70 and 0.59 for ZSU23. Although the WVV-based similarity alone did not show good recognition results, the performance was considerably improved using the proposed weights.

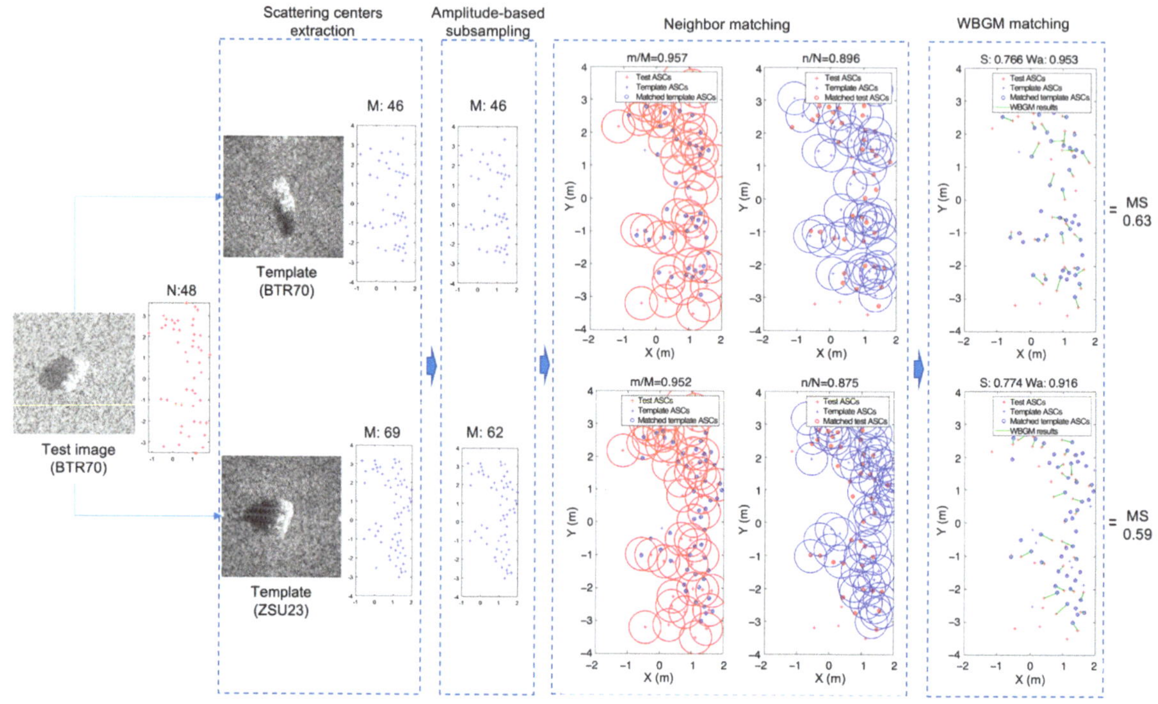

Figure 9. Example of recognition results obtained using the proposed algorithm.

3.2. Standard Operating Condition

The proposed method was first evaluated under SOC on 10 classes of targets for overall classification accuracy. An actual image was used as the test target chip and a simulated image was used as the template. In the identification performance experiment using a synthetic image as a template, a simulated image can be used under the same observation conditions (depression angle) as the real image. However, owing to the characteristics of the SAR image, the distribution of scattering centers may vary depending on the imaging parameters, such as the depression angle or subtle changes in the pose of the target. In addition, because there is a difference between synthetic and real images, it is more advantageous for target identification to use not only the same observation angle data, but also adjacent observation angle data, as the template image. The merged training data were generated by combining the synthetic data with depression angles of 16° and 17° for 1052 chips.

Tables 2 and 3 show the confusion matrix of the proposed method using a real 17° and 16° as the test image and a synthetic 16°/17° as the template. The performance was expressed by the percentage of correct classifications (PCCs). The average PCCs of all 10 targets in Table 2 were 90.7%, whereas the M2 target was recognized with PCCs under 80%, and the remaining targets were over 86%. For the result of real 16° data (Table 3), not only M2 but also BMP2 and M548 showed PCCs under 80%. Meanwhile, T72 and ZSU23 showed higher PCCs than the real 17° data. For real 16° and 17° data identifications, when only synthetic 16° and 17° data corresponding to each other were used as template data, the recognition rates were 87.0% and 88.0%, respectively. When synthetic 16°/17° data were used to identify real 16° and 17° data, the identification rates were improved to 87.3% and 90.7%, respectively. With the merged data from synthetic 16° and 17° angles, the real 17° data improved the recognition performance by 2.7%, while the real 16° data improved by only 0.3%.

Table 2. Recognition results obtained using the proposed method within 10 target classification tests (test image: real 17°, template image: synthetic 16°/17°).

Type	2S1	BMP2	BTR70	M1	M2	M35	M548	M60	T72	ZSU23	PCC (%)
2S1	54	0	1	0	1	0	0	0	0	2	93.10
BMP2	4	47	1	0	0	0	0	0	0	0	90.38
BTR70	2	0	47	0	0	0	0	0	0	0	95.92
M1	0	0	0	50	1	0	0	0	0	0	98.04
M2	4	4	0	0	41	0	0	0	1	3	77.36
M35	0	0	0	0	0	52	1	0	0	0	98.11
M548	0	0	0	0	0	3	49	0	0	1	92.45
M60	3	0	0	4	0	0	0	52	1	0	86.67
T72	1	3	0	0	1	0	0	2	45	0	86.54
ZSU23	6	0	0	0	1	0	0	0	0	51	87.93
Average											90.65

Table 3. Recognition results obtained using the proposed method within 10 target classification tests (test image: real 16°; template image: synthetic 16°/17°).

Type	2S1	BMP2	BTR70	M1	M2	M35	M548	M60	T72	ZSU23	PCC (%)
2S1	44	0	2	0	0	0	0	0	0	4	88.00
BMP2	10	42	1	0	2	0	0	0	0	0	76.36
BTR70	0	0	41	0	1	0	0	0	0	1	95.35
M1	0	0	0	52	0	0	0	0	0	0	100.00
M2	4	3	0	0	40	0	0	0	1	4	76.92
M35	0	0	7	0	0	43	0	0	0	2	82.69
M548	0	0	7	0	0	2	41	0	0	2	78.85
M60	3	0	0	5	0	0	0	42	1	0	82.35
T72	0	0	0	0	0	0	0	1	55	0	98.21
ZSU23	2	0	1	0	0	0	0	0	0	47	94.00
Average											87.27

We attempted to prove the effectiveness of our proposed method using a subset of ASCs. There are three types of ASCs: ASC_{all} means that all the test and template ASCs are used in WVV-based feature reconstruction as in Ref. [26], $ASC_{subset(neighbor)}$ indicates that the subset of ASCs is selected by neighbor matching, and $ASC_{subset(amp,neighbor)}$ indicates that the subset of ASCs is selected by amplitude-based sub-sampling followed by neighbor matching. Figure 10 shows the recognition rate of the 10 targets according to the ASC types used for classification. As described above, the merged data from synthetic 16° and 17° angles were used as a single template. This is the result of averaging the respective recognition performance obtained by using real 16° and 17° data as the test images. The overall performance of ASC_{all} using all extracted ASCs suggested by Ref. [26] was lower than the results achieved by our proposed methods. When using the ASC_{all}, the recognition rate of BMP2 was very low, at about 70%. The $ASC_{subset(amp,neighbor)}$ showed a high recognition rate of about 85% or more in all targets except M2, and the performance variance for the 10 targets was the smallest among the three types of ASC. The results of $ASC_{subset(amp,neighbor)}$ were appropriate for the 10-class target classification, and in particular, the recognition rate of BMP2, BTR70, and T72 was greatly improved, by more than 5% compared to ASC_{all}.

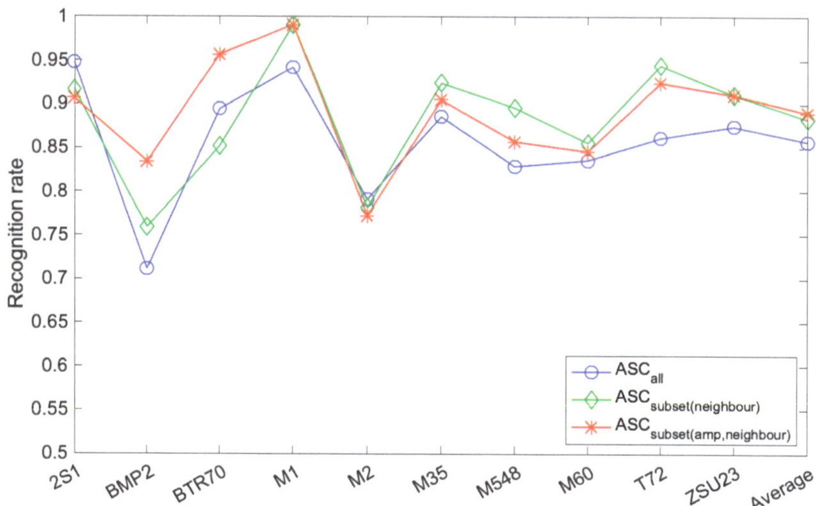

Figure 10. Recognition rate of each target according to different sets of scattering centers.

3.3. Occlusion and Random Missed Pixels

In the real world, the occlusion of a target by the external environment, such as artificial or natural objects, can always occur. The performance of the proposed method was investigated using directional occlusions. Similarly, the test samples in Table 1 were simulated to obtain samples with different levels of directional occlusion for classification. Figure 11 shows the recognition rate for each method with varying levels of directional occlusions. When the target was partially occluded, only part of the local structure was discriminative for the target. The WVV-based reconstruction with ASCs selected by neighbor matching made it possible to consider the local similarity between the test and the template.

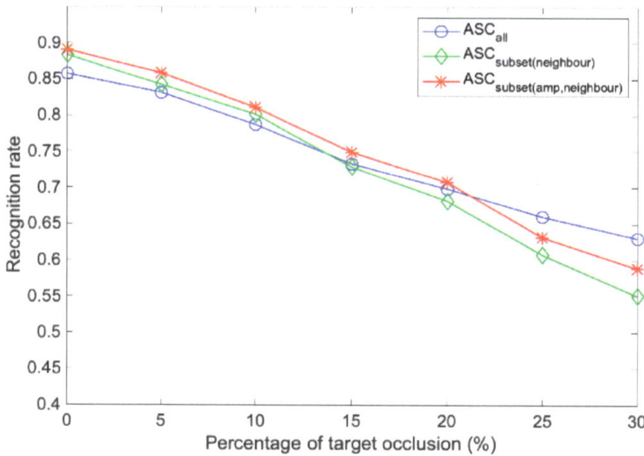

Figure 11. Performance comparison of different subsets by the degree of partial occlusion.

When the test was occluded randomly from different directions, the performance of the proposed method, $ASC_{subset(amp,neighbor)}$, remained at a high level of over 70% until the occlusion level was 20%. Compared to the overall recognition rate of ASC_{all}, it was improved by 1–4%. When the WVV was reconstructed with a part selected as a neighbor rather than the entire scattering center, better results were obtained with less than 20%

occlusion. In addition, the method applying $ASC_{subset(amp,neighbor)}$ was maintained at approximately 3% higher than the neighbor subset, although there was no considerable difference in value. However, when >25% occlusion occurred, it was better to use ASC_{all}. In the case of occlusion of 20% or more, the average recognition rate was lowered to 70% or less. That is, the overall reliability of the performance was too low to be used in practice. Therefore, it is better to use a subset instead of all ASCs, assuming a low occlusion rate, for a high recognition rate.

The ASCs could be interrupted by noise and differences in image resolution. A sensitivity analysis for randomly missed points was also performed. We randomly removed them according to the percentage of missed points. The remaining SCs were used to reconstruct WVV-based features that were matched with the templates. The percentage of missed points varied from 0 to 50%, and 10 Monte Carlo simulations were conducted for each percentage.

As shown in Figure 12, $ASC_{subset(neighbor)}$ always showed a higher recognition rate than ASC_{all}, verifying its robustness to random occlusion. The high performance of over 80% in $ASC_{subset(neighbor)}$ was maintained until a removal level of 45%. On the other hand, $ASC_{subset(amp,neighbor)}$ showed better results than ASC_{all} only up to 15% random occlusion. This is because the ASCs in the test image were randomly removed, while the ASCs in the template image were removed based on the amplitude of ASCs, so the relevance between the remaining pixels decreased significantly as the percentage of random occlusion increased.

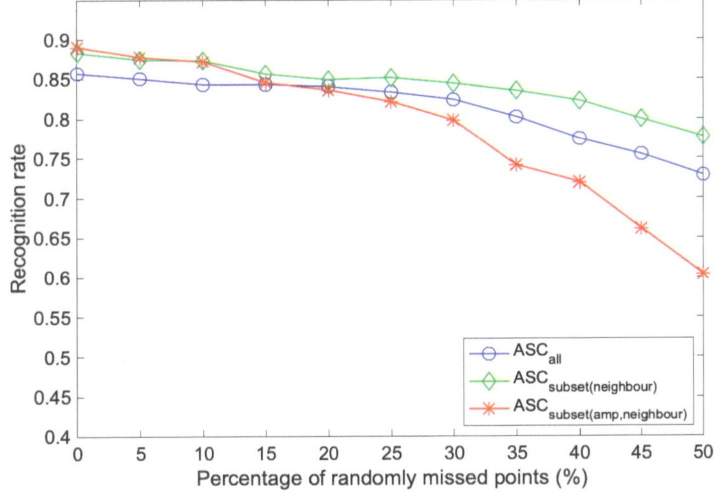

Figure 12. Performance comparison of the subsets by randomly missed points.

We achieved high performances using the SAMPLE dataset, with 100% synthetic data in the training set. When the experiments were conducted by changing the types of ASCs used, the overall recognition rate under SOC (no occlusion and no random missed pixels) was improved about 4% compared to ASC_{all}. Additionally, the proposed method was less sensitive to a small amount of partial occlusion and random pixel removal.

We compared the proposed method with existing methods [39] to illustrate its performance. The experimental results, where the average recognition rate was 24.97% when solely synthetic data were used in the training batch, were initially presented by Ref. [31], the creators of the SAMPLE dataset. Their algorithm was based on a convolutional neural network (CNN). They then achieved average accuracies of 51.58% by training DenseNet with the assistance of a generative adversarial network (GAN) in their most recent work [25]. Their deep learning-based performances seem to be superior to our proposed method in

Refs. [51,52]. However, the accuracy dropped notably below 85% when they used 100% synthetic data in training. Consequently, our proposed method had a higher performance, which was also more stable when using 100% synthetic data as the training dataset.

In terms of the validity under EOCs, our performances were also compared to other previous works where they used the SAMPLE dataset for experiments, but the organization of data setting (number of classes, depression angle, etc.) was not same as our dataset. For example, the target recognition rate decreased by 30% during 50% random occlusion in Ref. [17], when they use a measured dataset (MSTAR) for both testing and training, but only by 10.6% in our study using $ASC_{subset(neighbor)}$. This was the best performance under random occlusion among the previous works. Compared to the results of other methods, our similarity measure can effectively overcome the recognition difficulties caused by random occlusion when $ASC_{subset(neighbor)}$ is used, though the same condition with our dataset was not applied in the methods. We believe that the advantage of the proposed algorithm is that it mostly focuses on the local features related to the intersected target parts to improve the feasibility of the SAR ATR system in response to a realistic scenario. However, there are several limitations resulting from human intervention. One is the existence of an empirical threshold for amplitude-based subsampling, and the need to set a specific radius in neighbor matching, which can cause significant performance degradation in our algorithm when most parts of the target are occluded. To resolve these limitations, we will also consider how to integrate global similarities with the proposed method in the case of target occlusion.

4. Conclusions

A robust algorithm is required to identify partial differences between real and synthetic images to perform target identification based on a dataset of synthetic images, such as SAMPLE. We proposed an improved WVV-based ATR method using a subset of the template's ASCs, using the ASCs' amplitude and the proximity of the scattering center location, which is less susceptible to the partial differences between two ASC sets. The SAMPLE dataset, with 10 classes of military targets, was used in the experiments. With the merged template of synthetic $16°/17°$ images, the recognition rates were 87.3% for the $16°$ real images and 90.7% for the $17°$ real images. The performance with occlusion remained above 70% until the occlusion level reached 20%. In addition, the subset of ASCs selected by neighbor matching achieved recognition rates great than 80% until the proportion of randomly missed points reached 45%. Therefore, we expect that the proposed method will be useful in practical SAR ATR systems using synthetic images with partial and random differences in ASCs due to occlusion.

Author Contributions: Conceptualization, methodology, formal analysis, investigation, resources, data curation, visualization, S.L. and S.-W.K.; writing—original draft preparation, S.L.; writing—review and editing, supervision, project administration, funding acquisition, S.-W.K. All authors have read and agreed to the published version of the manuscript.

Funding: This research was supported in part by the Agency for Defense Development (ADD) in South Korea (No. UD200004FD).

Institutional Review Board Statement: Not applicable.

Informed Consent Statement: Not applicable.

Data Availability Statement: The SAMPLE dataset was obtained in accordance with the instructions contained in Ref. [31] and is available online: https://github.com/benjaminlewis-afrl/SAMPLE_dataset_public (accessed on 25 August 2022).

Acknowledgments: The authors would like to thank the US Air Force Research Lab for providing the public SAMPLE datasets.

Conflicts of Interest: The authors declare no conflict of interest.

References

1. Wang, L.; Bai, X.; Zhou, F. SAR ATR of ground vehicles based on ESENet. *Remote Sens.* **2019**, *11*, 1316. [CrossRef]
2. Brusch, S.; Lehner, S.; Fritz, T.; Soccorsi, M.; Soloviev, A.; van Schie, B. Ship surveillance with TerraSAR-X. *IEEE Trans. Geosci. Remote Sens.* **2010**, *49*, 1092–1103. [CrossRef]
3. Song, D.; Zhen, Z.; Wang, B.; Li, X.; Gao, L.; Wang, N.; Xie, T.; Zhang, T. A novel marine oil spillage identification scheme based on convolution neural network feature extraction from fully polarimetric SAR imagery. *IEEE Access* **2020**, *8*, 59801–59820. [CrossRef]
4. Fu, K.; Dou, F.-Z.; Li, H.-C.; Diao, W.-H.; Sun, X.; Xu, G.-L. Aircraft recognition in SAR images based on scattering structure feature and template matching. *IEEE J. Sel. Top. Appl. Earth Obs. Remote Sens.* **2018**, *11*, 4206–4217. [CrossRef]
5. El-Darymli, K.; Gill, E.W.; Mcguire, P.; Power, D.; Moloney, C. Automatic target recognition in synthetic aperture radar imagery: A state-of-the-art review. *IEEE Access* **2016**, *4*, 6014–6058. [CrossRef]
6. Moreira, A.; Prats-Iraola, P.; Younis, M.; Krieger, G.; Hajnsek, I.; Papathanassiou, K.P. A tutorial on synthetic aperture radar. *IEEE Geosci. Remote Sens. Mag.* **2013**, *1*, 6–43. [CrossRef]
7. Park, J.-H.; Seo, S.-M.; Yoo, J.-H. SAR ATR for limited training data using DS-AE network. *Sensors* **2021**, *21*, 4538. [CrossRef]
8. Paulson, C.; Nolan, A.; Goley, S.; Nehrbass, S.; Zelnio, E. Articulation study for SAR ATR baseline algorithm. In *Algorithms for Synthetic Aperture Radar Imagery XXVI*; SPIE: Bellingham, WA, USA, 2019; Volume 10987, pp. 73–89.
9. Kechagias-Stamatis, O.; Aouf, N. Automatic target recognition on synthetic aperture radar imagery: A survey. *IEEE Aerosp. Electron. Syst. Mag.* **2021**, *36*, 56–81. [CrossRef]
10. Dogaru, T.; Phelan, B.; Liao, D. Imaging of buried targets using UAV-based, ground penetrating, synthetic aperture radar. In *Radar Sensor Technology XXIII*; SPIE: Bellingham, WA, USA, 2019; Volume 11003, pp. 18–35.
11. Ross, T.D.; Bradley, J.J.; Hudson, L.J.; O'connor, M.P. SAR ATR: So what's the problem? An MSTAR perspective. In *Algorithms for Synthetic Aperture Radar Imagery VI*; SPIE: Bellingham, WA, USA, 1999; Volume 3721, pp. 662–672.
12. Keydel, E.R.; Lee, S.W.; Moore, J.T. MSTAR extended operating conditions: A tutorial. *Algorithms Synth. Aperture Radar Imag. III* **1996**, *2757*, 228–242.
13. Chen, S.; Wang, H.; Xu, F.; Jin, Y.-Q. Target classification using the deep convolutional networks for SAR images. *IEEE Trans. Geosci. Remote Sens.* **2016**, *54*, 4806–4817. [CrossRef]
14. Dudgeon, D.E.; Lacoss, R.T. An Overview of Automatic Target Recognition. *Linc. Lab. J.* **1993**, *6*, 3–9.
15. Liang, W.; Zhang, T.; Diao, W.; Sun, X.; Zhao, L.; Fu, K.; Wu, Y. SAR target classification based on sample spectral regularization. *Remote Sens.* **2020**, *12*, 3628. [CrossRef]
16. Feng, S.; Ji, K.; Zhang, L.; Ma, X.; Kuang, G. SAR target classification based on integration of ASC parts model and deep learning algorithm. *IEEE J. Sel. Top. Appl. Earth Obs. Remote Sens.* **2021**, *14*, 10213–10225. [CrossRef]
17. Lu, C.; Fu, X.; Lu, Y. Recognition of occluded targets in SAR images based on matching of attributed scattering centers. *Remote Sens. Lett.* **2021**, *12*, 932–943. [CrossRef]
18. Potter, L.C.; Moses, R.L. Attributed scattering centers for SAR ATR. *IEEE Trans. Image Process.* **1997**, *6*, 79–91. [CrossRef]
19. Fan, J.; Tomas, A. Target Reconstruction Based on Attributed Scattering Centers with Application to Robust SAR ATR. *Remote Sens.* **2018**, *10*, 655. [CrossRef]
20. Liu, Z.; Wang, L.; Wen, Z.; Li, K.; Pan, Q. Multi-Level Scattering Center and Deep Feature Fusion Learning Framework for SAR Target Recognition. *IEEE Trans. Geosci. Remote Sens.* **2022**, *60*, 5227914.
21. Ding, B.; Wen, G.; Ma, C.; Yang, X. An efficient and robust framework for SAR target recognition by hierarchically fusing global and local features. *IEEE Trans. Image Process.* **2018**, *27*, 5983–5995. [CrossRef]
22. Lv, J.; Liu, Y. Data augmentation based on attributed scattering centers to train robust CNN for SAR ATR. *IEEE Access* **2019**, *7*, 25459–25473. [CrossRef]
23. Feng, S.; Ji, K.; Wang, F.; Zhang, L.; Ma, X.; Kuang, G. Electromagnetic Scattering Feature (ESF) Module Embedded Network Based on ASC Model for Robust and Interpretable SAR ATR. *IEEE Trans. Geosci. Remote Sens.* **2022**, *60*, 5235415. [CrossRef]
24. Ding, B.; Wen, G.; Zhong, J.; Ma, C.; Yang, X. Robust method for the matching of attributed scattering centers with application to synthetic aperture radar automatic target recognition. *J. Appl. Remote Sens.* **2016**, *10*, 016010. [CrossRef]
25. Lewis, B.; DeGuchy, O.; Sebastian, J.; Kaminski, J. Realistic SAR data augmentation using machine learning techniques. In *Algorithms for Synthetic Aperture Radar Imagery XXVI*; SPIE: Bellingham, WA, USA, 2019; Volume 10987, pp. 12–28.
26. Tian, S.; Yin, K.; Wang, C.; Zhang, H. An SAR ATR method based on scattering centre feature and bipartite graph matching. *IETE Tech. Rev.* **2015**, *32*, 364–375. [CrossRef]
27. Ding, B.; Wen, G. Target reconstruction based on 3-D scattering center model for robust SAR ATR. *IEEE Trans. Geosci. Remote Sens.* **2018**, *56*, 3772–3785. [CrossRef]
28. Diemunsch, J.R.; Wissinger, J. Moving and stationary target acquisition and recognition (MSTAR) model-based automatic target recognition: Search technology for a robust ATR. In *Algorithms for synthetic aperture radar Imagery V*; SPIE: Bellingham, WA, USA, 1998; Volume 3370, pp. 481–492.
29. Camus, B.; Barbu, C.L.; Monteux, E. Robust SAR ATR on MSTAR with Deep Learning Models trained on Full Synthetic MOCEM data. *arXiv* **2022**, arXiv:2206.07352.
30. Vernetti, A.; Scarnati, T.; Mulligan, M.; Paulson, C.; Vela, R. Target Pose Estimation using Fused Radio Frequency Data within Ensembled Neural Networks. In Proceedings of the 2022 IEEE Radar Conference (RadarConf22), New York, NY, USA, 21–25 March 2022; pp. 1–6.

31. Lewis, B.; Scarnati, T.; Sudkamp, E.; Nehrbass, J.; Rosencrantz, S.; Zelnio, E. A SAR dataset for ATR development: The Synthetic and Measured Paired Labeled Experiment (SAMPLE). In *Algorithms for Synthetic Aperture Radar Imagery XXVI*; SPIE: Bellingham, WA, USA, 2019; Volume 10987, pp. 39–54.

32. Arnold, J.M.; Moore, L.J.; Zelnio, E.G. Blending synthetic and measured data using transfer learning for synthetic aperture radar (SAR) target classification. In *Algorithms for Synthetic Aperture Radar Imagery XXV*; SPIE: Bellingham, WA, USA, 2018; Volume 10647, pp. 48–57.

33. Inkawhich, N.; Inkawhich, M.J.; Davis, E.K.; Majumder, U.K.; Tripp, E.; Capraro, C.; Chen, Y. Bridging a gap in SAR-ATR: Training on fully synthetic and testing on measured data. *IEEE J. Sel. Top. Appl. Earth Obs. Remote Sens.* **2021**, *14*, 2942–2955. [CrossRef]

34. Bhanu, B.; Jones, G. Target recognition for articulated and occluded objects in synthetic aperture radar imagery. In Proceedings of the Proceedings of the 1998 IEEE Radar Conference, RADARCON'98. Challenges in Radar Systems and Solutions (Cat. No.98CH36197), Dallas, TX, USA, 14 May 1998; pp. 245–250.

35. Murtagh, F. A new approach to point pattern matching. *Publ. Astron. Soc. Pac.* **1992**, *104*, 301. [CrossRef]

36. Chen, Y.; Huang, D.; Xu, S.; Liu, J.; Liu, Y. Guide Local Feature Matching by Overlap Estimation. *arXiv* **2022**, arXiv:2202.09050. [CrossRef]

37. Sarlin, P.-E.; DeTone, D.; Malisiewicz, T.; Rabinovich, A. Superglue: Learning feature matching with graph neural networks. In Proceedings of the IEEE/CVF Conference on Computer Vision and Pattern Recognition, Seattle, WA, USA, 13–19 June 2020; pp. 4938–4947.

38. Yun, D.-J.; Lee, J.-I.; Bae, K.-U.; Song, W.-Y.; Myung, N.-H. Accurate Three-Dimensional Scattering Center Extraction for ISAR Image Using the Matched Filter-Based CLEAN Algorithm. *IEICE Trans. Commun.* **2018**, *101*, 418–425. [CrossRef]

39. Araujo, G.F.; Machado, R.; Pettersson, M.I. Non-Cooperative SAR Automatic Target Recognition Based on Scattering Centers Models. *Sensors* **2022**, *22*, 1293. [CrossRef]

40. Högbom, J. Aperture synthesis with a non-regular distribution of interferometer baselines. *Astron. Astrophys. Suppl. Ser.* **1974**, *15*, 417.

41. Ding, B.; Wen, G. Combination of global and local filters for robust SAR target recognition under various extended operating conditions. *Inf. Sci.* **2019**, *476*, 48–63. [CrossRef]

42. Ding, B.; Wen, G.; Zhong, J.; Ma, C.; Yang, X. A robust similarity measure for attributed scattering center sets with application to SAR ATR. *Neurocomputing* **2017**, *219*, 130–143. [CrossRef]

43. Tang, T.; Su, Y. Object recognition based on feature matching of scattering centers in SAR imagery. In Proceedings of the 2012 5th International Congress on Image and Signal Processing, Chongqing, China, 16–18 October 2012; pp. 1073–1076.

44. Ozdemir, C. *Inverse Synthetic Aperture Radar Imaging with MATLAB Algorithms*; John Wiley & Sons: Hoboken, NJ, USA, 2012.

45. Demirkaya, O.; Asyali, M.H.; Sahoo, P.K. *Image Processing with MATLAB: Applications in Medicine and Biology*; CRC Press: Boca Raton, FL, USA, 2008.

46. Novak, L.M.; Owirka, G.J.; Brower, W.S. Performance of 10-and 20-target MSE classifiers. *IEEE Trans. Aerosp. Electron. Syst.* **2000**, *36*, 1279–1289.

47. Karine, A.; Toumi, A.; Khenchaf, A.; El Hassouni, M. Radar target recognition using salient keypoint descriptors and multitask sparse representation. *Remote Sens.* **2018**, *10*, 843. [CrossRef]

48. Yu, J.; Zhou, G.; Zhou, S.; Yin, J. A Lightweight Fully Convolutional Neural Network for SAR Automatic Target Recognition. *Remote Sens.* **2021**, *13*, 3029. [CrossRef]

49. Gao, F.; Huang, T.; Sun, J.; Wang, J.; Hussain, A.; Yang, E. A new algorithm for SAR image target recognition based on an improved deep convolutional neural network. *Cogn. Comput.* **2019**, *11*, 809–824. [CrossRef]

50. Choi, Y. Simulated SAR Target Recognition using Image-to-Image Translation Based on Complex-Valued CycleGAN. *J. Korean Inst. Inf. Technol.* **2022**, *20*, 19–28.

51. Inkawhich, N.A.; Davis, E.K.; Inkawhich, M.J.; Majumder, U.K.; Chen, Y. Training SAR-ATR models for reliable operation in open-world environments. *IEEE J. Sel. Top. Appl. Earth Obs. Remote Sens.* **2021**, *14*, 3954–3966. [CrossRef]

52. Sellers, S.R.; Collins, P.J.; Jackson, J.A. Augmenting simulations for SAR ATR neural network training. In Proceedings of the 2020 IEEE International Radar Conference (RADAR), Washington, DC, USA, 28–30 April 2020; pp. 309–314.

Article

Estimation and Classification of NLFM Signals Based on the Time–Chirp Representation

Ewa Swiercz *, Dariusz Janczak and Krzysztof Konopko

Faculty of Electrical Engineering, Bialystok University of Technology, 15-351 Bialystok, Poland
* Correspondence: e.swiercz@pb.edu.pl

Abstract: A new approach to the estimation and classification of nonlinear frequency modulated (NLFM) signals is presented in the paper. These problems are crucial in electronic reconnaissance systems whose role is to indicate what signals are being received and recognized by the intercepting receiver. NLFM signals offer a variety of useful properties not available for signals with linear frequency modulation (LFM). In particular, NLFM signals can ensure the desired reduction of side-lobes of an autocorrelation (AC) function and desired power spectral density (PSD); therefore, such signals are more frequently used in modern radar and echolocation systems. Due to their nonlinear properties, the discussed signals are difficult to recognize and therefore require sophisticated methods of analysis, estimation and classification. NLFM signals with frequency content varying with time are mainly analyzed by time–frequency algorithms. However, the methods presented in the paper belong to time–chirp domain, which is relatively rarely cited in the literature. It is proposed to use polynomial approximations of nonlinear frequency and phase functions describing signals. This allows for applying the cubic phase function (CPF) as an estimator of phase polynomial coefficients. Originally, the CPF involved only third-order nonlinearities of the phase function. The extension of the CPF using nonuniform sampling is used to analyse the higher order polynomial phase. In this paper, a sixth order polynomial is considered. It is proposed to estimate the instantaneous frequency using a polynomial with coefficients calculated from the coefficients of the phase polynomial obtained by CPF. The determined coefficients also constitute the set of distinctive features for a classification task. The proposed CPF-based classification method was examined for three common NLFM signals and one LFM signal. Two types of neural network classifiers: learning vector quantization (LVQ) and multilayer perceptron (MLP) are considered for such defined classification problem. The performance of both the estimation and classification processes was analyzed using Monte Carlo simulation studies for different SNRs. The results of the simulation research revealed good estimation performance and error-free classification for the SNR range encountered in practical applications.

Keywords: NLFM signal classification; NLFM signal estimation; cubic phase function; multiclass classification; instantaneous frequency rate

Citation: Swiercz, E.; Janczak, D.; Konopko, K. Estimation and Classification of NLFM Signals Based on the Time–Chirp Representation. *Sensors* **2022**, *22*, 8104. https://doi.org/10.3390/s22218104

Academic Editors: Janusz Dudczyk and Piotr Samczyński

Received: 22 September 2022
Accepted: 18 October 2022
Published: 22 October 2022

1. Introduction

In this paper, an approach based on the time–chirp $(T - \Omega)$ transform used for the estimation and classification of signals with nonlinear frequency modulation (NLFM) has been presented. The chirp rate (Ω), also called the instantaneous frequency rate (IFR), is the signal phase acceleration and can be calculated as the time derivative of a frequency function. The analysis and processing of NLFM signals are exploited in a wide range of applications for example in Electronic Support Measures/Electronic Intelligence (ESM/ELINT), Electronic Warfare (EW), Electronic Reconnaissance (ER) systems, as well as in passive bistatic radar (PBR) [1]. Modern electronic intelligence and electronic support are designed to automatically distinguish the modulation type of an intercepted radar signal, which can be utilized in early warning systems or give more information about hostile radars [2–7]. Passive bistatic radar uses emissions from communications, broadcast, or radionavigation transmitters instead of dedicated, cooperative radar transmitters. The transmitted waveforms are not explicitly designed for passive radar purposes. Therefore knowledge about the received signal is crucial in the ability of waveform recognition and reconstruction. New sources of target illumination in passive radars are constantly being searched. Solutions with using the 5G cellular network as a source of illumination in a passive radar system have recently appeared [8]. The NLFM waveform for synthetic aperture radar (SAR) applications is important for improving spaceborne SAR image quality and reducing system costs [9,10]. The NLFM waveform was also proposed for active sonars [11].

The NLFM signal can be synthesized in different ways in order to obtain desired properties by shaping the power spectral density (PSD). Due to nonlinear frequency modulation, such signals can achieve the desired PSD and desired autocorrelation function with reduced sidelobes compared to LFM signals [12–14]. In the case of narrowband signals the Doppler offset may be miscalculated in the narrowband ambiguity function [15,16]. The potential advantage of NLFM is its Doppler shift tolerance. These properties make NLFM signals very attractive especially for radar applications. Generally, synthesis of signals with limitation of the required level of AC-sidelobes and desired power spectral density is a great challenge. These conditions require multi-objective optimization approach with strong constraints and high computational load and may result in inconsistent requirements [11,15,17–20]. Appropriate NLFM chirps achieving the desired shape of the power spectrum have been suggested as a solution to this problem. Signals for which parameterized nonlinear frequency modulations satisfy the desired spectral properties achieving low level sidelobes, have been proposed by a lot of authors. Signal models presented by Collins and Atkins [11], Pirce [21] or Yue and Zhang [22] are very popular and are also the subject of analysis in this paper.

A special class of NLFM signals, which enables the shaping of the PSD and obtains the desired AC function, is polynomial phased signals (PPS). This kind of nonlinear signal plays a significant role in modern radar systems, including SAR, ISAR and OTHR systems, as well as in sonars, biomedicine, machine engine testing, etc., especially in the ISAR, where due to target movement or extreme target maneuvers, the radar returns contain mainly PPS components, possibly with high-order phase terms [23,24]. The simplest type of a signal with a polynomial phase is a signal with linear frequency modulation. Unfortunately, the PSD of an LFM signal is approximately rectangular and after matched filtering (MF) the peak-to-sidelobe ratio (PSLR) is rather low, reaching about 13.3 [dB]. Therefore, NLFM signals are considered to be a good alternative to LFM signals.

Modern radar systems, especially surveillance systems, can emit a pulse train with inter-pulse and intra-pulse complex modulation, including both linear and nonlinear frequency modulation. If such radars operate in the dense, hostile electromagnetic environment, intercepted signals should be recognized or classified by means of spectrum-sensing systems such as ELINT, ER and EW. The main problem of recognition and estimation of intercepted signals is the determination of the modulation type and its parameters. If any information is not available, it is reasonable to assume a phase polynomial for such signals with a sufficiently high order of polynomial nonlinearity. Even parameterized nonlinearities, which usually have a complex analytical description, can be approximated by polynomial form with a sufficiently high order to obtain a simpler description for further analysis. In this paper, the polynomial approximation of the NLFM models has been evaluated for estimation and classification purposes. The set of selected coefficients of polynomial approximation is suggested as distinctive features allowing the identification of a type of unknown emission by means of classification. This approach requires a database containing a set of nonlinearity types. In this paper, the estimation of the PPS parameters as well as identification by classification of other type of nonlinearities are considered.

The general description of nonstationary signals embedded in noise can be presented as follows:

$$s(t) = A(t)e^{j\phi(t)} + s_n(t) \tag{1}$$

where $\phi(t)$ is the phase of the signal, $A(t)$ is the amplitude of the signal and $s_n(t)$ is white Gaussian noise.

There is an unambiguous relationship between frequency function $f(t)$ (the instantaneous frequency (IF)) and the instantaneous phase function $\phi(t)$ of a nonstationary signal $s(t)$ through the differentiation operation:

$$f(t) = \frac{1}{2\pi} \frac{d\phi(t)}{dt} \tag{2}$$

Knowledge of the IF function automatically determines the phase function and vice versa. It seems natural to use time–frequency (T-F) distributions for IF analysis and estimation of NLFM signals. However, the most known quadratic T-F distributions, such as the pseudo Wigner–Ville distribution and the Choi–Williams distribution, contain cross-term components, and the estimation of the instantaneous frequency modulation is performed with unacceptable accuracy. Therefore, they are practically useless [2]. For analysis of the PPS, the high order ambiguity function (HAF) or the product HAF (PHAF) seems to be attractive. In these distributions, the phase-differentiation (PD) operation is repeated many times until a single complex sinusoidal signal is obtained [25,26]. The frequency of the obtained sinusoidal signal indicates the highest order PPS coefficient. Next, the original signal is dechirped with the use of this PPS coefficient. The remaining parameters are estimated by repeating the same procedure. Although these methods provide good accuracy, they suffer from high computational burden and error propagation during dechirping operations. Therefore, they seem to be useless for NLFM.

This paper deals with the effective estimation of the PPS of higher order. The proposed method is based on the concept of nonlinear sampling and the cubic phase function distribution (CPF) developed on the time–chirp $(T - \Omega)$ plane [27]. The CPF distribution turned out to be effective in parameter estimation of the quadratic frequency modulation signals [28]. In this paper, the CPF is proposed to extract distinctive features of parameterized nonlinearities approximated by a polynomial useful in classification. Nonlinearities within the paper are replaced by the polynomial form containing sufficient information to classify NLFM signals. Although the CPF was originally designed for the estimation of the third order PPS, in this paper, the CPF method is used for estimation of the sixth order PPS.

The classification of NLFM using the set of distinctive properties obtained from the CPF method on the $(T - \Omega)$ plane is presented in the next part of the paper. The classification process comprises three types of nonlinearities typical for radar applications. The classification of NLFM signals is not trivial, especially for signals with abrupt frequency changes, and requires advanced systems such as neural networks to achieve high classification efficiency [29–32]. Generally, due to the ability of self-learning and adaptability, neural networks can outperform other classification approaches. Two neural classifiers based on learning vector quantization (LVQ) and multilayer perceptron (MLP) networks have been used and compared. Other concepts of LFM and NLFM classification based on neural networks can be found in [33–35]. The neural network classification, considered in this paper, also takes into account the classification between NLFM and LFM signals [36].

The paper is organized as follows: Section 2 presents selected NLFM signals used in radar and sonar and discusses the use of their polynomial approximations. Section 3 discusses various methods of IF estimation and describes the proposed IF estimator based on the cubic phase function, while Section 4 presents the results of simulation studies on the quality of the proposed IF estimator. Section 5 proposes the use of the CPF-based estimator of phase polynomial coefficients in the classification of the NLFM signals and presents the results of simulation tests. Section 6 presents the conclusions.

2. Nlfm Signals and Their Polynomial Approximations

In this section, some selected examples of nonlinear functions used for the generation of NLFM signals for radar and echolocation systems are presented. Then, using the Taylor expansion of these functions, some aspects related to the accuracy of the polynomial approximation are illustrated and discussed.

Currently, NLFM signals are mainly developed for use in radar and sonar technologies. To present the proposed estimation and classification methods, three representative NLFM functions have been selected and presented by Formulas (3)–(7). The functions and their parameters were designed by their authors to minimize the sidelobes of the autocorrelation function.

The NLFM signal developed by Collins and Atkins [11] consists of a linear and nonlinear part. The nonlinearity impact is determined by two parameters α and γ:

$$f_1(t, \alpha, \gamma) = \frac{B}{2}\left[\frac{\alpha \tan(2\gamma t/T)}{\tan(\gamma)} + \frac{2(1-\alpha)t}{T}\right]; \quad -\frac{T}{2} \le t \le \frac{T}{2} \tag{3}$$

where B is the signal bandwidth, T is the time duration of the pulse, and α is the parameter that defines the weight between the linear and nonlinear part, while the parameter γ affects the intensity of the nonlinear part.

Parameter values that minimize the sidelobes of the autocorrelation function are $\alpha = 0.52$ and $\gamma = 1.47$ [11].

The NLFM signal proposed by Price [21] also consists of a linear part represented by the B_L parameter and a nonlinear part controlled by B_C:

$$f_2(t, B_L, B_C) = \frac{t}{T}\left(B_L + \frac{B_C}{\sqrt{1 - 4t^2/T^2}}\right). \tag{4}$$

In the paper, the following form of Formula (4), explicitly showing the signal bandwidth B [12], is used:

$$f_2(t, B_l, B_c) = B\frac{t}{T}\left(B_l + \frac{B_c}{\sqrt{1 - 4t^2/T^2}}\right), \tag{5}$$

where the parameters: $B_l = B_L/B$ and $B_c = B_C/B$, with values $B_l = 0.561105$ and $B_c = 0.23799$ minimizing sidelobes [12], and the time interval $-0.45T \le t \le 0.45T$ covering bandwidth B.

The NLFM signal proposed by Yue and Zhang [22] is given by the formula:

$$f_3(t, k_1, k_2) = Bk_1 \tan\left(k_2\frac{t}{T}\right); \quad -\frac{T}{2} \le t \le \frac{T}{2}. \tag{6}$$

where the parameters k_1 and k_2 that minimize the sidelobes take values: $k_1 = 0.1171$ and $k_2 = 2.607$.

In the presented analyses, the *LFM* signal is also used. It has the same bandwidth B and the pulse duration T as NLFM signals. The frequency of the *LFM* signal is described by the following formula:

$$f_{LFM}(t) = \frac{B}{T}t; \quad -\frac{T}{2} \le t \le \frac{T}{2}. \tag{7}$$

Nonlinear, continuous functions $f_1(t, \alpha, \gamma)$, $f_2(t, B_l, B_c)$, $f_3(t, k_1, k_2)$ determined on the closed interval: $-\frac{T}{2} \le t \le \frac{T}{2}$, according to the Weierstrass approximation theorem, can be approximated with desired accuracy using polynomial functions of a sufficiently high order. The approximation error for a particular nonlinear function depends on the order of the polynomial.

The proposed estimation and classification method is based on simplified polynomial models of nonlinear functions that describe the frequency and phase of the NLFM signals. An important issue is the selection of the order of the approximating polynomial. In the paper, this issue is analyzed using the Taylor series approximation, which makes it possible to obtain an analytical description of the signal frequency and phase as a function of time.

Examples of the approximation of the nonlinear functions $f_1(t, \alpha, \gamma)$, $f_2(t, B_l, B_c)$, $f_3(t, k_1, k_2)$ are presented by (8)–(10). Polynomials of the fifth order were calculated using the Taylor expansion around the time point $t = 0$:

$$f_1^{Ta}(t, \alpha, \gamma) = \frac{B}{T}\left(1 - \alpha + \frac{\alpha\gamma}{tan(\gamma)}\right)t + \frac{4B\alpha\gamma^3}{3T^3 tan(\gamma)}t^3 + \frac{32B\alpha\gamma^5}{15T^5 tan(\gamma)}t^5 + R_{f1}(t, \alpha, \gamma) \quad (8)$$

$$f_2^{Ta}(t, B_l, B_c) = \frac{B(B_c + B_l)}{T}t + \frac{2BB_c}{T^3}t^3 + \frac{6BB_c}{T^5}t^5 + R_{f2}(t, B_c, B_l) \quad (9)$$

$$f_3^{Ta}(t, k_1, k_2) = \frac{Bk_1k_2}{T}t + \frac{Bk_1k_2^3}{3T^3}t^3 + \frac{2Bk_1k_2^5}{15T^5}t^5 + R_{f3}(t, k_1, k_2) \quad (10)$$

where $R_{f1}(t, \alpha, \gamma)$, $R_{f2}(t, B_c, B_l)$ and $R_{f3}(t, k_1, k_2)$ are reminder terms.

Figure 1 illustrates a change in the accuracy of $f_1^{Ta}(t, \alpha, \gamma)$ approximation depending on the polynomial order. Figure 1a shows the function $f_1(t, \alpha, \gamma)$ and its polynomial approximations of order $M \in \{5, 9, 13\}$. Figure 1b shows the error $\Delta f_1(t, \alpha, \gamma)$ of these approximations.

$$\Delta f_1(t, \alpha, \gamma) = f_1(t, \alpha, \gamma) - f_1^{Ta}(t, \alpha, \gamma) \quad (11)$$

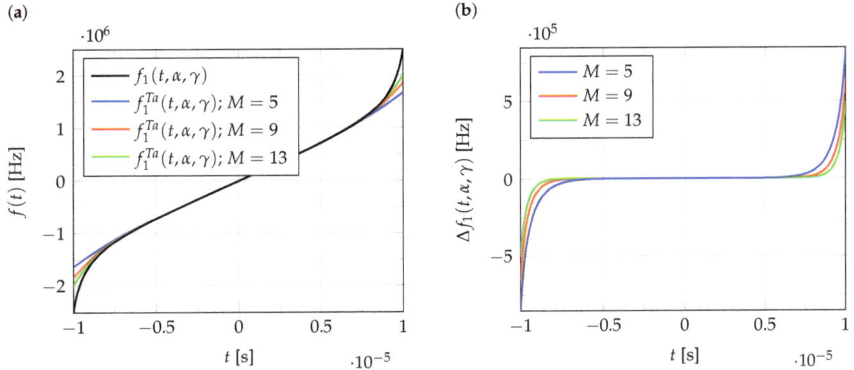

Figure 1. The accuracy of nonlinear frequency approximations, (**a**) the function $f_1(t, \alpha, \gamma)$ and its Taylor polynomial approximations of the order $M \in \{5, 9, 13\}$, (**b**) the error of approximations.

As can be seen from the analysis of the results presented in Figure 1, the error of polynomial approximation for the order $M > 5$ slowly decreases. Similar results are obtained for the approximation of the functions $f_2(t, B_l, B_c)$ and $f_3(t, k_1, k_2)$.

Based on the relationship (2), it is possible to calculate the analytical functions (12)–(15) that describe the instantaneous phase of the NLFM and LFM signals:

$$\phi_1(t, \alpha, \gamma) = \pi B\left[\frac{\alpha T \ln(\tan(2\gamma t/T)^2 + 1)}{4\gamma \tan(\gamma)} + \frac{(1 - \alpha)t^2}{T}\right] + \phi_1^0, \quad (12)$$

$$\phi_2(t, B_l, B_c) = \frac{\pi}{2}B\left(\frac{2B_l}{T}t^2 - B_c\sqrt{T^2 - 4t^2}\right) + \phi_2^0, \quad (13)$$

$$\phi_3(t, k_1, k_2) = \frac{\pi BTk_1}{k_2}\ln\left(\tan(k_2\frac{t}{T})^2 + 1\right) + \phi_3^0, \quad (14)$$

$$\phi_{LFM}(t) = \frac{\pi B}{T}t^2 + \phi^0, \quad (15)$$

where ϕ_1^0, ϕ_2^0, ϕ_3^0, ϕ^0 are initial phase values, while the time interval is is $-\frac{T}{2} \leq t \leq \frac{T}{2}$.

The Taylor polynomial approximation is calculated around the time moment $t = 0$ nd phase value: $\phi(0)$, as the phase function is symmetrical with respect to this point. Taylor approximations of the phase with polynomials of the sixth order are presented by (16)–(18):

$$\phi_1^{Ta}(t, \alpha, \gamma) = \frac{\pi B}{T}\left(1 - \alpha + \frac{\alpha\gamma}{\tan(\gamma)}\right)t^2 + \frac{2\pi B\alpha\gamma^3}{3T^3\tan(\gamma)}t^4 + \frac{32\pi B\alpha\gamma^5}{45T^5\tan(\gamma)}t^6 + R_{\phi1}(t, \alpha, \gamma) \quad (16)$$

$$\phi_2^{Ta}(t, B_l, B_c) = -\frac{\pi}{2}BB_cT + \frac{\pi B(B_c + B_l)}{T}t^2 + \frac{\pi BB_c}{T^3}t^4 + \frac{2\pi BB_c}{T^5}t^6 + R_{\phi2}(t, B_c, B_l) \quad (17)$$

$$\phi_3^{Ta}(t, k_1, k_2) = \frac{\pi Bk_1k_2}{T}t^2 + \frac{\pi Bk_1k_2^3}{6T^3}t^4 + \frac{2\pi Bk_1k_2^5}{45T^5}t^6 + R_{\phi3}(t, k_1, k_2) \quad (18)$$

where $R_{\phi1}(t, \alpha, \gamma)$, $R_{\phi2}(t, B_c, B_l)$, $R_{\phi3}(t, k_1, k_2)$ are reminder terms.

The influence of polynomial order on the accuracy of the polynomial approximation of the phase function $\phi_1(t, \alpha, \gamma)$ is presented in Figure 2. Figure 2a shows the function $\phi_1(t, \alpha, \gamma)$, and its polynomial approximations of order $M \in \{6, 10, 14\}$. Figure 2b shows the error $\Delta\phi_1(t, \alpha, \gamma)$ of these approximations:

$$\Delta\phi_1(t, \alpha, \gamma) = \phi_1(t, \alpha, \gamma) - \phi_1^{Ta}(t, \alpha, \gamma) \quad (19)$$

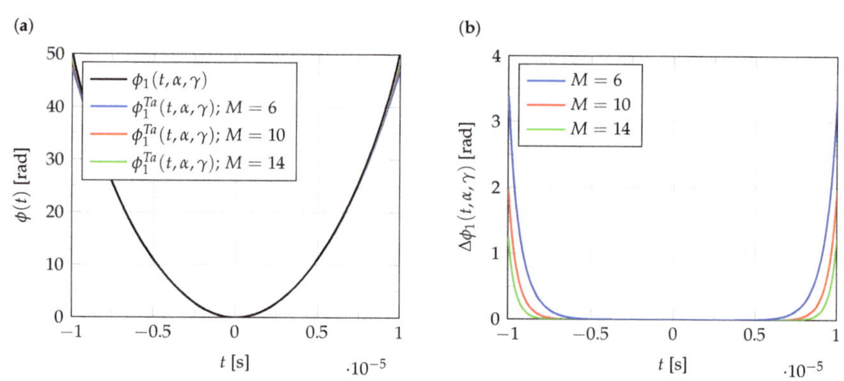

Figure 2. The accuracy of phase function approximations, (**a**) the function $\phi_1(t, \alpha, \gamma)$ and its Taylor polynomial approximations of order $M \in \{6, 10, 14\}$; (**b**) the error of approximations.

As can be seen from the analysis of the results presented in Figure 2, the error of polynomial approximation of phase function is quite small compared to phase value. Similar results are obtained for the approximation of the functions $\phi_2(t, B_l, B_c)$ and $\phi_3(t, k_1, k_2)$.

In the paper, the polynomial approximation of a phase function has been analyzed to classify NLFM signals. The nonlinear change in the frequency of analyzed signals is described by the odd function, which results in an odd order of approximating polynomial. Taking into account this feature and the relationship (2), the polynomial approximation of the phase function is of an even order. In our case, after preliminary investigation, we decided that the order of the polynomial approximation of the phase function can be limited to $M = 6$. The set of selected coefficients of the approximating polynomial is chosen as a set of distinctive features in the proposed method of classification of NLFM signals. The sixth order of the approximating polynomial seems sufficient to constitute a set of distinctive features. Although the selected order of the approximating polynomial does not guarantee a perfect polynomial fitting, especially on signal parts with abrupt nonlinear changes, it allows for effective classification with the low computational load.

3. The CPF-Based Estimator of the IFR

The recognition of signals with nonlinear frequency modulation is usually performed with the use of estimation of the instantaneous frequency of the signal. There are many methods for analyzing nonlinear frequency functions, including all distributions of the group belonging to the Cohen class. These are, among others, time–frequency distributions such as the Wigner–Ville distribution, the Choi–Wiliams distribution and the short-time Fourier transform (STFT). A large group of signals with a nonlinear frequency are PPS. Very good estimation results of the higher order PPS are obtained by the quasi-maximum likelihood (QML) method, which is an extended version of the STFT transformation [37–39]. Estimation of the IF is performed by the STFT. Coefficients of the PPS are obtained from the IF estimates using the classical polynomial regression. The QML method requires an additional refining procedure to improve the quality of coarse estimates of polynomial coefficients. The refining process consists of four steps: dechirping the received signal using coarse initial estimates of the polynomial phase parameters provided by the STFT, filtering through an M-point moving average (MA) filter combined with an M-fold decimation, polynomial phase estimation of the obtained signal by phase unwrapping and least squares estimation. The final estimates are calculated as a combination of estimates obtained in step 3 with the initial coarse estimates [40]. In the last step of the QML method, the optimal STFT window is searched by maximizing the quasi-ML function. Rather than directly searching through all parameters of phase polynomial, the maximum QML function is calculated for the estimates provided by STFT and polynomial regression. The QML is computationally exhaustive for higher order PPS because it requires multiple STFT calculations and multiple polynomial regression calculations, which are slow but precise processes. Therefore, it is desirable to search for a new transform with results comparable to the QML transform, but with less computational effort. A proposed alternative method of estimating the parameters of approximating polynomials is the CPF distribution defined for discrete signals.

Discrete signals of interest (i.e., PPS) $z_r(n)$ are characterized by a constant amplitude b_0 and phase $\phi(n)$ and are defined as:

$$z_r(n) = z_s(n) + z_w(n) = b_0 e^{j\phi(n)} + z_w(n), \quad -\frac{N-1}{2} \le n \le \frac{N-1}{2}, \tag{20}$$

where the phase function $\phi(n)$ is described by the $M-th$ order polynomial with coefficients a_m and $z_w(n)$ is Gaussian white noise with variance σ^2. The discrete phase function is specified by the following formula:

$$\phi(n) = \sum_{m=0}^{M} a_{\phi m} n^m \tag{21}$$

The CPF is defined as follows:

$$CPF_{z_r}(n, \Omega) = \sum_{m=0}^{\frac{N-1}{2}} z_r(n+m) z_r(n-m) e^{-j\Omega m^2} \tag{22}$$

where Ω is the frequency rate.

The discrete time $t_n = nT_s$ resulting from sampling with the period T_s and the discrete frequency rate Ω define a discrete grid on the time—frequency rate plane (n, Ω). Therefore, the discrete results of the estimation of phase polynomial coefficients may differ from the continuous case if the discrete grid is sparse. The estimate of IFR [28,41] for each point in time is obtained as follows:

$$\widehat{IFR}(n) = \underset{\Omega}{\mathrm{argmax}}\, CPF_{z_r}(n, \Omega) \tag{23}$$

The CPF presented in the literature is mainly used to estimate the parameters of a signal phase polynomial up to the third order [28,41,42]. However, according to the analysis presented in Section 2, the proposed classification approach requires estimation of the coefficients of the sixth order phase polynomial. In this paper, the extension of the CPF to analysis of a sixth order polynomial is considered. The proposed approach assumes that only one run of the CPF-based method is used to estimate coefficients of the phase polynomial of the required order. Having had the set of estimated phase polynomial coefficients, the set of frequency polynomial coefficients can also be calculated according to the relationship (2). The proposed method based on the CPF has a lower computational load than the QML and other commonly known methods dedicated to the analysis of signals in the frequency or time–frequency domains.

NLFM signals are most often defined by frequency functions, such as functions $f_1(t, \alpha, \gamma)$, $f_2(t, B_L, B_C)$ and $f_3(t, k_1, k_2)$ given in Section 2 described by (3)–(6). Generating signals by means of Equation (1) requires the knowledge of the corresponding phase functions, which for the analyzed signals are represented by the functions $\phi_1(t, \alpha, \gamma)$, $\phi_2(t, B_L, B_C)$ and $\phi_3(t, k_1, k_2)$ described by (12)–(14). The CPF algorithm, which processes the received noise-disturbed signal (20), estimates the coefficients of the discrete phase polynomial model. In the proposed classification method, these coefficients of the phase polynomial are used as a set of distinctive features. Classification methods are described in more detail in Section 5. Moreover, the estimation of the instantaneous frequency with the use of a discrete polynomial approximating the continuous frequency function is proposed. The coefficients of the approximating frequency polynomial are calculated from the coefficients of the phase polynomial obtained using CPF.

The proposed $M - th$ order polynomial approximating frequency function $f(n)$ is of the form:

$$\widehat{f(n)} = \sum_{m=0}^{M} a_{fm} n^m, \tag{24}$$

where a_{fm} are the coefficients of the frequency polynomial.

The coefficients of the frequency polynomial (24) can be obtained from phase polynomial coefficients using relationships (2) and take the following value:

$$a_{fm} = \frac{(m+1)a_{\phi m+1}}{2\pi T_s} \tag{25}$$

where the coefficients $a_{\phi m}$ of the phase polynomial are obtained by the CPF.

Modification of the kernel of the classical CPF distribution (22) with the use of nonuniform sampling allows for higher order PPS decomposition. By sampling the signal at nonuniformly spaced time moments, the order of the PPS estimator can be lowered [43]. The kernel of the original CPF distribution (22) is as follows:

$$K(z_r, n) = z_r(n - m)z_r(n + m) \tag{26}$$

and the modified kernel takes the following form:

$$K(z_r, n) = z_r\left(n - \sqrt{Cm}\right)z_r\left(n + \sqrt{Cm}\right), \quad m = 0, 1, \ldots, \frac{N-2}{2} - |n| \tag{27}$$

where $\sqrt{Cm}$ defines the nonlinear sampling [44].

The proposed sampling allows the calculation of the CPF using the FFT method, which results in a significant reduction of the computational load.

If we consider the modified kernel for $n = 0$ and the signal model (2) processed by this kernel, a third order polynomial form is obtained with only even coefficients from the set of all coefficients of the sixth order phase polynomial, which is proposed as a polynomial approximation of considered nonlinearities:

$$K(z_r, 0) = z_{r1}(m) = e^{j2\left(Ca_2 m + C^2 a_4 m + C^3 a_6 m^3\right)} \tag{28}$$

where the parameter $C = (N - 1)/2$ controls the sampling process, and a_2, a_4, a_6 are polynomial parameters related to phase polynomial coefficients from (21).

The kernel $K(z_r, 0)$ creates the signal with the polynomial phase of the 3rd order. Coefficients of such a polynomial can be efficiently computed using the CPF distribution (22). Parameters $2Ca_2$, $2C^2 a_4$ and $2C^3 a_6$ are related to $a_{\phi 2}$, $a_{\phi 4}$, $a_{\phi 6}$, respectively. Therefore, the estimation of the polynomial parameters can be performed directly by the CPF procedure:

$$\widehat{\Omega}_1 = arg \max_{\Omega} CPF_{z_r}(n_1, \Omega) \tag{29}$$

$$\widehat{\Omega}_2 = arg \max_{\Omega} CPF_{z_r}(n_2, \Omega) \tag{30}$$

It is natural to assume the parameter $n_1 = 0$, but parameter n_2 should be chosen to obtain a statistically optimal estimate of parameters. It strongly depends on the properties of analyzed signals. The $\widehat{\Omega}_1$ and $\widehat{\Omega}_2$ allow for calculating $\widehat{a_{\phi 4}}$ and $\widehat{a_{\phi 6}}$ according to Equations (31) and (32) [28]:

$$\widehat{a_{\phi 4}} = \frac{\Omega_1 n_2 - \Omega_2 n_1}{4C^2(n_2 - n_1)} \tag{31}$$

$$\widehat{a_{\phi 6}} = \frac{\Omega_2 - \Omega_1}{12C^3(n_2 - n_1)} \tag{32}$$

An estimate of a_2 from (28) is obtained by dechirping and finding the Fourier transform peak and $\widehat{a_{\phi 2}}$ can be calculated as follows:

$$\widehat{a_{\phi 2}} = \frac{\widehat{a_2}}{2C} \tag{33}$$

The kernel (28) is independent of the remaining parameters $a_{\phi 1}$, $a_{\phi 3}$, $a_{\phi 5}$ of the phase polynomial. The accuracy of the estimation process depends on the SNR, and operations above a certain SNR threshold are performed with acceptable accuracy.

The estimated parameters $a_{\phi 2}$, $a_{\phi 4}$ and $a_{\phi 6}$ are used for the estimation of the nonlinear frequency function and the signal classification process. The proposed method including estimation and classification is summarized in the flow diagram presented in Figure 3.

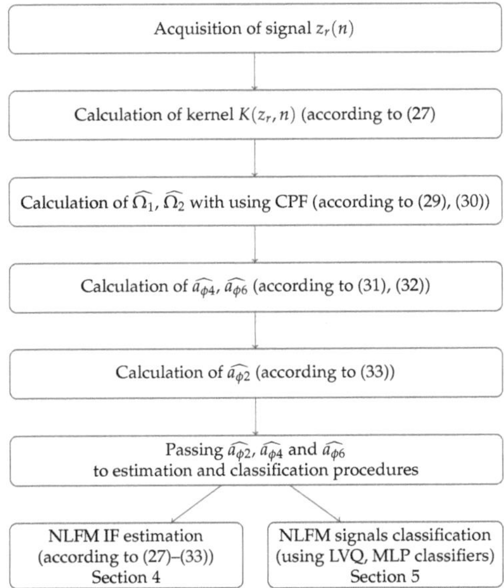

Figure 3. Flow diagram of the proposed method including estimation and classification.

4. Frequency Estimation of NLFM Signals Based on CPF

Simulation investigations of the proposed algorithms were carried out for three waveforms NLFM and one LFM waveforms. The NLFM functions, marked as $f_1(t, \alpha, \gamma)$, $f_2(t, B_L, B_C)$ and $f_3(t, k_1, k_2)$, are presented in Section 2 and are described by (3)–(6). Their specific parameters ensuring minimal sidelobes are presented there. The up-chirp LFM waveform $f_{LFM}(t)$ is described by Formula (15). The following time and bandwidth parameters for all signals were assumed: pulse duration $T = 20 \cdot 10^{-6}$ s. (with time frame $-T/2 \leq t \leq T/2$); sampling frequency $f_s = 100$ MHz; signal bandwidth $B = 5$ MHz. The signal noise was assumed to be complex Gaussian with variance depending on the SNR. To evaluate the proposed methods, $N_{sr} = 500$ Monte Carlo simulations were carried out for each case.

For assumed sixth order phase polynomial and odd frequency functions, the estimation function takes the form:

$$\widehat{f(n)} = a_{f1}n + a_{f3}n^3 + a_{f5}n^5, \tag{34}$$

where a_{f1}, a_{f3} and a_{f5} are polynomial coefficients that can be calculated as follows:

$$a_{f1} = \frac{a_{\phi 2}}{\pi T_s} \tag{35}$$

$$a_{f3} = \frac{a_{\phi 4}}{\pi T_s} \tag{36}$$

$$a_{f5} = \frac{a_{\phi 6}}{\pi T_s} \tag{37}$$

where $a_{\phi 2}$, $a_{\phi 4}$ and $a_{\phi 6}$ are the coefficients of the polynomial approximating phase obtained from the CPF.

The quality of the estimation of the instantaneous frequency of the signal can be assessed by the root mean square error (RMSE) determined according to the relationship:

$$RMSE(k) = \sqrt{\frac{1}{N_{sr}} \left[\sum_{n=1}^{N_{sr}} \left(\widehat{f_n(k)} - f_{NLFM}(k) \right)^2 \right]} \tag{38}$$

where N_{sr} is the number of simulation runs.

Figures 4 and 5 show the RMSE of the estimation of the instantaneous value of the frequency of NLFM signals obtained using the proposed CPF-based method. For comparison, Figure 5b shows the RMSE for the LFM signal.

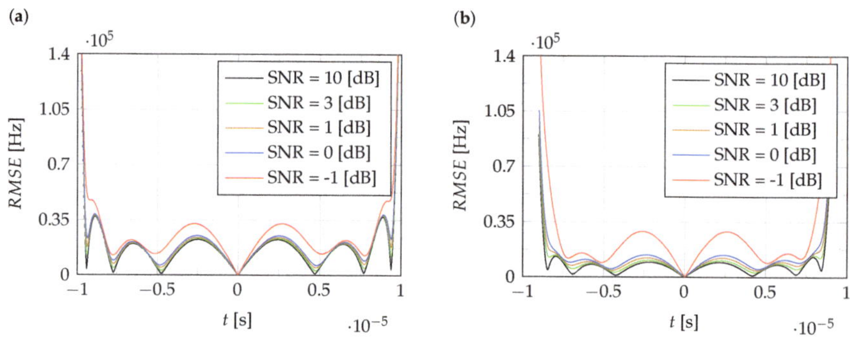

Figure 4. The RMSEof estimation of the instantaneous value of the frequency: (**a**) $f_1(t, \alpha, \gamma)$, (**b**) $f_2(t, B_l, B_c)$.

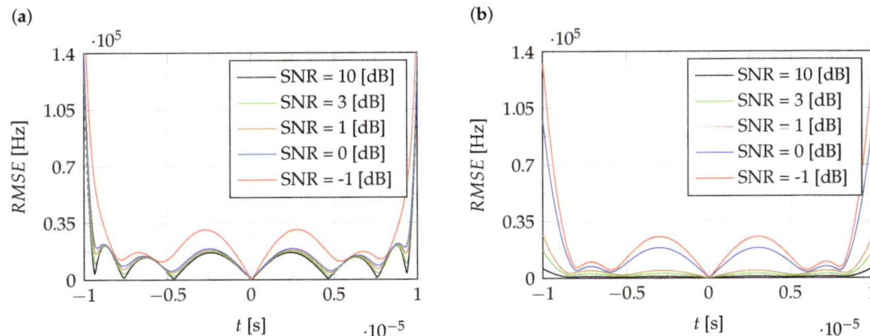

Figure 5. The RMSE of estimation of the instantaneous value of the frequency: (**a**) $f_3(t, k_1, k_2)$; (**b**) $f_{LFM}(t)$.

The analysis of RMSE presented in Figures 4 and 5 shows that the estimation error level is mainly around 0.5% of the bandwidth B, except for the edges where it increases to 5%.

Then, the mean square error (MSE) dependence on the SNR was determined for the estimates of the individual NLFM and LFM signals. The MSE was defined as follows:

$$MSE(SNR) = \log_{10} \frac{1}{N_{sr}} \frac{1}{N_{sl}} \sum_{n=1}^{N_{sr}} \sum_{n=1}^{N_{sr}} \left(\widehat{f_n(k)} - f_{NLFM}(k) \right)^2 \tag{39}$$

where N_{sr} is the number of simulation runs, and Nsl is the number of signal samples (signal length).

The MSE defined in this way allows for determining the estimation error for the entire signal (the entire pulse) and enabling the comparison of the obtained results with the results presented in the publication [2]. Figure 6 presents the MSE of estimation of the instantaneous frequency in various SNR conditions.

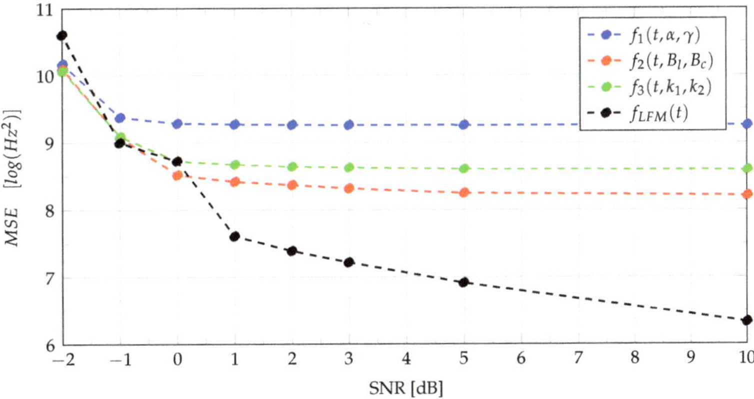

Figure 6. The MSE of estimation of the instantaneous frequency under various SNR conditions.

The analysis of the MSE of NLFM signals estimation presented in Figure 6 shows that the estimation error level of the NLFM signal depends on its type. The error remains constant at $SNR \geq 0$ dB and increases slightly for $SNR = -1$ dB. However, in the case of $SNR \leq -2$ dB, the MSE increases significantly. On the other hand, the MSE for LFM is much lower than for NLFM, and the error decreases as the noise level decreases. Compared to other methods of estimation of the IF, the method based on CPF shows a similar estimation quality. For example, comparing the obtained results with those presented in [2], it can be noticed that, for LFM, the MSE obtained using the CPF method for $SNR \geq 2$ dB is smaller than for the QML method, which, according to [2], shows lower estimation errors than methods such as the backward finite difference (BFD) method, central finite difference (CFD) estimator, Kay estimator, estimators based on the Choi–Williams distribution (CWD) and pseudo-Wigner distribution (PWD). However, for a higher noise level, QML provides better results. In the case of the estimation of NLFM signals, the QML method comes up with a slightly lower MSE than the CPF method over the entire SNR range.

The complexity of the QML algorithm, in terms of the number of operations performed, depends on the assumed instantaneous frequency (IF) resolution in the procedure of searching the STFT maximum. Similarly, the complexity of the CPF algorithm strongly depends on the instantaneous frequency rate (IFR) resolution in the procedure of searching for the CPF maximum. These maximization operations are performed on two different planes i.e., time–frequency (T-F) and time–chirp rate (T-FR). This imposes different resolution requirements that must be applied when we calculate the maximum points for the STFT and the CPF in successive identical time moments. This affects the accuracy of the estimation, as well as the execution time of the algorithms. In the comparison, the experiments with the typical values of the procedure parameters in both algorithms showed the lower computational load for the CPF algorithm compared to the QML method several dozen times.

Despite the slightly lower quality of the estimation, considering the much lower computational load of the CPF method compared to the QML, it can be concluded that the CPF may be preferred in real-time applications.

5. Classification of Signals Based on Phase Polynomial Coefficiencies Obtained from CPF

The classification procedure presented in this paper concerns the problem of recognizing signal types with nonlinear frequency modulation. The main problem is to find a set of distinctive features that allow the received signals to be distinguished and classified into a class related to a specific emitter. Generally, three kinds of classification tasks are mainly used:

1. Binary classification—in this case, there are only two classes;
2. Multiclass classification—in this case, there are more than two classes, and the classifier can only report one of them as output;
3. Multilabel classification—in this case, the classifier is allowed to choose many answers. This type of classification can be simply considered as a combination of multiple independent binary classifiers.

The classification task considered in the paper can be associated with a multiclass classification, in which the class is defined by a specific type of nonlinearity of the frequency function. Therefore, it is the type of classification indicated in item 2 of the above three-point list.

Many different modifications of the multiclass classification method have been proposed in the literature [45–47]. The proposed method uses multiclass classification with a vector of features. The classification between three types of NLFM signals and one LFM signal described in Section 2 is considered. The feature vector is formed by aggregation of selected CPF coefficients describing the polynomial approximation of the considered phase functions.

The extracted features are processed by the classifier to select the most probable class. A supervised classification is considered, where the classes are known in advance and samples of the features describing each class are available. The feature vectors for individual nonlinearities can be treated as a pattern in the feature space. Therefore, classification carried out, especially by a neural network, is a problem of recognizing patterns [33,34]. In this paper, two types of neural network classifiers: learning vector quantization (LVQ) and multilayer perceptron (MLP) are considered. The LVQ neural network has been chosen because of its high ability to learn data classification, where similar input vectors are grouped into a region represented by the so-called coded vector (CV). LVQ can be applied directly to multiclass classification problems. LVQ is a supervised version of vector quantization. LVQ uses known target output classifications for each input pattern in supervised learning of the neural network. The input space of samples is covered by the "codebook vectors" (CVs) determined during the neural network learning stage. The LVQ neural network is built as a feedforward net with one hidden layer of neurons (the Kohonen layer), fully connected with the input layer and one output layer. During the training stage, the values of weights used to form the coded vectors are adjusted, according to the previously predefined input patterns. The distance d_i of an input vector signed x to the weight vector w_i of each node in the Kohonen layer is computed. The node of a particular class, which has the smallest distance to the presented input vector (for example the Euclidean distance), is declared to be the winner:

$$d_i = \|w_i - x\| = \sqrt{\sum_{j=1}^{M}(w_{ij} - x_j)^2} \tag{40}$$

The weights will be moved closer to that class, which is expected as the winning class. Otherwise, they will be moved away. The classification after learning is relied on finding a Voronoi cell, specified by the CV with the smallest distance to the input vector and assigning it to a particular class.

The designed LVQ classifier contains $4N$ competing neurons with the logistic sigmoid function as an activation function, where N is the number of classes. The MLP is a fundamental type of neural network architecture with the ability to learn nonlinear models. The multi-layer perceptron (MLP) is a type of artificial neural network organized in several layers in which the flow of information takes place from the input layer to the output layer; therefore, it is a feedforward network. Each layer is made up of a variable number of neurons, and the neurons of the last layer (called the "output") are the outputs of the entire system. A network of such perceptrons is termed a neural network of perceptrons. A perceptron with only an input and output layer is called a simple perceptron. A single layer feed-forward network consists of one or more output neurons o, each of which is connected with a weighting factor w_{io} to all of the inputs i. The input of the neuron is the weighted sum of the inputs plus the bias term θ. A example of a single layer network with n inputs and one output is shown in Figure 7.

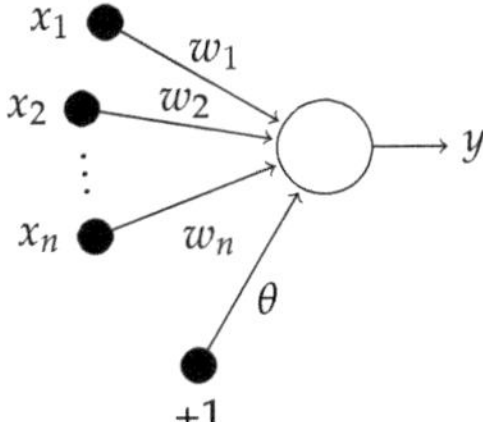

Figure 7. Single layer network with one output and n inputs.

The output of the network is formed by the activation of the output neuron, which is some function of inputs:

$$y = f\left(\sum_{i=1}^{N} w_i x_i + \theta\right) \tag{41}$$

The training of a neural network is the procedure of setting its weights. If there is one hidden layer, this one is a two layer perceptron. The aim of the supervised MLP network training is to achieve an appropriately small mean square error obtained in the Levenberg–Marquardt backpropagation procedure by adjusting weights. The complexity of the neural network classifier strongly depends on a number of neurons, which require an adjustment of their weighs at the learning stage. Therefore, the neural architecture should be as simple as possible.

For the assumed classification task, a simple single hidden layer MLP classifier with the number of neurons equal to $2N$ has been chosen.

Evaluation of the classification process requires appropriate quality criteria. The classification performance is usually visualized using the confusion matrix, which is a table summarizing true and false decisions. The matrix compares the actual types of objects with those predicted by the classifier. This issue can be easily presented on the example of binary classification. In this case, the 2×2 confusion matrix is formulated as shown in Figure 8.

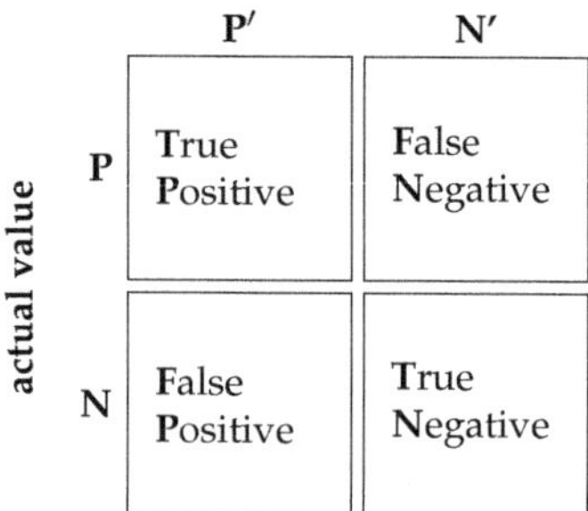

Figure 8. Interpretation of the binary confusion matrix.

The abbreviations TP, TN, FP and FN shown in Figure 8 denote:

- true positives (TP): the actual value is positive and the prediction is also positive;
- true negatives (TN): the actual value is negative and the prediction is also negative;
- false positives (FP): the actual value is negative, but the prediction is positive (type I error);
- false negatives (FN): the actual value is positive, but the prediction is negative (type II error).

It is difficult to compare the properties of different classifiers based on the confusion matrix alone. Therefore, a simpler description of the classification can be obtained using metrics calculated on the basis of data from the confusion matrix. Some common metrics [48] can be calculated as follows:

1. Accuracy (ACC)

$$ACC = \frac{TP + TN}{TP + TN + FP + FN} \tag{42}$$

2. Precision also known as positive predictive value (PPV)

$$PPV = \frac{TP}{TP + FP} \tag{43}$$

3. Sensitivity also known as recall, hit rate or true positive rate (TPR)

$$TPR = \frac{TP}{TP + FN} \tag{44}$$

The above metrics are defined similarly for multiclass classifiers. To evaluate the performance of the proposed multiclass model, the $N \times N$ confusion matrix is used, where N is the number of classes that describe the NLFM or LFM signals. The quality of the selected classifiers can be assessed by comparing the metrics calculated from their confusion matrix.

The performance of the proposed method was evaluated with the use of the simulation experiment. The proposed CPF-based classification method was tested for three NLFM

$(f_1(t, \alpha, \gamma), f_2(t, B_l, B_l), f_3(t, k_1, k_2))$ and one LFM $(f_{LFM}(t))$ signal. Their specific parameters are presented in Section 2. The simulation parameters were assumed to be the same as in the case of simulation presented in Section 4, namely: pulse duration $T = 20 \cdot 10^{-6}$ s.; sampling frequency $f_s = 100$ MHz; signal bandwidth $B = 5$ MHz and Gaussian noise with variance depending on SNR. Thus, having four signals to recognize, the problem of 4-class classification is studied. The feature vector used in the classification process was formed by the coefficients $(a_{\phi 2}, a_{\phi 4}, a_{\phi 6})$ of the polynomial which is an approximate signal phase. The coefficients are determined by the proposed CPF method. Figure 9 shows an example of the realizations of the coefficients $a_{\phi 2}, a_{\phi 4}$ and $a_{\phi 6}$ obtained by CPF for four signal classes in the case of $SNR = 5$ dB, $SNR = 1$ dB, $SNR = -1$ dB and $SNR = -5$ dB. Each signal is marked with a different colour. The points in the figure represent the estimates of the coefficients $a_{\phi 2}, a_{\phi 4}$ and $a_{\phi 6}$ obtained in $N_{sr} = 500$ realizations of individual signals.

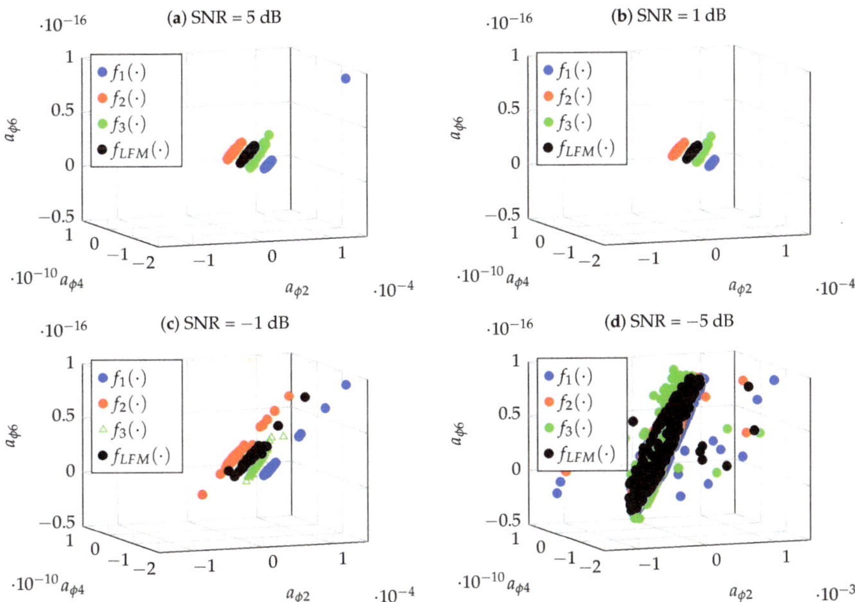

Figure 9. Example of the realizations of the coefficients $a_{\phi 2}, a_{\phi 4}$ and $a_{\phi 6}$ obtained by CPF for four signal classes in the case of (**a**) $SNR = 5$ dB; (**b**) $SNR = 1$ dB; (**c**) $SNR = -1$ dB; and (**d**) $SNR = -5$ dB.

The analysis of the spatial position of the coefficients $a_{\phi 2}, a_{\phi 4}$ and $a_{\phi 6}$ shown in Figure 9 shows that, in the case of $SNR = 5$ dB and $SNR = 1$ dB, the constellations of points corresponding to individual signals are well separable, which should result in the effective operation of classification algorithms. In the case of $SNR = -1$ dB, a significant relocation of several points can be observed, which may cause deterioration of the separation and thus of the quality of the classification. For $SNR = -5$ dB, the spaces of the individual coefficients overlap, which may result in the incorrect classification.

The classification task was addressed using LVQ and MLP neural networks. The tests were carried out at different SNRs. In each case, $N_{sr} = 500$ Monte Carlo simulations were carried out for each signal. The $N_l = 1400$ realizations constituted a training set and the remaining $N_t = 600$ realizations were used to test the classifiers. Figures 10 and 11 show the confusion matrices for MLP and LVQ classifiers obtained for $SNR = -1$ dB and $SNR = -5$ dB, respectively.

According to the analysis of the confusion matrix presented in Figure 10, for $SNR = -1$ dB, the classification is correct in the case of the MLP classifier and slightly worse for the LVQ. This means that even the significant relocation of points corresponding to the individual signals, which is visible in Figure 9, does not affect the proper classification performed by classifiers. However, as it results from the analysis of the confusion matrix shown in Figure 11, in the case of $SNR = -5$ dB, the classification is not correct for both MLP and LVQ. This result corresponds to the results shown in Figure 9, where for this SNR the constellations of points overlap.

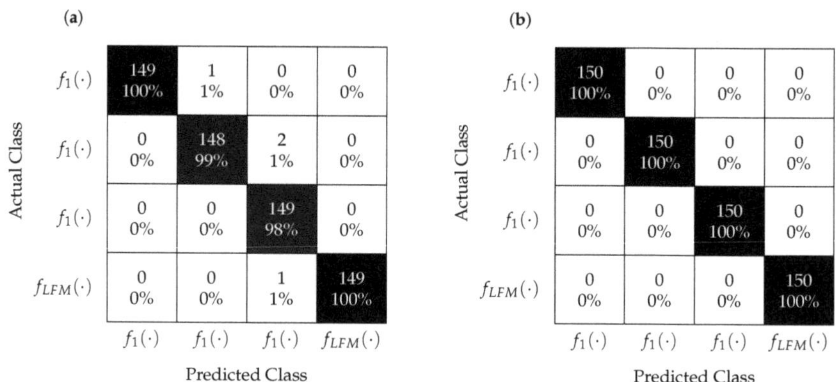

Figure 10. Confusion matrices for the LVQ (**a**) and MLP (**b**) classifiers obtained for $SNR = -1$ dB.

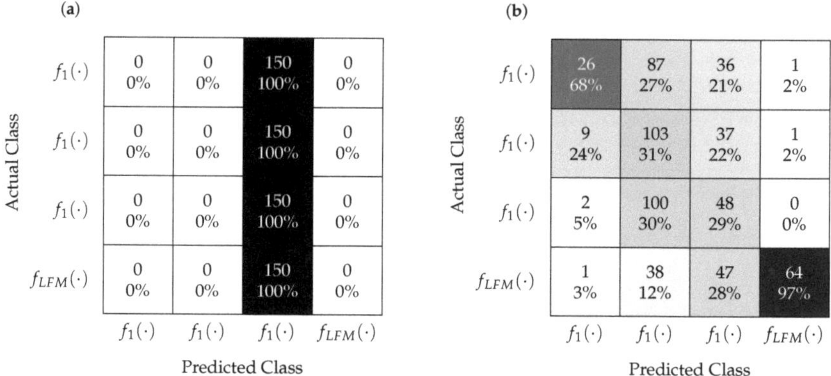

Figure 11. Confusion matrices for the LVQ (**a**) and MLP (**b**) classifiers obtained for $SNR = -5$ dB.

The quality of the MLP and LVQ classifiers was also analyzed with the use of ACC metrics. The results for -5 dB$\leq SNR \leq 5$ dB are shown in Figure 12.

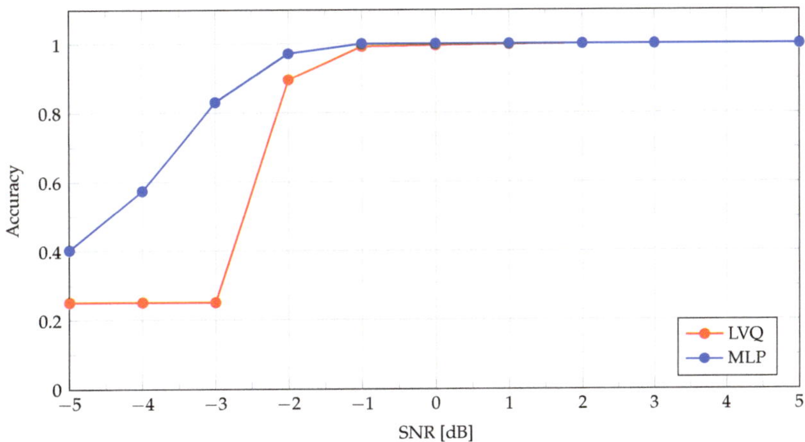

Figure 12. Accuracy (42) metrics of MLP and LVQ classifiers.

As can be seen in Figure 12, both classifiers for $SNR \geq -1$ dB provide almost 100% correct identifications of signals. In the case of $SNR = -2$ dB, the classification accuracy slightly decreases, with MLP being a more effective classifier. For SNR below -2 dB, the classification quality for the MLP algorithm gradually decreases, while in the case of LVQ, the decisions are made completely randomly.

The ACC metrics assess the overall quality of classification without being able to evaluate the identification ability of each class. In this case, PPV and TPR metrics can be used. For multiclass classification, the PPV for each class is the ratio of a correctly predicted class to all predicted classes, while TPR is defined as the ratio of a correctly predicted class to all true class values. Figures 13 and 14 show the PPV and TPR metrics for four classes for MLP and LVQ classifiers. When analyzing the results of the PPV metrics, it should be kept in mind that, in case of the absence of recognition of a given class (both TP and FP), according to definition (43), the PPV value is undefined, and therefore there are missing points in the figure.

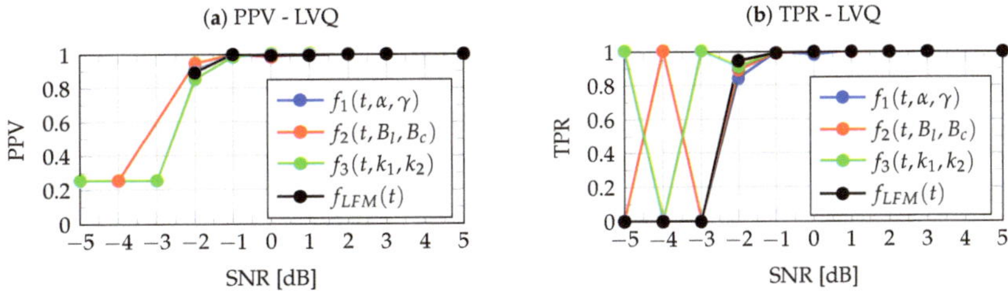

Figure 13. PPV (**a**) and TPR (**b**) metrics of the LVQ classifier.

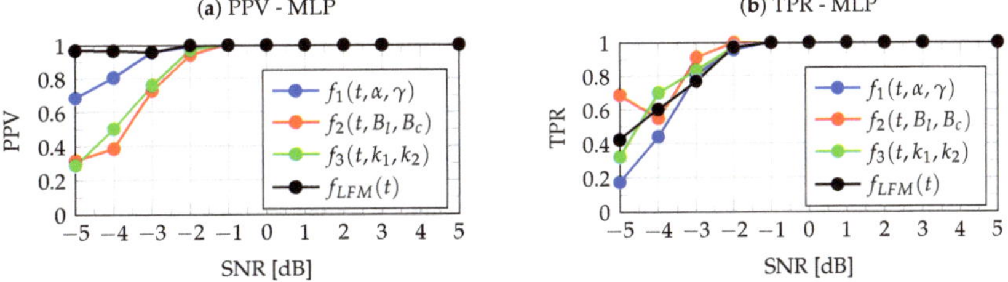

Figure 14. PPV (**a**) and TPR (**b**) metrics of the MLP classifier.

The analysis of Figures 13 and 14 allows for assessing the quality of the MLP and LVQ classifiers. The quality of the classifier is indicated by the combined analysis of PPV and TPR values. As can be seen in these figures, in the case of $SNR \geq -1$ dB, both PPV and TPR for all classes take values approximately equal to one, which means that errors of type I and type II are at a minimum level. This means that the identification of each class for $SNR \geq -1$ dB is very good for both MLP and LVQ classifiers. As can be seen in the figures, for the $SNR = -2$ dB, the classifiers' performance breaks down. In this case, for the MLP classifier, the PPV and TPR metrics take values in the range of $< 0.937, 1 >$ while for LVQ slightly less, i.e., $< 0.84, 0.947 >$. This means a slight decrease in the quality of the classification, while the MLP classifier is slightly better. In the case of $SNR \leq -3$ dB, the quality of both classifiers is slightly degraded. In this case, as can be seen in Figure 8 for the LVQ classifier, there is a significant dispersion of type I and type II errors, which means favoring certain classes. However, in the case of MLP (Figure 14), type I and type II errors are at a similar level, and the predictions of classes are more uniform. The incorrect classification results from errors in the estimation of parameters $a_{\phi 2}$, $a_{\phi 4}$ and $a_{\phi 6}$ the determination of which depends on the accuracy of the IFR estimation (23). Due to noise, the location of the maximum CPF obtained for IFR_0 for noiseless NLFM is shifted to a new random location:

$$\widehat{IFR} = IFR_0 + \delta IFR \tag{45}$$

It should be emphasized that, for $SNR \geq -1$ dB, which is important for practical applications, the classification is error-free for the analyzed classifiers. Classification errors appear for $SNR \leq -2$ dB. Due to the gradual shifting and overlapping of the $a_{\phi 2}$, $a_{\phi 4}$ and $a_{\phi 6}$ parameter space, visible in Figure 9, classification becomes problematic. An additional source of error is the use of the polynomial approximation of the NLFM functions proposed for classification, as discussed in Section 3.

6. Conclusions

In this paper, estimation and classification of NLFM signals based on the time–chirp representation have been presented. The majority of useful nonlinearities can be defined or successfully approximated by the polynomial form. However, NLFM radar signals, with the desired PSD and desired AC, require a sufficiently high order of the polynomial for representation of phase and frequency functions. Simulation experiments have shown that CPF is an efficient method of obtaining a polynomial approximation of a phase function.

Having had the coefficients of the phase polynomial from the CPF, the coefficients of the frequency polynomial can be calculated. Based on this relation, the IF estimation of considered NLFM signals is proposed. The simulation results revealed a slightly higher estimation MSE for the proposed CPF-based compared to the QML for NLFM signals. Due to much lower computational load of the proposed approach, the CPF-based method might be preferred in real-time applications.

The promising idea for classification of signals with nonlinear frequency modulations is to use the CPF distribution for the extraction of distinctive features. The modification of the CPF algorithm based on the nonuniform sampling is used for estimation of phase polynomial coefficients for a polynomial of the sixth order. Such calculated coefficients constitute distinctive features for classification.

The classification process has been successfully performed by the LVQ and MLP classifiers, which have the ability to process nonlinear models. Simulations were carried out for different SNRs. The training and classification tasks were performed without the preliminary noise reduction process. This means that classifiers were trained on data with noise that significantly distorted a classified signal, especially for $SNR < -1$. This results in an unacceptable decrease in the effectiveness of the classification. PPV and TPR measures were successfully used to assess classification results. The results of the research showed that the proposed approach allows for classifying four classes: three considered NLFM signals and an LFM signal. It should be emphasized that, for $SNR \geq -1$ dB, which is important for practical applications, the obtained classification is error-free for the analyzed classifiers. The proposed classification process can be easily extended to more than four classes.

Author Contributions: Conceptualization, E.S., D.J. and K.K.; methodology, E.S., D.J. and K.K.; software, E.S. and D.J.; validation, E.S., D.J. and K.K.; formal analysis, E.S., D.J. and K.K.; investigation, E.S. and D.J.; data curation, E.S. and D.J.; writing—original draft preparation, E.S., D.J. and K.K.; writing—review and editing, E.S., D.J. and K.K.; visualization, E.S., D.J. and K.K.; supervision, E.S. All authors have read and agreed to the published version of the manuscript.

Funding: The researchwas conducted at a Bialystok University of Technology within the project WZ/WE-IA/1/2020, financially supported by the Polish Ministry of Science and Higher Education.

Institutional Review Board Statement: Not applicable.

Informed Consent Statement: Not applicable.

Data Availability Statement: Not applicable.

Conflicts of Interest: The authors declare no conflict of interest.

References

1. Samczynski, P.; Krysik, P.; Kulpa, K. Passive radars utilizing pulse radars as illuminators of opportunity. In Proceedings of the 2015 IEEE Radar Conference, Johannesburg, South Africa, 27–30 October 2015; pp. 168–173. [CrossRef]
2. Milczarek, H.; Leśnik, C.; Djurović, I.; Kawalec, A. Estimating the Instantaneous Frequency of Linear and Nonlinear Frequency Modulated Radar Signals—A Comparative Study. *Sensors* **2021**, *21*, 2840. [CrossRef]
3. Robertson, S. *Practical ESM Analysis*; The Artech House Electronic Warfare Library, Artech House: Boston, MA, USA, 2019.
4. Schroer, R. Cockpit Instruments [A century of powered flight: 1903–2003]. *IEEE Aerosp. Electron. Syst. Mag.* **2003**, *18*, 13–18. [CrossRef]
5. Grishin, Y.; Janczak, D. Computer-aided methods of the LPI radar signal detection and classification. *Photon. Appl. Astron. Commun. Ind. High-Energy Phys. Exp.* **2007**, *6937*, 69373N. [CrossRef]
6. Konopko, K. A detection algorithm of LPI radar signals. In Proceedings of the Signal Processing Algorithms, Architectures, Arrangements, and Applications SPA 2007, Poznan, Poland, 7 September 2007; pp. 103–108. [CrossRef]
7. Konopko, K.; Grishin, Y.P.; Janczak, D. Radar signal recognition based on time–frequency representations and multidimensional probability density function estimator. In Proceedings of the 2015 Signal Processing Symposium (SPSympo), Debe, Poland, 10–12 June 2015; pp. 1–6. [CrossRef]
8. Samczynski, P.; Abratkiewicz, K.; Plotka, M.; Zielinski, T.P.; Wszolek, J.; Hausman, S.; Korbel, P.; Ksiezyk, A. 5G Network-Based Passive Radar. *IEEE Trans. Geosci. Remote Sens.* **2022**, *60*, 1–9. [CrossRef]
9. Jin, G.; Liu, K.; Deng, Y.; Sha, Y.; Wang, R.; Liu, D.; Wang, W.; Long, Y.; Zhang, Y. Nonlinear Frequency Modulation Signal Generator in LT-1. *IEEE Geosci. Remote Sens. Lett.* **2019**, *16*, 1570–1574. [CrossRef]
10. Jin, G.; Deng, Y.; Wang, R.; Wang, W.; Wang, P.; Long, Y.; Zhang, Z.M.; Zhang, Y. An Advanced Nonlinear Frequency Modulation Waveform for Radar Imaging With Low Sidelobe. *IEEE Trans. Geosci. Remote Sens.* **2019**, *57*, 6155–6168. [CrossRef]
11. Collins, T.; Atkins, P. Nonlinear frequency modulation chirps for active sonar. *IEE Proc.—Radar Sonar Navig.* **1999**, *146*, 312. [CrossRef]
12. Alphonse, S.; Williamson, G.A. Evaluation of a class of NLFM radar signals. *Eurasip J. Adv. Signal Process.* **2019**, *2019*, 62. [CrossRef]
13. Leśnik, C.; Kawalec, A. Modification of a Weighting Function for NLFM Radar Signal Designing. *Acta Phys. Pol.* **2008**, *114*, A-143–A-149. [CrossRef]
14. Leśnik, C. Nonlinear Frequency Modulated Signal Design. *Acta Phys. Pol.* **2009**, *116*, 351–354. [CrossRef]
15. Milczarek, H.; Lesnik, C.; Kawalec, A. Doppler-tolerant NLFM Radar Signal Synthesis Method. In Proceedings of the 2020 IEEE Radar Conference (RadarConf20), Florence, Italy, 21–25 September 2020; pp. 1–5. [CrossRef]
16. Vespe, M.; Jones, G.; Baker, C. Lessons for radar: Waveform diversity in echolocating mammals. *IEEE Signal Process. Mag.* **2009**, *26*, 65–75. [CrossRef]
17. Gladkova, I. Zak transform and a new approach to waveform design. *IEEE Trans. Aerosp. Electron. Syst.* **2001**, *37*, 1458–1464. [CrossRef]
18. Ghavamirad, R.; Sebt, M.; Babashah, H. Phase improvement algorithm for NLFM waveform design to reduction of sidelobe level in autocorrelation function. *Electron. Lett.* **2018**, *54*, 1091–1093. [CrossRef]
19. Pan, Y.; Peng, S.; Yang, K.; Dong, W. Optimization design of NLFM signal and its pulse compression simulation. In Proceedings of the IEEE International Radar Conference, Arlington, VA, USA, 9–12 May 2005; pp. 383–386. [CrossRef]
20. Kurdzo, J.M.; Cheong, B.L.; Palmer, R.D.; Zhang, G. Optimized NLFM pulse compression waveforms for high-sensitivity radar observations. In Proceedings of the 2014 International Radar Conference, Lille, France, 13–17 October 2014; pp. 1–6. [CrossRef]

21. Price, R. *Chebyshev Low Pulse Compression Sidelobes via a Nonlinear FM*; National Radio Science Meeting of URSI: PortSaid, Egypt, 1979.
22. Yue, W.; Zhang, Y. A novel nonlinear frequency modulation waveform design aimed at side-lobe reduction. In Proceedings of the 2014 IEEE International Conference on Signal Processing, Communications and Computing (ICSPCC), Guilin, China, 5–8 August 2014; pp. 613–618. [CrossRef]
23. Zheng, J.; Su, T.; Zhang, L.; Zhu, W.; Liu, Q.H. ISAR Imaging of Targets With Complex Motion Based on the Chirp Rate–Quadratic Chirp Rate Distribution. *IEEE Trans. Geosci. Remote Sens.* **2014**, *52*, 7276–7289. [CrossRef]
24. Djurović, I.; Thayaparan, T.; Stanković, L. Adaptive Local Polynomial Fourier Transform in ISAR. *Eurasip J. Adv. Signal Process.* **2006**, *2006*, 036093. [CrossRef]
25. Porat, B.; Friedlander, B. Asymptotic statistical analysis of the high-order ambiguity function for parameter estimation of polynomial-phase signals. *IEEE Trans. Inf. Theory* **1996**, *42*, 995–1001. [CrossRef]
26. Barbarossa, S.; Scaglione, A.; Giannakis, G. Product high-order ambiguity function for multicomponent polynomial-phase signal modeling. *IEEE Trans. Signal Process.* **1998**, *46*, 691–708. [CrossRef]
27. Swiercz, E.; Janczak, D.; Konopko, K. Identification of Parameters of High Order Polynomial Phase Signals. In Proceedings of the 2021 21st International Radar Symposium (IRS), Berlin, Germany, 21–22 June 2021; pp. 1–10, ISSN: 2155-5753. [CrossRef]
28. O'Shea, P. A Fast Algorithm for Estimating the Parameters of a Quadratic FM Signal. *IEEE Trans. Signal Process.* **2004**, *52*, 385–393. [CrossRef]
29. Liu, H. Book review: Machine Learning, Neural and Statistical Classification Edited by D. Michie, D.J. Spiegelhalter and C.C. Taylor (Ellis Horwood Limited, 1994). *ACM Sigart Bull.* **1996**, *7*, 16–17. [CrossRef]
30. Ding, J.; Yan, Y.; Liu, Y. Radar signals recognition based on attention and denoising residual network. *Itm Web Conf.* **2022**, *45*, 02012. [CrossRef]
31. Niranjan, R.; Rama Rao, C.; Singh, A. FPGA based Identification of Frequency and Phase Modulated Signals by Time Domain Digital Techniques for ELINT Systems. *Def. Sci. J.* **2021**, *71*, 79–86. [CrossRef]
32. Ma, Z.; Yu, W.; Zhang, P.; Huang, Z.; Lin, A.; Xia, Y. LPI Radar Waveform Recognition Based on Neural Architecture Search. *Comput. Intell. Neurosci.* **2022**, *2022*, 1–15. [CrossRef]
33. Zhu, B.; Jin, W.d. Radar Emitter Signal Recognition Based on EMD and Neural Network. *J. Comput.* **2012**, *7*, 1413–1420. [CrossRef]
34. Wang, C.; Wang, J.; Zhang, X. Automatic radar waveform recognition based on time–frequency analysis and convolutional neural network. In Proceedings of the 2017 IEEE International Conference on Acoustics, Speech and Signal Processing (ICASSP), New Orleans, LA, USA, 5–9 March 2017; pp. 2437–2441, ISSN: 2379-190X. [CrossRef]
35. Świercz, E. Automatic Classification of LFM Signals for Radar Emitter Recognition Using Wavelet Decomposition and LVQ Classifier. *Acta Phys. Pol.* **2011**, *119*, 488–494. [CrossRef]
36. Swiercz, E.; Konopko, K.; Janczak, D. Time-chirp Distribution for Detection and Estimation of LPI Radar Signals. In Proceedings of the 2020 21st International Radar Symposium (IRS), Warsaw, Pakistan, 5–8 October 2020; pp. 362–367. [CrossRef]
37. Djurović, I.; Stanković, L. STFT-based estimator of polynomial phase signals. *Signal Process.* **2012**, *92*, 2769–2774. [CrossRef]
38. Cheng, H.; Zeng, D.; Zhu, J.; Tang, B. Maximum Likelihood Estimation of Co-channel multicomponent polynomial phase signals using IMPORTANCE sampling. *Prog. Electromagn. Res.* **2011**, *23*, 111–122. [CrossRef]
39. Djurović, I.; Stanković, L. Quasi-maximum-likelihood estimator of polynomial phase signals. *IET Signal Process.* **2014**, *8*, 347–359. [CrossRef]
40. O'Shea, P. On Refining Polynomial Phase Signal Parameter Estimates. *IEEE Trans. Aerosp. Electron. Syst.* **2010**, *46*, 978–987. [CrossRef]
41. Swiercz, E.; Janczak, D.; Konopko, K. Detection of LFM Radar Signals and Chirp Rate Estimation Based on Time-Frequency Rate Distribution. *Sensors* **2021**, *21*, 5415. [CrossRef]
42. Djurović, I.; Simeunović, M.; Wang, P. Cubic phase function: A simple solution to polynomial phase signal analysis. *Signal Process.* **2017**, *135*, 48–66. [CrossRef]
43. O'Shea, P.; Wiltshire, R. A New Class of Multilinear Functions for Polynomial Phase Signal Analysis. *IEEE Trans. Signal Process.* **2009**, *57*, 2096–2109. [CrossRef]
44. O'Shea, P. Improving Polynomial Phase Parameter Estimation by Using Nonuniformly Spaced Signal Sample Methods. *IEEE Trans. Signal Process.* **2012**, *60*, 3405–3414. [CrossRef]
45. Hastie, T.; Rosset, S.; Zhu, J.; Zou, H. Multi-class AdaBoost. *Stat. Its Interface* **2009**, *2*, 349–360. [CrossRef]
46. Hosmer, D.W.; Lemeshow, S.; Sturdivant, R.X. *Applied Logistic Regression*, 3rd ed.; Number 398 in Wiley Series in Probability and Statistics; Wiley: Hoboken, NJ, USA, 2013.
47. Hsu, C.-W.; Lin, C.-J. A comparison of methods for multiclass support vector machines. *IEEE Trans. Neural Netw.* **2002**, *13*, 415–425. [CrossRef] [PubMed]
48. Tharwat, A. Classification assessment methods. *Appl. Comput. Inform.* **2021**, *17*, 168–192. [CrossRef]

Communication

Radar Detection-Inspired Signal Retrieval from the Short-Time Fourier Transform

Karol Abratkiewicz

Institute of Electronic Systems, Warsaw University of Technology, 00-665 Warsaw, Poland; karol.abratkiewicz@pw.edu.pl

Abstract: This paper presents a novel adaptive algorithm for multicomponent signal decomposition from the time–frequency (TF) plane using the short-time Fourier transform (STFT). The approach is inspired by a common technique used within radar detection called constant false alarm rate (CFAR). The areas with the strongest magnitude are detected and clustered, allowing for TF mask creation and filtering only those signal modes that contribute the most. As a result, one can extract a particular component void of noise and interference regardless of the signal character. The superiority understood as an improved reconstructed waveform quality of the proposed method is shown using both simulated and real-life radar signals.

Keywords: signal extraction; radar signal processing; time–frequency analysis

Citation: Abratkiewicz, K. Radar Detection-Inspired Signal Retrieval from the Short-Time Fourier Transform. *Sensors* **2022**, *22*, 5954. https://doi.org/10.3390/s22165954

Academic Editor: Ram M. Narayanan

Received: 11 July 2022
Accepted: 6 August 2022
Published: 8 August 2022

Publisher's Note: MDPI stays neutral with regard to jurisdictional claims in published maps and institutional affiliations.

1. Introduction

Non-stationary and multicomponent signals are present in a variety of applications such as radar [1,2], sonar [3], biomedical engineering [4], and others [5–8]. In all practical cases, signals are corrupted by noise, interference, disturbances, and multipath propagation (the latter is meaningfully observable for wireless systems, radar, and sonar) [9,10]. The difficulties in analysis resulting from signal quality degradation are challenging since particular component extraction is not trivial. The problem with multicomponent signal decomposition stems from several issues. Firstly, the noise is a common problem preventing unequivocal signal analysis and spoils the properties of many techniques for mode extraction. Secondly, real-life signals may have components of a different nature, e.g., harmonic terms, pulse-shape transients, amplitude and frequency modulation (AM-FM), and many others. Thirdly, common relations between terms can also be obstructed, e.g., components can be entangled or occupy the same bandwidth, time frame, or both simultaneously.

A typical approach to component extracting is the TF processing due to the informative character of the result, the possibility to distinguish particular signal terms, and invertibility [11]. In the literature, several attempts in signal retrieval from different TF distributions can be identified concerning the STFT [12–15], but also other techniques such as the wavelet transform [16], empirical mode decomposition [17,18], and the Wigner–Ville distribution [19]. In the considered problem, the main goal is to detect the component and create a mask around it. After that, a given component can be retrieved using an inverse transformation. However, one cannot extract and reconstruct the component without any information about the signal character, location on a time–frequency plane, and duration. One of the most popular approaches is signal decomposition using vertical synchrosqueezing (VSS), which is based on relocating transform values along the frequency axes to the local instantaneous frequency ridge [20,21]. As a result, one can obtain a sharp and concentrated distribution, facilitating dominant component extraction. However, the principle of the VSS operation relies on shifting the transform values vertically; hence, impulsive signals and spikes are poorly concentrated or even smeared over the plane. The problem was addressed in [22], where the classical, non-concentrated spectrogram was used. In this approach, the idea was to connect spectrogram zeros using Delaunay triangulation. Then only those triangles were used in signal decomposition whose edge length was in a given range. On this basis, the mask around the detected component was created. The advantage of the method was the possibility of extracting both vertical and horizontal components, which is superior to VSS. However, the triangulation-based approach works moderately with amplitude modulation in low noise levels. In practice, the edge length of the triangle including a valid signal, can vary; hence, its discrimination based on a strict length range is insufficient [21,23].

This paper uses the 2D CFAR algorithm to adaptively detect TF distribution local maxima. The proposed approach is similar to that presented in [24], where the CFAR algorithm was also used. However, the components were extracted after parametric morphological operations, which may deteriorate the quality of the reconstruction due to noise extraction or the removal of a fragment of the useful signal. The optimal morphological operation type and the structuring element size may have a crucial influence on the reconstructed signal quality and depend on the propagation environment and the signal itself. Therefore, the processing may need to tune the algorithm to a specific scenario to detect the signal under interest. In the proposed approach, detected regions are grouped using the density-based spatial clustering of applications with noise (DBSCAN), allowing the TF mask to be defined. In order to extract the entire signal content, the mask is modified regarding surrounding spectrogram zeros, which ensures precise signal reconstruction for any waveform character and local disturbances.

The paper has the following structure: Section 2 presents the proposed algorithm. In Section 3, the method is validated using numerical experiments and compared to two techniques known from the literature: VSS [21] and the triangulation-based method [22]. Section 4 presents the application of the real-life radar pulse filtering, and Section 5 concludes the presented findings.

2. Algorithm Description

For the complex and continuous signal $x(t)$ and the even and real Gaussian window $h(t) = \frac{1}{\sqrt{2\pi}\sigma}e^{-\frac{t^2}{2\sigma^2}}$ with a standard deviation σ, the STFT is defined as follows:

$$F_x^h(t, \omega) = \int_{\mathbb{R}} x(\tau)h(\tau - t)e^{-j\omega\tau}d\tau, \tag{1}$$

where $j = \sqrt{-1}$. The energy distribution, referred to as a spectrogram, is given as

$$S_x^h(t, \omega) = |F_x^h(t, \omega)|^2. \tag{2}$$

In the discrete-time domain, the complex signal is defined as $x[n] = x\left(\frac{n}{N}\right)$ with $n = 0, ..., N - 1$, and the Gaussian window $h[n]$ is truncated to be supported on $[-M, ..., M]$ and obeys $2M + 1 = K$, where K is the number of frequency bins. This work assumes that the window length equals the Fourier transform size. For the Gaussian window, which theoretically has an infinite length, the assumption amounts to elongation of the window with samples very close to 0, which can be regarded as zero-padding. The discrete STFT variant of (1) is given as [25]

$$F_x^h[m, k] = \sum_{n \in \mathbb{Z}} x[n]h[n - m]e^{-j2\pi\frac{k(n-m)}{N}}, \tag{3}$$

where $k = 0, ..., K - 1$, and $m = 0, ..., N - 1$. The analogous distribution for (2) follows as

$$S_x^h[m, k] = |F_x^h[m, k]|^2. \tag{4}$$

In this paper, the TF distribution obtained using (1) is processed in the same way as in radar. In short, a radar system is a device that sends an electromagnetic signal into space and then receives signal reflections from obstacles. The transmitted and received signals are compared, which allows the radar to estimate a target's range (comprising the delay between the transmitted and received signal) and its radial velocity (by analyzing the Doppler shift presentin the received signal). A simplified diagram of the radar operation is shown in Figure 1.

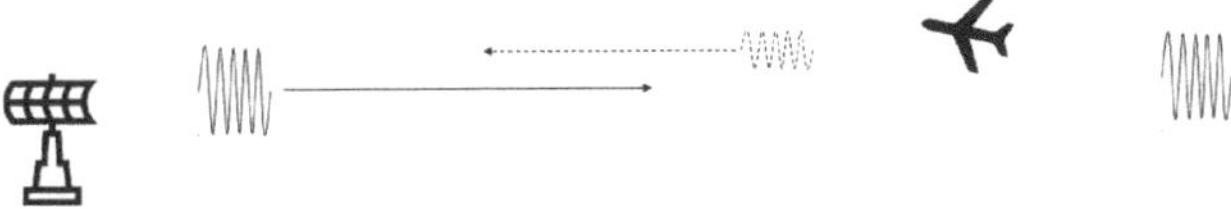

Figure 1. A simplified idea of the radar system. Solid arrow—transmitted signal, dotted arrow—reflected signal.

In practice, the receiver gathers multiple reflections not only from the target, but also from stationary objects such as buildings, trees, etc. Thus, spikes resulting from signal reflection from moving targets in the presence of noise are extracted using the CFAR algorithm, allowing an adaptive threshold to be obtained [26,27], and then, a binary decision is made [26]. The main advantage of the CFAR algorithm is the ability to adapt the threshold level to different kinds of noise, such as Gaussian and alpha stable, among others [28]. For the distribution $X(t)$, one may write the detection condition as

$$D(|X(t)|) = \begin{cases} 1 & \text{if } |X(t)| \geq T, \\ 0 & \text{otherwise,} \end{cases} \tag{5}$$

where T is the regularized threshold governed by

$$T = RC, \tag{6}$$

where C is the noise estimate depending on the CFAR algorithm variant (in the literature, one can find a wide range of descriptions of different CFAR techniques [26,27], and for the sake of clarity and consistency, they are not duplicated in this work). R is the factor manipulated by the probability of false alarm P_f, so that

$$R = N_T \left(P_f^{-\frac{1}{N_T}} - 1 \right), \tag{7}$$

where N_T is the number of points used to estimate the clutter level. An example of adaptive target detection in a real-life radar is shown in Figure 2. The way in which the CFAR algorithm was applied is typical in radar; namely, after the so-called range compression, the target echo can be extracted from the noise. The result shows a one-dimensional CFAR detection performance to demonstrate its main properties. The principle is the same for a bivariate case; however, it would be much more difficult to illustrate. As can be observed, the threshold is adaptively adjusted based on the local noise estimate. The threshold rapidly grows and then drops for the apparent target around the 900 range bin, which yields the target detection. The same procedure is applied to the spectrogram in the proposed algorithm.

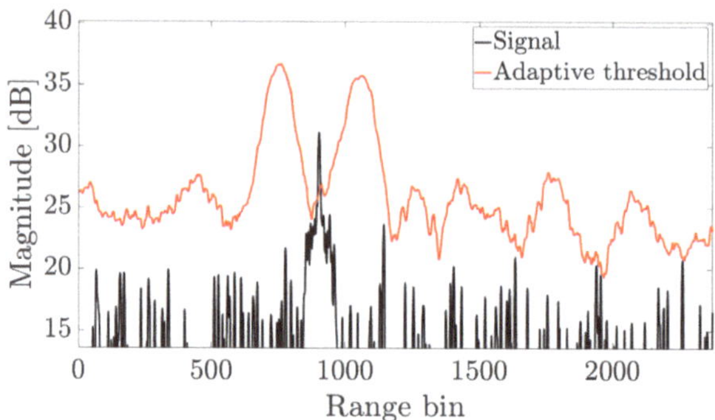

Figure 2. An example of a CFAR detection result using a real-life radar signal after range compression.

Since the spectrogram is a two-dimensional distribution, the CFAR algorithm is two-dimensional as well. In this work, vertical and horizontal training cells are denoted as N_T^V and N_T^H, respectively, $N_T = N_T^V + N_T^H$, and vertical and horizontal guard cells are given by N_G^V and N_G^H, respectively. The idea behind the CFAR algorithm is illustrated in Figure 3.

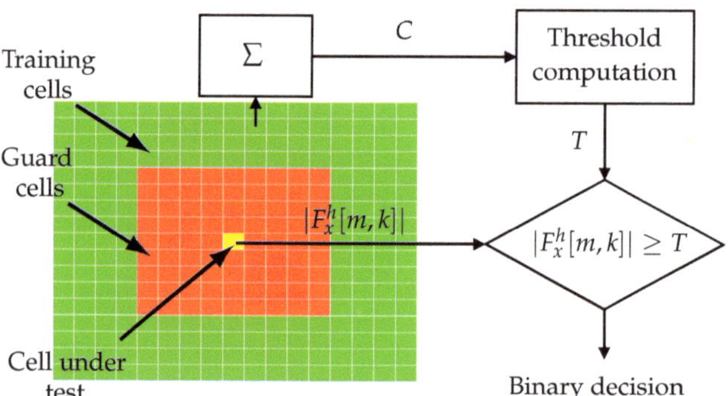

Figure 3. The principle of the CFAR algorithm's operation.

In this paper, for the STFT analysis, two sets are created around each point of the $|F_x^h[m,k]|$ distribution (yellow cell under test). The first one is composed of so-called guard cells marked in the diagram in red. They are not considered during processing since they may influence noise estimates when the strong component occurs. The second group (in green) is built of the training cells. They are used for noise level estimation, usually by averaging or statistical operations. The process is carried out for each $[m,k]$ of the STFT, and as a result, a binary detection map for the whole distribution is obtained according to (5). In this work, the greatest of cell averaging (GOCA) CFAR is used due to its better properties with heterogeneous distributions [26]. However, any other detector can be applied.

Detected points are clustered using the DBSCAN method. In short, this is an efficient, non-parametric technique of grouping points that are closely located and labeling as outliers those groups that lie outside [29]. The algorithm divides points (in this work, points resulting from the CFAR detection) into three sets: core point (which the clustering process starts with); border point (when no more points can be grouped into it); noise point (does not belong to any cluster and two other groups). STFT regions where any coherent structure occurs provide closely packaged points after CFAR detection that can be easily clustered using the DBSCAN algorithm. Detected regions (sub-domains) are sorted concerning the signal energy on the spectrogram. In the CFAR algorithm, P_f selection comprises a trade-off between weak signal extraction and incorrect detection and is usually in the range of $P_f \in (10^{-6}, 10^{-3})$. Thanks to signal clustering using the DBSAN method, false detections are not destructive for component retrieval. Since component energy is taken into account when sorting, all modes stronger than the noise appear higher than the clutter in the sorted list of sub-domains. To ensure a component's extraction, P_f can be greater than in radar, which is a meaningful advantage and results in robustness when joined with the DBSCAN method. The selection of appropriate processing parameters (e.g., the number of guard and training cells, the probability of false alarm) depends on the application, the number of time and frequency bins of the STFT, and the noise level.

The selected sub-domains allow TF masks to be created. As mentioned in this section, the CFAR algorithm assumes a given regularized threshold to detect the spikes. On the other hand, a part of the signal below the threshold is pruned off. Thus, some part of the signal information is lost, since, in practice, the threshold is non-zero. The problem does not matter in radar since the precise signal reconstruction is not needed in most situations, and only the target range and/or velocity are estimated. In signal extraction from the TF plane, the cut-off part of the distribution results in imprecise reconstruction due to the energy pruning. However, as shown in [22,30], the STFT is fully characterized by spectrogram zeros. Following the idea Flandrin and Bardent et al., the STFT given by (1) can be expressed as

$$F_x^h(t, \omega) = e^{-\frac{|z|^2}{4}} \mathcal{F}_x(z),$$ (8)

where

$$\mathcal{F}_x(z) = \int_{\mathbb{R}} A(z, u) x(u) \, du,$$ (9)

is the Bargmann transform [31], whose kernel takes on the following form:

$$A(z, u) = \pi^{-\frac{1}{4}} e^{-\frac{u^2}{2} - juz + \frac{z^2}{4}}.$$ (10)

Equation (9) is a function of positive order and admits the Weierstrass–Hadamard factorization:

$$\mathcal{F}_x(z) \propto \prod_{n=1}^{\infty} \left(1 - \frac{z}{z_n}\right) e^{\frac{z}{z_n} + \frac{1}{2}\left(\frac{z}{z_n}\right)^2},$$ (11)

where $z_n = \omega_n + jt_n$ are zeros of the Bargmann transform (9). These properties were used for component retrieval through the connection of spectrogram zeros using the Delaunay triangulation. However, sensitivity to noise, clutter, and amplitude disturbances make the approach difficult to apply in, e.g., radar systems [21,23]. However, a zero distribution is used in this work to spread the TF mask and extract the entire signal from the STFT. Doing so for the detected signal component clustered using the DBSCAN method, one has to find surrounding spectrogram zeros and extend the mask in order to precisely distinguish the specific signal mode. After that, the STFT can be multiplied by the mask, and the inverse STFT can be applied, obtaining a precise signal reconstruction.

The proposed algorithm can be divided into the following steps:

1. Compute the STFT $F_x^h[m, k]$.
2. Detect protruding regions (sub-domains) from $|F_x^h[m, k]|$ using the adaptive CFAR thresholding.
3. Group the detected points into clusters using the DBSCAN algorithm.
4. Sort clusters comprising their energy.
5. For each desired component, spread the mask to the nearest zeros of the spectrogram $S_x^h[m, k]$.
6. Apply the STFT masking.
7. Retrieve the signal back in the time domain using the inverse STFT.

The outcomes for the exemplified multicomponent signal with vertical, horizontal , and amplitude-modulated terms are presented in Figure 4. In Figure 4a, the spectrogram is illustrated. Three entirely different modes contaminated by noise are apparent. After the adaptive thresholding, detected points are presented as black regions in Figure 4b. Next, they are clustered as shown in Figure 4c, where a different color specifies each sub-domain. Next, the masks are spread to their adjacent spectrogram zeros for a defined number of components. The masks (in black) with detected zeros (green points) are shown in Figure 4d, and their final form is depicted in Figure 4e. The spectrogram after applying the mask looks like the one shown in Figure 4f, and the real and imaginary part of the retrieved components is shown in Figure 4g and 4h, respectively. As shown, the method allows for precise signal retrieval from the TF plane regardless of its character. A typical problem known in the literature is extracting various types of signals, e.g., horizontal, vertical, and mixed terms. In the proposed approach, the signal character does not matter, and the extracted signal can be of any nature, e.g., impulsive spike, chirp, harmonic term, or any non-defined TF structure such as a telecommunication signal. The main limitation is the difficulty of separating the intersecting components, and such signals are not analyzed in this work. The issue requires further in-depth analysis. In the next section, the method is compared to the techniques known from the literature.

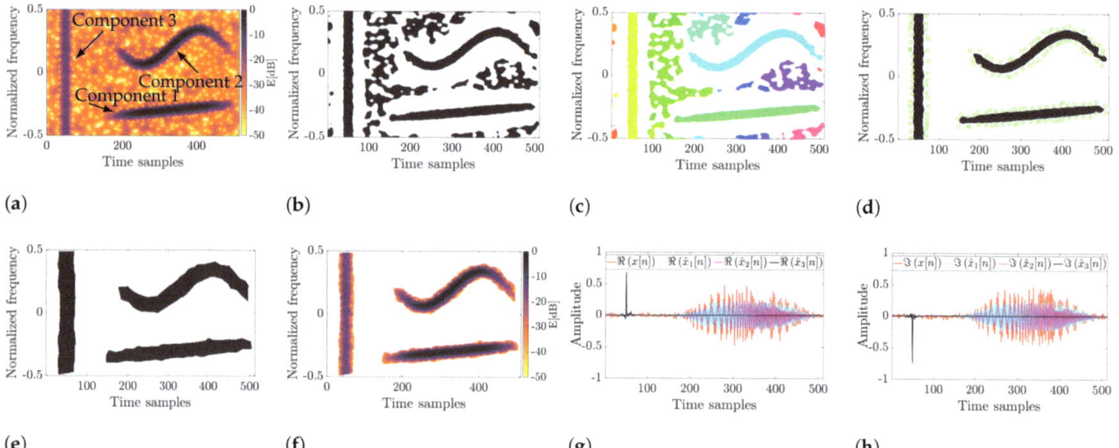

Figure 4. Simulation results for the multicomponent amplitude- and frequency-modulated signal. (**a**) Simulated signal; (**b**) Detected components; (**c**) Clustered components; (**d**) Initial masks with zeros; (**e**) Spectrogram after masking; (**f**) Final TF masks; (**g**) Extracted modes (real part); (**h**) Extracted modes (imaginary part).

3. Simulations

The proposed method was compared to two other well-defined techniques in the literature. The first is the one proposed by Flandrin [22] with a default range of the triangle edge length for noise $e_l \in [0.2, 2.2]$. The second one was VSS, whose definition is as follows [20]:

$$S_x^h(t, \omega) = \int_{\mathbb{R}} F_x^h(t, \Omega)\delta(\omega - \hat{\omega}_x(t, \Omega))\mathrm{d}\Omega, \tag{12}$$

where $\hat{\omega}_x(t, \omega)$ is the frequency reassignment operator. In this work, one of the most precise and efficient approaches to VSS was used, namely the technique called the enhanced first-order VSS (EVSS1) proposed in [21]. For mode extraction from the synchrosqueezed STFT, the technique initially proposed in [32] and implemented in [33] was applied. The approach is based on the local minimum computation of the formula:

$$E_x(\psi_1, ..., \psi_K) = \sum_{k=1}^{K} -\int_{\mathbb{R}} |\mathrm{TF}_x(t, \psi_k(t)|^2 + \lambda\psi_k'(t)^2 + \varepsilon\psi_k''(t)^2\mathrm{d}t, \tag{13}$$

where K stands for the known number of components $\psi_k(t)$ extracted from the distribution TF_x. According to [34], λ and ε were set to 0 in the analysis due to their irrelevant influence on ridge detection. Additionally, in all analyzed cases (simulated and real-life) and for all methods in question, the width of the analysis window was selected to minimize the Rényi entropy [35] or, for signals with several completely different modes, to ensure a relatively constant resolution for all components. Furthermore, in all experiments, the sliding step of the window was assumed to be unitary since it streamlines the inverse STFT computation; however, this value is not fixed and can be easily manipulated.

A comparison of the results for the known techniques and the proposed algorithm is shown in Figure 5. As can be seen, none of the techniques from the literature allow for extracting all components sufficiently from the spectrogram depicted in Figure 5a. For example, vertical synchrosqueezing (Figure 5b) concentrated the harmonic term and sinusoidal chirp well. However, the pulse at the beginning was smeared, and its reconstruction is impossible. Known algorithms are unable to distinguish incorrectly concentrated components. Thus, the common techniques are of limited usability. The vertical component was not even detected for the approach based on spectrogram zero triangulation, as shown in Figure 5c, where different colors represent detected signal modes. As presented in Figure 5d, the proposed method can precisely retrieve all of these components, which is the superiority among the methods known in the literature. As already mentioned, the proposed technique does not assume a signal model; thus, any structure can be extracted regardless of its nature.

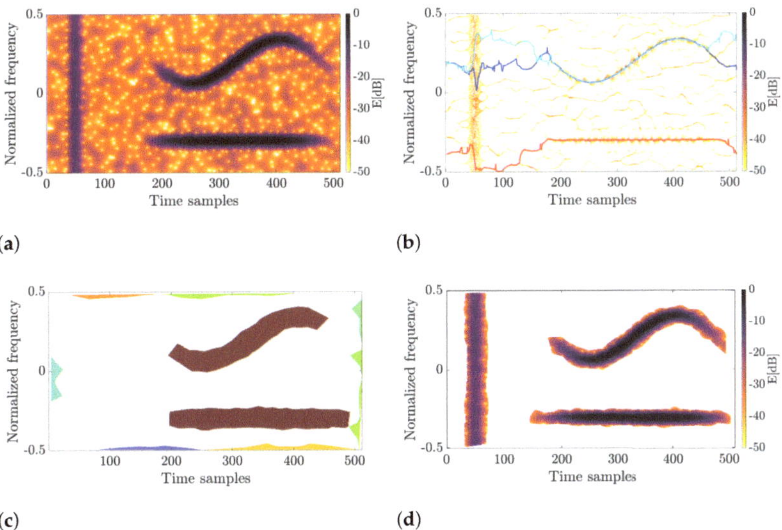

Figure 5. Comparison of different methods for signal retrieval from the time–frequency distribution. For vertical synchrosqueezing and the triangulation-based method, lines and regions with different colors express detected components. (**a**) Spectrogram; (**b**) Vertical synchrosqueezing; (**c**) Triangulation; (**d**) Proposed method.

To assess the proposed method's usability quantitatively, let us consider a nonlinear chirp with sharp nonlinearity at its ends. Such a kind of signal is commonly used in military and civilian radar systems, e.g., air traffic control (ATC) [21]. The simulated signal $x[n]$ of a length N and sampled with a rate $f_s = 12.5$ MSa/s is defined as

$$x[n] = A_x \exp\left(j2\pi \left[\alpha \frac{(n - \frac{N}{2})^2}{2} + \gamma \frac{(n - \frac{N}{2})^{10}}{10} \right] \right) + w[n], \tag{14}$$

where $n \in [0, ..., N-1]$, $\alpha = 1.5 \cdot 10^{11} \frac{\text{Hz}}{\text{s}}$ and $\gamma = 2 \cdot 10^{50} \frac{\text{Hz}}{\text{s}^9}$ are the chirp rate and the nonlinear frequency modulation term, respectively, $\mathcal{A}_x$ is the unitary amplitude, and $w[n]$ is the additive white Gaussian noise. The pulse was multiplied by the Tukey window to simulate a non-zero duration of the rising and falling edges at the beginning and the end of the waveform (amplitude modulation), and after this operation, the signal duration was $T = 30$ µs.

In order to define the reconstructed signal quality, the reconstruction quality factor (RQF) was used for all considered methods [23]:

$$\text{RQF} = 10 \log_{10} \frac{\sum_{n=0}^{N-1} |x[n]|^2}{\sum_{n=0}^{N-1} |x[n] - \hat{x}[n]|^2}, \tag{15}$$

where $x[n]$ is the known and pure waveform and $\hat{x}[n]$ is its retrieved form. The calculations were carried out for six signal-to-noise ratio (SNR) values SNR $= \{5, 10, 15, 20, 25, 30\}$ dB and averaged over 100 Monte Carlo simulations for each method. The STFT was computed using the fast Fourier transform algorithm with 512 points and the Gaussian window with a normalized standard deviation $\sigma = 8$. In this case, $P_f = 0.4$ results in noise detection, but allows for the extraction of amplitude-varying and weak components' extraction. The remaining parameters are as follows: training cells $N_T^V = 16$, $N_T^H = 16$, guard cells $N_G^V = 16$, $N_G^H = 16$.

Figure 6 presents an example of a spectrogram and the results of signal retrieval. As shown, the proposed algorithm allows for the best signal reconstruction over the validated techniques. One can even extract vertical components thanks to operating on a non-concentrated distribution. Furthermore, the independence of the length between spectrogram zeros makes the introduced technique efficient and robust in noise and amplitude modulation. The RQF is higher even by 15 dB compared to the triangulation-based approach and 10 dB compared to VSS (differences for SNR $= 30$ dB). The signal reconstruction results were improved mainly thanks to the non-parametric analysis of the signal distribution on the TF plane. Therefore, the signal model is not used in the extraction process, as is usually the case with methods from the literature. In the proposed solution, the signal may contain any component, and the proposed technique does not assume a specific function defining the phase.

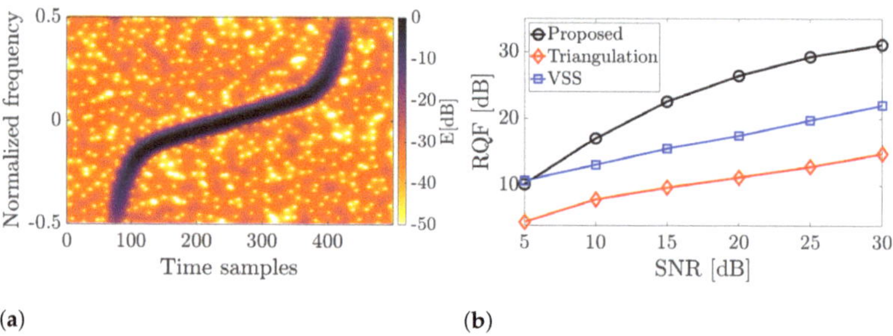

(a)   (b)

Figure 6. Simulation results for the nonlinear chirp signal. (**a**) Spectrogram; (**b**) RQF.

4. Real-Life Signal Analysis

Signal decomposition is an issue in various applications, and the proposed algorithm can be used widely. As a representative example, frequency-modulated radar chirps were considered. The main goal was to extract only the direct pulse, whilst maintaining amplitude, phase, and frequency dependencies unaltered. In this case, a simple inverse STFT is useless. As a result, one would obtain the unmodified signal (the signal was recorded in the time–domain so that the inverse STFT will result in the same signal). Thus, the proposed algorithm can be practically used in such an application.

4.1. Nonlinear Frequency Modulated Pulse

The first signal originates from the ATC radar system located at Warsaw's Chopin Airport, Poland. The sampling rate during the data collection was $f_s = 40$ MHz; then, the signal was downsampled so that the final $f_s = 5$ MHz. The selected pulse (shown in Figure 7a) was processed in the same way as during the simulations. The window standard deviation, in this case, was $\sigma = 10$, and the processing was carried out for 512 points of the FFT. In this case, the CFAR parameters were as follows: $P_f = 0.4$, $N_T^V = 12$, $N_T^H = 8$, $N_G^V = 24$, and $N_G^H = 16$. Since the transmitting radar was non-cooperative and the waveform signature is unknown, the RQF cannot be computed in this case. The signal under consideration and the processing outcomes are presented in Figure 7.

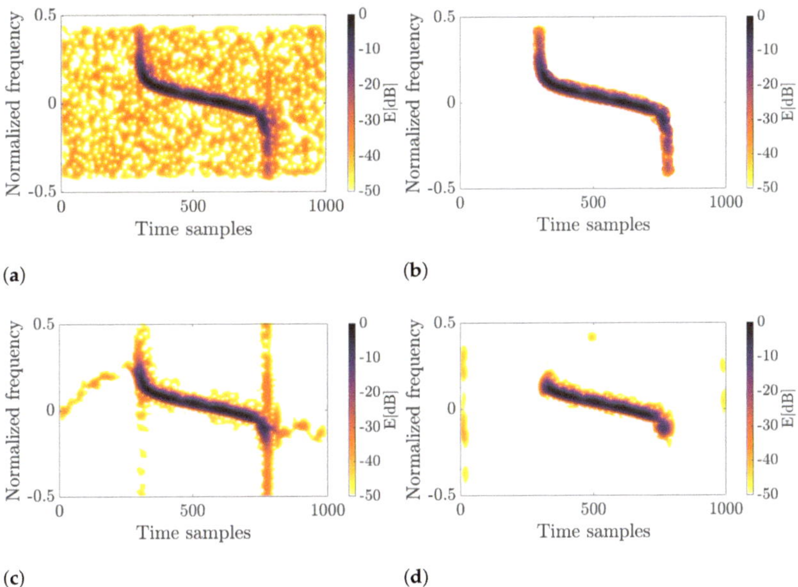

Figure 7. Input real-life radar pulse and the retrieval results for three methods under consideration. (**a**) Input signal; (**b**) Proposed method; (**c**) Vertical synchrosqueezing; (**d**) Triangulation method.

The results clearly show differences between the retrieved pulses. After processing using the method proposed in this work, the entire waveform was extracted, including its linear and nonlinear terms, as shown in Figure 7b. With the VSS-based approach, vertical components (transitions and highly modulated terms) are lost because they cannot be concentrated, and therefore, energy is spread across the plane instead of being focused on the instantaneous frequency ridge. Therefore, the major part of the reconstructed pulse consists of the linear chirp, and the whole signal information was not captured. A similar outcome was achieved for the spectrogram zero triangulation method. In this case, pruning of the pulse ends results from the assumption on the triangle edge length. As can be observed, the edge length for noise $e_l \in [0.2, 2.2]$ does not allow for dealing with all components, especially those with rapid frequency and amplitude modulation.

The reconstructed pulse can be used in a passive radar that uses another active radar as a source of illumination [36] or for specific emitter identification in electronic warfare. The quality and correlation properties of the pulse have a key influence on the radar's detection capability. The better the reconstruction quality, the greater the radar's ability to detect the target. The results of the auto-correlation of the reconstructed pulses are shown in Figure 8.

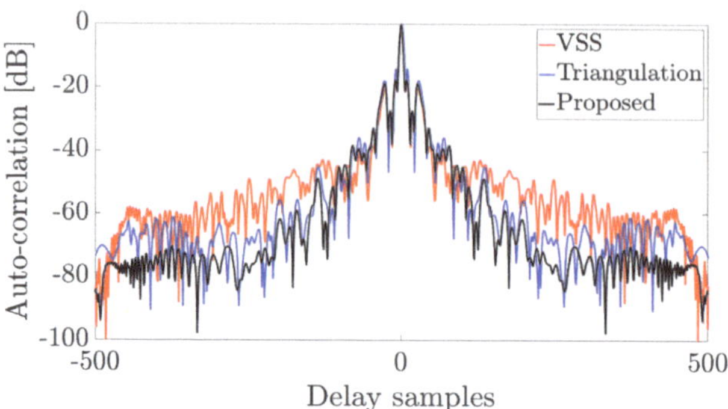

Figure 8. Auto-correlation of the reconstructed NLFM pulses.

As can be seen, the best capabilities were obtained for the proposed method since the side-lobes have the lowest value, while the peak is high and narrow. The main factor influencing the quality of the recovered pulse is the possibility of extracting vertical components at the ends of the transmitted signal. Only the proposed method allows complete pulse extraction, including the part with linear and nonlinear frequency modulation. Furthermore, resistance to local amplitude variation is achievable with the proposed approach. It should be noted that local amplitude deviations resulting from various factors such as the nonlinearity of the radio frequency chain, noise, and multipath propagation influence the amplitude distribution on the spectrogram. That has a crucial impact on the efficiency of methods known from the literature [21,22], which are characterized by inferior properties to those of the presented algorithm. This shows the effectiveness of the proposed method and the possibility of its use in practical systems.

4.2. Linear Frequency-Modulated Pulse with Strong Multipath Interference

The second signal was transmitted by the medium-range radar and recorded with a similar setup as in the NLFM case. The results are illustrated in Figure 9. Apart from the strong direct signal in Figure 9a, delayed copies of the pulse are also apparent in the receiver. For the proposed method, the entire pulse was correctly retrieved with preserved amplitude, phase, and frequency. For the VSS-based approach, the pulse was recovered, but the surrounding noise disturbed its quality. In addition, the component extraction method contributed to the extraction of a certain amount of noise across the entire time axis. The result is shown in Figure 9c. In the method using the triangulation of spectrogram zeros, the results of which are presented in Figure 9d, the reconstruction is also worse than for the proposed method. The impulse was shortened, and some multipath disturbances were also extracted. The best result was obtained for the new method presented in this article, as shown in Figure 9b, where the precisely retrieved signal is shown. It is worth noting that the precise pulse extraction was achievable thanks to the use of the CFAR algorithm, which estimates the noise level and defines the threshold adaptively. The nature of the disturbance may have a different distribution than Gaussian for the multipath effect just after the direct pulse. As can be seen, the CFAR algorithm correctly assessed its level giving a correct threshold value. For the proposed algorithm, the normalized standard deviation of the window was $\sigma = 10$, and the processing was performed for 512 points of the FFT and the following CFAR parameters $P_f = 0.4$, $N_T^V = 8$, $N_T^H = 8$, $N_G^V = 8$, and $N_G^H = 8$. As in the previous case, the pure waveform character was unknown to the author; thus, the RQF cannot be calculated. Therefore, the auto-correlation function was computed and is shown in Figure 10. It is clear that the pulse retrieved for the proposed technique is characterized by the best performance from the radar point of view. Apart from the lowest side-lobe level, the peak is the narrowest, allowing target detection.

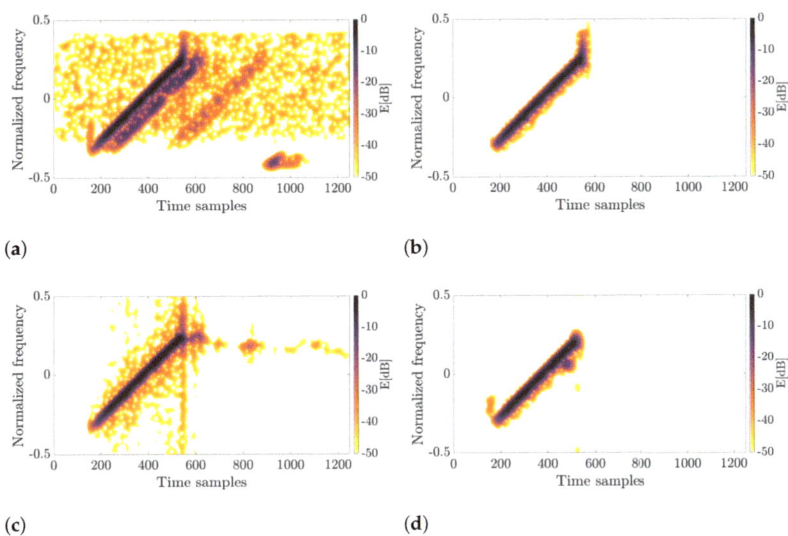

Figure 9. Results for the second signal. (**a**) Input signal; (**b**) Proposed method; (**c**) Vertical synchrosqueezing; (**d**) Triangulation method.

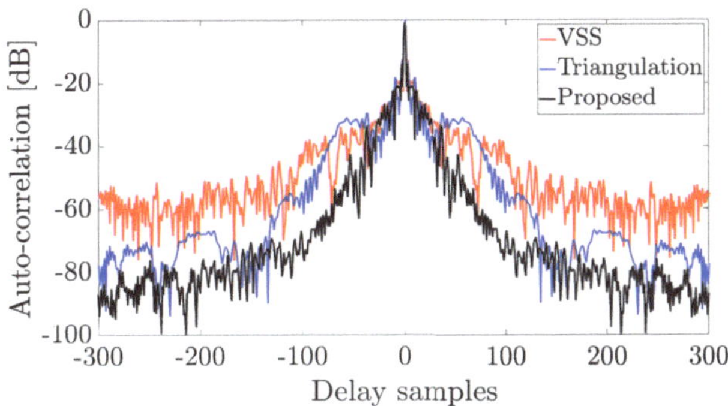

Figure 10. Auto-correlation of the reconstructed LFM pulses.

5. Conclusions

This paper presented a novel signal decomposition approach to component retrieval from the STFT with examples supporting its effectiveness. The superiority of the presented finding relies on the unsupervised ability to extract differently oriented waveforms, e.g., bursts, transient, nonlinear chirps, harmonic terms, and signals with amplitude modulation. The proposed approach's properties were confirmed by thorough numerical experiments using simulated signals and real-life radar pulses. Future research should cover the proposed algorithm's extension to the analysis of multicomponent signals with crossing modes and rapidly oscillating instantaneous frequency. In another direction, work should involve CFAR algorithm parametrization comprising signal and processing parameters.

Funding: This research received no external funding.

Institutional Review Board Statement: Not applicable.

Informed Consent Statement: Not applicable.

Data Availability Statement: Not applicable.

Conflicts of Interest: The author declares no conflict of interest.

References

1. Hanif, A.; Muaz, M.; Hasan, A.; Adeel, M. Micro-Doppler Based Target Recognition With Radars: A Review. *IEEE Sensors J.* **2022**, *22*, 2948–2961. [CrossRef]
2. Uysal, F.; Selesnick, I.; Isom, B.M. Mitigation of Wind Turbine Clutter for Weather Radar by Signal Separation. *IEEE Trans. Geosci. Remote. Sens.* **2016**, *54*, 2925–2934. [CrossRef]
3. Czarnecki, K.; Fourer, D.; Auger, F.; Rojewski, M. A fast time-frequency multi-window analysis using a tuning directional kernel *Signal Process.* **2018**, *147*, 110–119. [CrossRef]
4. Khan, N.A.; Ali, S.; Choi, K. Modified Time-Frequency Marginal Features for Detection of Seizures in Newborns. *Sensors* **2022**, *22*, 3036. [CrossRef]
5. Jin, S.; Johansson, P.; Kim, H.; Hong, S. Enhancing Time-Frequency Analysis with Zero-Mean Preprocessing. *Sensors* **2022**, *22*, 2477. [CrossRef]
6. Maciusowicz, M.; Psuj, G. Time-Frequency Analysis of Barkhausen Noise for the Needs of Anisotropy Evaluation of Grain-Oriented Steels. *Sensors* **2020**, *20*, 768. [CrossRef]
7. Kim, B.S.; Jin, Y.; Lee, J.; Kim, S. FMCW Radar Estimation Algorithm with High Resolution and Low Complexity Based on Reduced Search Area. *Sensors* **2022**, *22*, 1202. [CrossRef]
8. Sahoh, B.; Kliangkhlao, M.; Kittiphattanabawon, N. Design and Development of Internet of Things-Driven Fault Detection of Indoor Thermal Comfort: HVAC System Problems Case Study. *Sensors* **2022**, *22*, 1925. [CrossRef]
9. Xuebo, Z.; Wenwei, Y.; Bo, Y. Parameter Estimation for Class a Modeled Ocean Ambient Noise. *J. Eng. Technol. Sci.* **2018**, *50*, 330–345. [CrossRef]
10. Mahmood, A.; Chitre, M. Modeling colored impulsive noise by Markov chains and alpha-stable processes. In Proceedings of the OCEANS 2015, Genova, Italy, 18–21 May 2015; pp. 1–7. [CrossRef]
11. Fourer, D.; Harmouche, J.; Schmitt, J.; Oberlin, T.; Meignen, S.; Auger, F.; Flandrin, P. The ASTRES toolbox for mode extraction of non-stationary multicomponent signals. In Proceedings of the 2017 25th European Signal Processing Conference (EUSIPCO), Kos Island, Greece, 28 August–2 September 2017; pp. 1130–1134. [CrossRef]
12. Laurent, N.; Meignen, S. A Novel Time-Frequency Technique for Mode Retrieval Based on Linear Chirp Approximation. *IEEE Signal Process. Lett.* **2020**, *27*, 935–939. [CrossRef]
13. Colominas, M.A.; Meignen, S.; Pham, D.H. Fully Adaptive Ridge Detection Based on STFT Phase Information. *IEEE Signal Process. Lett.* **2020**, *27*, 620–624. [CrossRef]
14. Colominas, M.A.; Meignen, S.; Pham, D.H. Time-Frequency Filtering Based on Model Fitting in the Time-Frequency Plane. *IEEE Signal Process. Lett.* **2019**, *26*, 660–664. [CrossRef]
15. Legros, Q.; Fourer, D. A Novel Pseudo-Bayesian Approach for Robust Multi-Ridge Detection and Mode Retrieval. In Proceedings of the 2021 29th European Signal Processing Conference (EUSIPCO), Dublin, Ireland, 23–27 August 2021; pp. 1925–1929. [CrossRef]
16. Daubechies, I.; Lu, J.; Wu, H.T. Synchrosqueezed wavelet transforms: An empirical mode decomposition-like tool. *Appl. Comput. Harmon. Anal.* **2011**, *30*, 243–261. [CrossRef]
17. Huang, N.E.; Shen, Z.; Long, S.R.; Wu, M.C.; Shih, H.H.; Zheng, Q.; Yen, N.C.; Tung, C.C.; Liu, H.H. The empirical mode decomposition and the Hilbert spectrum for nonlinear and non-stationary time series analysis. *Proc. R. Soc. Lond. Ser.* **1998**, *454*, 903–998. [CrossRef]
18. Li, H.; Xu, B.; Zhou, F.; Yan, B.; Zhou, F. Empirical Variational Mode Decomposition Based on Binary Tree Algorithm. *Sensors* **2022**, *22*, 4961. [CrossRef]
19. Wu, X.; Liu, T. Spectral decomposition of seismic data with reassigned smoothed pseudo Wigner–Ville distribution. *J. Appl. Geophys.* **2009**, *68*, 386–393. [CrossRef]
20. Pham, D.H.; Meignen, S. High-Order Synchrosqueezing Transform for Multicomponent Signals Analysis—With an Application to Gravitational-Wave Signal. *IEEE Trans. Signal Process.* **2017**, *65*, 3168–3178. [CrossRef]
21. Abratkiewicz, K.; Gambrych, J. Real-Time Variants of Vertical Synchrosqueezing: Application to Radar Remote Sensing. *IEEE J. Sel. Top. Appl. Earth Obs. Remote. Sens.* **2022**, *15*, 1760–1774. [CrossRef]
22. Flandrin, P. Time–Frequency Filtering Based on Spectrogram Zeros. *IEEE Signal Process. Lett.* **2015**, *22*, 2137–2141. [CrossRef]
23. Abratkiewicz, K.; Samczyński, P.J.; Rytel-Andrianik, R.; Gajo, Z. Multipath Interference Removal in Receivers of Linear Frequency Modulated Radar Pulses. *IEEE Sens. J.* **2021**, *21*, 19000–19012. [CrossRef]
24. Wang, J. CFAR-Based Interference Mitigation for FMCW Automotive Radar Systems. *IEEE Trans. Intell. Transp. Syst.* **2021**, 1–10. [CrossRef]
25. Quatieri, T. *Discrete-Time Speech Signal Processing: Principles and Practice*, 1st ed.; Prentice Hall Press: Hoboken, NJ, USA, 2001.
26. Rohling, H. Radar CFAR Thresholding in Clutter and Multiple Target Situations. *IEEE Trans. Aerosp. Electron. Syst.* **1983**, *AES-19*, 608–621. [CrossRef]
27. Rohling, H. Ordered statistic CFAR technique—An overview. In Proceedings of the 2011 12th International Radar Symposium (IRS), Leipzig, Germany, 7–9 September 2011; pp. 631–638.

8.   Abu, A.; Diamant, R. CFAR detection algorithm for objects in sonar images. *IET Radar, Sonar Navig.* **2020**, *14*, 1757–1766. [CrossRef]

9.   Ester, M.; Kriegel, H.P.; Sander, J.; Xu, X. A Density-Based Algorithm for Discovering Clusters in Large Spatial Databases with Noise. In Proceedings of the Second International Conference on Knowledge Discovery and Data Mining, KDD'96, Portland, Oregon, 2–4 August 1996; AAAI Press: Palo Alto, CA, USA, 1996; pp. 226–231.

30.   Bardenet, R.; Flamant, J.; Chainais, P. On the zeros of the spectrogram of white noise. *Appl. Comput. Harmon. Anal.* **2020**, *48*, 682–705. [CrossRef]

31.   Bargmann, V. On a Hilbert space of analytic functions and an associated integral transform part I. *Commun. Pure Appl. Math.* **1961**, *14*, 187–214. [CrossRef]

32.   Carmona, R.; Hwang, W.; Torresani, B. Characterization of signals by the ridges of their wavelet transforms. *IEEE Trans. Signal Process.* **1997**, *45*, 2586–2590. [CrossRef]

33.   Thakur, G.; Brevdo, E.; Fučkar, N.S.; Wu, H. The Synchrosqueezing algorithm for time-varying spectral analysis: Robustness properties and new paleoclimate applications. *Signal Process.* **2013**, *93*, 1079–1094. [CrossRef]

34.   Meignen, S.; Pham, D.; McLaughlin, S. On Demodulation, Ridge Detection, and Synchrosqueezing for Multicomponent Signals. *IEEE Trans. Signal Process.* **2017**, *65*, 2093–2103. [CrossRef]

35.   Meignen, S.; Colominas, M.; Pham, D.H. On the Use of Rényi Entropy for Optimal Window Size Computation in the Short-Time Fourier Transform. In Proceedings of the ICASSP 2020—2020 IEEE International Conference on Acoustics, Speech and Signal Processing (ICASSP), Barcelona, Spain, 4–8 May 2020; pp. 5830–5834. [CrossRef]

36.   Samczynski, P.; Krysik, P.; Kulpa, K. Passive radars utilizing pulse radars as illuminators of opportunity. In Proceedings of the 2015 IEEE Radar Conference, Arlington, VA, USA, 10–15 May 2015; pp. 168–173. [CrossRef]

Article

Multi-Sensory Data Fusion in Terms of UAV Detection in 3D Space

Janusz Dudczyk [1,*], Roman Czyba [2] and Krzysztof Skrzypczyk [2]

[1] Institute of Information Technology and Technical Sciences, Stefan Batory State University,
96-100 Skierniewice, Poland
[2] Department of Automatic Control and Robotics, Silesian University of Technology, 44-100 Gliwice, Poland;
roman.czyba@polsl.pl (R.C.); krzysztof.skrzypczyk@polsl.pl (K.S.)
* Correspondence: jdudczyk@pusb.pl

Abstract: The paper focuses on the problem of detecting unmanned aerial vehicles that violate restricted airspace. The main purpose of the research is to develop an algorithm that enables the detection, identification and recognition in 3D space of a UAV violating restricted airspace. The proposed method consists of multi-sensory data fusion and is based on conditional complementary filtration and multi-stage clustering. On the basis of the review of the available UAV detection technologies, three sensory systems classified into the groups of passive and active methods are selected. The UAV detection algorithm is developed on the basis of data collected during field tests under real conditions, from three sensors: a radio system, an ADS-B transponder and a radar equipped with four antenna arrays. The efficiency of the proposed solution was tested on the basis of rapid prototyping in the MATLAB simulation environment with the use of data from the real sensory system obtained during controlled UAV flights. The obtained results of UAV detections confirmed the effectiveness of the proposed method and theoretical expectations.

Keywords: UAV; anti-drone system; data fusion; drone detection; identification; recognition; sensing technologies; tracking algorithm

Citation: Dudczyk, J.; Czyba, R.; Skrzypczyk, K. Multi-Sensory Data Fusion in Terms of UAV Detection in 3D Space. *Sensors* **2022**, *22*, 4323. https://doi.org/10.3390/s22124323

Academic Editor: Andrzej Stateczny

Received: 13 May 2022
Accepted: 3 June 2022
Published: 7 June 2022

Publisher's Note: MDPI stays neutral with regard to jurisdictional claims in published maps and institutional affiliations.

1. Introduction

During the last two decades, unmanned aerial vehicles have experienced enormous development [1,2]. Since unmanned aerial vehicles (UAVs) were released for general civil use, the number of incidents involving them have been constantly increasing. Unfortunately, the threats they pose may endanger public and personal safety.

This article comprehensively raises the issue of anti-drone systems technology. The current state of knowledge in this area is presented, as well as an overview of the existing market solutions in the field of anti-drone systems that enable counteracting UAVs [3–7]. Radar, visual, acoustic and radio technologies that are used for UAV detection are characterised. The fundamental research problem raised in the article concerns the data fusion from the AeroScope radio system, EchoGuard radar equipped with four antenna sets and ADS-B in order to develop a UAV detection algorithm in 3D space. The article elaborates on the process of pre-processing data from the afore-mentioned sensors, the data synchronisation process and radar data fusions in order to develop a drone detection algorithm. The algorithm was tested on actual data.

In general, the main goal of the paper was to develop an algorithm based on the data fusion from various sensors, enabling the detection, identification and recognition in 3D space of a UAV violating restricted airspace.

The structure of the paper is as follows. First, a brief review of anti-drone systems and a wide range of UAV detection technologies is provided. Then, the concept of the test procedure in real-word conditions, in the presence of both single and multiple drones, is presented. Section 5 contains a description of selected sensors along with data analysis, which has a significant impact on the choice of data fusion method and the development of an algorithm. Section 6 presents the data fusion algorithm along with a description of the individual stages of system design. Finally, Section 7 presents the results of UAVs detection, performed on the data of the actual experiment. The conclusions are discussed in the last section.

2. Review of Anti-Drone Systems

Currently, there are numerous solutions in the field of anti-drone systems available on the market. Their main task is to counteract UAVs by detecting and combating them using kinetic methods or energy directed at the UAV in the form of an electromagnetic pulse [8,9]. The most advanced technical solutions use full azimuth coverage of 360° and hemispherical elevation coverage in the range of −45° to 225°. The accuracy of computing the azimuth resolution for these types of systems can even reach the value of ±0.5°, and the accuracy of computing the elevation is ±0.7°. Other competitive systems of this type are available on the market, most often used for limited spatial coverage, both in azimuth and elevation. An example is the AARONIA system based on the AARTOS detector (RF detector), which provides an elevation coverage of 10°, and the SpotterRF system (3D-500 Radar), which provides an elevation coverage of 90°. In addition to the above-mentioned systems that are approved for operation in urban areas, there are also systems that, due to their emission parameters, are not approved for urban usage. These include the Blighter (A400 Radar), ROBIN (ELVIRA), ELTA (ELM2026) and Echodyne (EchoGuard) systems, which not only are not approved for urban usage, but also allow you to only cover the space of: 180° in azimuth and 20° in elevation, 360° in azimuth and 60° in elevation, 90° in azimuth and 60° in elevation and 120° in azimuth and 80° in elevation, respectively. In the latest technical solutions, a single radar sensor is able to provide complete hemispherical spatial coverage and semi-spherical elevation coverage, so there is no need to integrate several RF sensors. This approach minimises mutual electromagnetic interference and at the same time enables the efficient use of the allocated spectrum resources. Furthermore, due to the use of a single radar sensor, the weight of the anti-drone system is significantly minimised. An example is the RS800 solution (by ARTsys360), the weight of which is approx. 5 kg, which is incomparably smaller than other systems operating in urban areas, such as SpotterRF or AARONIA, whose weights are approx. 6 kg and 30 kg, respectively. A highly crucial parameter of anti-drone systems is the detection range. This parameter largely depends on the radar cross section (RCS) of the UAV and the speed at which the recognised UAV is moving. Furthermore, a UAV radar cross-section signature may be a highly efficient distinguishing feature in the process of drone detection and classification as shown in [10]. On average, the detection range of drones for this type of systems ranges from 800 m to 3500 m, moving at a speed of approx. 40 m/s. The exemplary EchoGuard radar by Echodyne is able to detect drones at a distance of 900 m, people at a distance of 2200 m and vehicles at a distance of 3500 m. The accuracy of detection is also a crucial parameter of anti-drone systems, the value of which should be as high as possible, especially when detecting and tracking quickly manoeuvring objects in low-altitude urban areas [11]. In addition to technical parameters, it is also worth mentioning the functional parameters of anti-drone systems, such as software-defined functionality or sensory multi-functionality. The software-defined functionality enables additional functions of the anti-drone system, such as automatic self-calibration, dynamic detection and frequency allocation, dynamic disturbance detection, the automatic launch of anti-disturbance modes, remote system operation and configuration and software updates [12]. The sensory multi-functionality of anti-drone systems is based on the application of multisensory sensor fusion (radar, optical, thermal imaging) enabling the detection and tracking of various types of objects, i.e., drones, people and vehicles in full azimuth and hemispheric coverage, and feature extraction for the classification process of detected objects [13].

For this reason, the data fusion process is a relevant technical and functional parameter that directly affects the effectiveness of UAV detection, recognition and identification by the anti-drone system. Therefore, a modern anti-drone system is a heterogeneous platform on which various sensors and effectors are installed. Only in this way is a modern anti-drone system able to effectively detect, recognise, classify and incapacitate UAVs. Commercial anti-drone systems on the market feature minimal sensory fusion. The mentioned systems are not optimised in terms of the integration of radar, optical or thermal imaging sensors on one platform. This leads to their functional degradation and selective use in the object detection process, while minimising the effectiveness and probability of object detection. At the moment, the analysed systems presented in Section 2 do not have the characteristics of the so-called 'data fusion'. Also, commercial solutions do not provide these features, contrary to the method proposed by the authors of this article. To conclude, it can be said that most of the anti-drone systems are dedicated to single purposes that prevent the fusion of many different sensors. By optimising the transmitted waveform and automatic frequency allocation to avoid mutual interference, the ability to transmit detected objects in order to increase the probability of detection and minimise false alarms and the possibility of fusing several sensors on a single display in the GUI, the synergistic coexistence of the anti-drone system is ensured.

3. UAV Detection Technologies

The methods of detecting unmanned aerial vehicle (UAV) interference can essentially be divided into passive and active methods. In [14], the authors present a comprehensive survey on anti-drone systems and anti-drone system analysis, investigating a wide range of anti-drone technologies and deriving proper system models for reliable drone defence. Each of these methods has its own typical advantages and disadvantages. The most relevant methods and their features are characterised below.

The main advantage of all passive methods is their undetectability by an intruder. This is highly important if you plan to detect and monitor the activity of an intruder in order to identify his intentions. Most passive methods are usually less capable of detecting threats, although this is not always the case. Passive methods are based on receiving signals sent by the drone in different frequency bands (it can be a radio, acoustic or optical signal). There are also methods of detecting magnetic anomalies caused by a moving unmanned vehicle, which by its nature causes a local disturbance of the magnetic field. Observation with the application of an optical camera system using visible light technology or thermal imaging is also a highly effective method. In general, optical observation systems are capable of detecting UAVs at a certain distance. This task is performed with the use of specialised imaging cameras and a mount that allows you to observe the entire protected area. As far as the automatic observation method is concerned, it most often uses image analysis algorithms that are based on the changes in the observed image. This allows moving objects to be distinguished in relation to the stationary background. Initial detection is followed by classification based on matching the observed image to the image pattern. This method requires using very high-resolution cameras to ensure that the image is scanned at a distance of several hundred metres from the camera. The required resolution increases with the square of the distance. In [15], the authors show how the performance of a UAV detection and tracking concept based on acousto-optical technology can be powerfully increased through active imaging. Of course, high resolution requires enormous computing power of the image analyser in order to be able to detect, classify and possibly identify the object in real time. In case of performing in the dark, a system of thermal imaging cameras is used. The advantage of this solution is the fact that the drone usually leaves a noticeable thermal trace during the flight due to the high power generated by its automation. In [16], the authors examined the CNN model that is suitable for visible camera-based drone identification. One of the passive methods of UAV detection is the acoustic method, which relies on tracking the object through listening and sensing and then analysing the sound. This method has been known practically since the beginning of aviation and was already used during World Wars I and II. Using a multi-microphone system, it is possible to pinpoint the direction with an accuracy of several dozen degrees. By using signal analysis with the application of appropriate patterns, it is possible to identify, with high probability, the sound emitted by the drone, which is quite distinctive. A significant drawback of the mentioned method is the influence of ambient sounds (the so-called acoustic background), which has a negative impact on the detection, classification and identification of UAVs. In addition, the strength and direction of the wind have a highly crucial bearing on the range of the acoustic method. Due to the highly limited range resulting from the high attenuation of the acoustic wave in the space of its propagation, acoustic waves are used sporadically for UAV detection. Another passive method is radio-electronic spectrum monitoring, which allows the source of the radio signal to be detected and located. When using this method to detect UAVs, it is necessary to know the radio band that is used to communicate with the drone and the signal structure, allowing for the approximate identification and possible exclusion of irrelevant signal sources. An antenna system with a dedicated receiver is required to detect such a radio-electronic signal. To determine the direction from which the signal is transmitted, it is necessary to provide a directional antenna or an array of antennas providing information to the multi-channel data analyser. This method can be used to determine the direction from which the signal is transmitted, but to pinpoint a specific position (dislocation), at least two such tracking units are required to pinpoint the position of the source by using the triangulation method [17]. Another desirable feature is the ability to identify the radio signal sent both by the drone and the operator itself in order to locate the position of said operator of the unmanned aerial vehicle. The ability to read the transmitted information allows you to obtain telemetry data, for instance, the data about the GPS position. This is possible with standard drones purchased on the market that do not use special encryption. A potential disadvantage of the system is the problem of the identification and interception of radio data in the case of specially developed non-standard structures, other than the so-called 'off-the-shelf technology' (COTS—commercial off-the-shelf). Another problem is the use of the drone structure for the so-called 'silent flight'. In this mode, the UAV turns off the two-way radio communication immediately after take-off and can only receive the signal from the operator (uplink). Then it is impossible to detect it in the standard radio band, and the drone flies along the programmed route and lands at a predetermined location.

Active methods use the signal they send to detect and locate unmanned aerial vehicles. Most often it is a radio (or radar) signal in the form of a directional electromagnetic beam. The afore-mentioned signal reflecting off the surface of an object or target is detected [18]. Knowing the delay of the reflected signal, the distance can be determined. The method is effective for purposes several times greater than the wavelength used in the device. Due to the fact that most often they are centimetre waves or shorter, it is possible to detect drone-type objects. The disadvantage of active methods is the easy detection of the 'pinpoint' attempt. In terms of operational activities, this enables an intruder to attempt to withdraw from the intended activities or to hide. Active UAV detection systems operating in the radio band are not dependent on the time of the day or weather conditions. They are widely used for detecting and locating sources of electromagnetic waves, and in case of an active radar beam, they are able to detect even a small object in the protected zone. In [19], the authors design a drone detection mechanism using the RF control signal exchanged between the drone and its remote controller. Fundamental techniques based on radar detection work very well, but they may often be insufficient when confronted with very small UAVs [20]. There are several studies on the analysis of radio signals emitted by UAVs and their controllers. The authors of [21] examine the distinctive features of the radio spectrum for some of the most popular UAV systems and propose an algorithm for detecting the presence of UAVs in the analysed radio spectrum. In [22], the authors propose a per-drone iterated algorithm that optimises drone-cell deployments for different drone-cell numbers and prevents the drawbacks of the pure particle swarm optimisation-based algorithm. The commercial UAV market is developing in an extremely swift manner and new models are based on newer and newer technologies. Therefore, it is vital to update the signature databases for given models, regardless of the selected detection method.

Nevertheless, the radio analysis approach seems to have the lowest variability. Currently, the most popular and the most flexible tool for radio spectrum analysis is software defined radio (SDR) platforms. These are software-tunable radio system platforms that allow for signal processing in digital form. This process is mainly based on the conversion of the recording of the radio signal from the time domain to the frequency domain by means of the Fourier transform and its appropriate analysis [23,24].

Active Methods of Disrupting and Directly Interfering with the UAV

Another way to prevent the drone from undesirable trespassing directly into the safety zone is active interference. This can be executed by disrupting the GPS signal that is necessary for the aircraft positioning and navigation, and also by disrupting or preventing the signal transmission between the operator and the aircraft. Jammers are used for this purpose. They can operate in the band of a specific device or system (e.g., GPS signal) or in a wideband, e.g., to interfere with data transmission. This type of device is most often equipped with a directional antenna that sends a beam aimed at the UAV that needs to be disabled. Sector or omnidirectional antennas are also used to operate in all directions. This solution is used in the defence of a large sector, when it is difficult to locate the target of the attack and precisely pinpoint the direction the UAV is attacking from. The methods of direct interference include active defence through the so-called 'kinetic' attack, which may take the form of firing dismantling missiles at the UAV (whose individual parts, after dismantling, hang on Kevlar lines and get entangled with the UAV rotors) or throwing a neutralising net at the hostile UAV. These types of methods can only be used to a limited extent due to the safety of the people in the vicinity. All the detection methods used, along with the sensors and the entire measurement infrastructure, must be connected to an IT system that supervises all activities. It is necessary to secure and organise the work of security services [25]. The structure of the system must enable data to be received from all sensors and security systems, archived and presented in order for appropriate actions to be taken by the system staff. In the case of large objects, it will usually be a distributed multi-station system with several levels of decision-making, both in the context of extracting the signatures of processed signals and building constant vectors (patterns) in the database for further recognition and identification [26].

In the next part of the paper, we will focus on the detection of UAVs that violate the restricted airspace, and not on methods of disrupting and directly interfering with UAVs. Our considerations concern the development of multisensory data fusion, which allows for the precise detection, identification and recognition in 3D space of UAVs or UAV formations with the distinction of individual platforms. Therefore, the review of anti-drone systems presented in Section 2 made it possible to assess the current state of the art in the studied area, while the available UAV detection technologies presented in Section 3, in particular in the field of sensory systems, allowed for the selection of specific sensors that are the subject of research in the next part of the paper.

4. Data Acquisition during Operational Activities

The main purpose of this operational procedure is to acquire data from a real sensory system operating in real conditions similar to the future operation of the system. In order to recreate the scenario of a monitored airspace violation, archiving necessary data is required, such as:

- Test start time;
- UAV take-off time;
- Time of violation of the observed airspace;
- Flight path (spatial coordinates associated with recording time);
- Flight parameters.

Figure 1 illustrates the concept of the test procedure. The flight scenario was defined so that the flight trajectory was inside the observation zone.

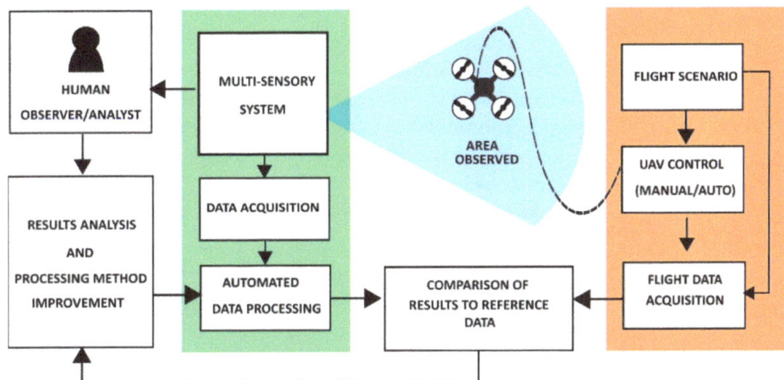

Figure 1. The concept of the system's tests in realistic conditions.

The data acquisition process took place in real conditions during operational activities under the supervision of qualified operators using unmanned aerial vehicles:

- DJI Matrice 600 with an ADS-B transponder on board;
- DJI Mavic 2.

On the basis of the review of the available UAV detection technologies carried out in the previous section, three sensory systems classified into the group of passive and active methods were selected. At this stage, a holistic approach was applied and the sensors were selected in order to obtain complementarity of the data. At the sensor selection stage, the acoustic method based on tracking and analysing the sound trace emitted by a flying object was abandoned. A significant disadvantage of the method mentioned above is the influence of ambient sounds (the so-called acoustic background), which has a negative impact on the detection, classification and identification of UAVs. Another disadvantage is the strength and direction of the wind, which have a significant impact on the range of the acoustic method.

Therefore, taking into account the above, the following sensors were selected, from which data were collected during the tests:

- ADS-B transponder;
- DJI AeroScope (the notation 'AEROSCOPE' and 'AeroScope' will be used interchangeably hereafter) radio system for tracking radio communication between the UAV and the operator;
- EchoGuard radar equipped with four antenna arrays covering a full angle of 360° horizontally and ±40° elevation (will be referred to as 'radar' hereafter).

As a result of the tests performed, three data sets were obtained each time, respectively, for each of the sensors. Further on in the paper, the symbols used in the results tables will be explained and the time plots of selected quantities will be presented in a graphical form.

5. Sensory Data Analysis

The sensors used in the system under construction determine the use of the following methods to develop effective data fusion algorithms ensuring UAV detection in a controlled airspace:

- Methods based on the ADS-B system;
- Passive radiolocation methods (RF—radio frequency);
- Active radiolocation methods (radars).

5.1. ADS-B Transponder

According to the amendment to the aviation law from 31 December 2020, each UAV should be equipped with an ADS-B transponder. The purpose of such an operation is to integrate the UAV with the controlled airspace in which no aircraft can move freely without being visible on the radar by flight controllers and other aircraft, e.g., a passenger equipped with a TCAS (Traffic Alert and Collision Avoidance System)—a collision warning and avoidance system that responds to signals from ADS-B transponders.

The system is event-based and provides data of all the aircraft currently in the airspace and within the range of the ADS-B receiver. Data are contained in an array of 39 columns Data logged by ADS-B receiver indicate the presence of various aircraft types, classified according to the ICAO classification (International Civil Aviation Organization):

- 0 = unidentified (no information on aircraft type);
- 2 = small aircraft (from 15,500 to 75,000 lb);
- 3 = large aircraft (from 75,000 to 300,000 lb);
- 5 = heavy > 300,000 lb;
- 14 = UAV (unmanned aerial vehicle).

Therefore, data analysis requires a preliminary filtration, narrowing down further considerations to unmanned aerial vehicles only. For the purposes of sensory information analysis and the synthesis of the data fusion algorithm, the following vector of measurement values describing the current state of the system was adopted:

$$X_1 = \left[ICAO, \ lat_1, \ long_1, \ h, \ y, \ V_{xy}, V_z, \ ET \right] \tag{1}$$

where:

$ICAO$ is the aircraft type code;
lat_1 is the latitude;
$long_1$ is the longitude;
h is the altitude;
y is the heading;
V_{xy} is the horizontal velocity;
V_z is the vertical velocity;
ET is the category of the aircraft emitting the signal.

The data obtained from the ADS-B transponder are reliable on the condition that the UAV is equipped with such a device. However, the problem is the mass of the ADS-B transponder. Even miniaturised devices are not light enough to be lifted by drones weighing less than 2 kg, and these currently fly the most in the airspace. Accordingly, as of today, there is no guarantee that all drones will be equipped with an ADS-B transponder, so this article proposes a fusion of data from several different sensors.

5.2. AeroScope Radio System

The essence of the method is the detection of the RF radio communication signal between the UAV and the ground operator. The flying platform communicates with the controller in a specified frequency band. Once this frequency band has been identified, there is a high probability that a UAV is within the detection range.

The system is event-based and provides pre-processed data in the form of a table with 19 columns, each of which contains temporal data of a specific physical quantity. For the purposes of further analysis of the sensory information and the synthesis of the data fusion algorithm, the following vector of measurement quantities describing the current state of the system has been assumed:

$$X_2 = [V, \ lat_2, \ long_2, d, \ h, \ y, \ DT, \ Did] \tag{2}$$

where:

V is the flight velocity;
lat_2 is the latitude;
$long_2$ is the longitude;
d is the distance of the UAV from the sensory system;
h is the altitude;

y is the heading;
DT is the UAV type;
Did is the UAV identifier.

Based on the obtained flight logs, the time diagrams of basic physical quantities and the reconstructed UAV flight altitude are presented below.

On the graph showing the changes in the distance of the UAV in relation to the AeroScope receiver (Figure 2) and the graph showing the flight altitude (Figure 3), there are noticeable disturbances in the form of large deviations from the regular flight path. Near the 100th and 175th sample, there are visible abrupt changes in distances of a large value, which are not realistic to achieve during the flight with a standard UAV. It can be concluded that these are disturbances that need to be filtered in the step of signal processing. The solution to this problem has been widely discussed in [27].

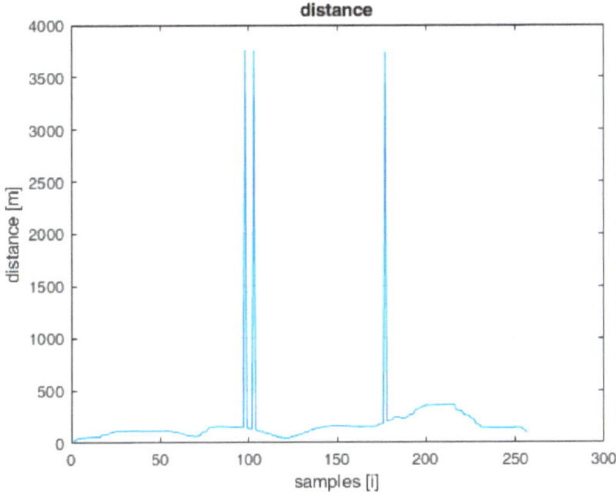

Figure 2. Distance to the UAV measured by AeroScope system.

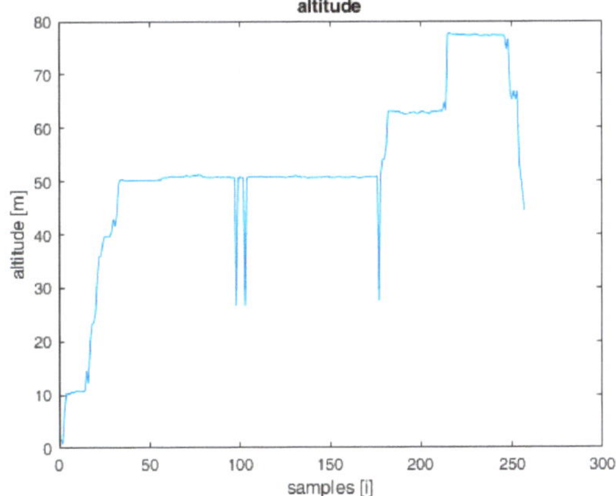

Figure 3. UAV flight altitude broadcasted by AeroScope.

5.3. EchoGuard Radar

The radar system is equipped with four antenna assemblies. The scanning range of a single radar/antenna is 120° in azimuth (±60°) and 80° in elevation (±40°). The antennas are placed every 90° covering the entire area of 360° in azimuth. Taking into account the scanning range of a single antenna of 120°, common scanning areas for adjacent antennas appear.

The system provides data in the form of a table containing 24 columns, each of which is the time data of individual state variables. For the purposes of further analysis of the sensory information and the synthesis of the data fusion algorithm, the following vector of measurement quantities describing the current state of the system has been assumed:

$$X_3 = \left[P_{\mathrm{UAV}}, \ CL, \ RCS, \ az, \ el, \ R, \ x, \ y, \ z, \ V_x, \ V_y, \ V_z \right] \tag{3}$$

where:

P_{UAV} is the probability of UAV detection;
CL is the confidence level;
RCS is the radar cross-section;
az is the estimated azimuth;
el is the estimated elevation;
R is the estimated distance between the UAV and radar;
x, y and z are the estimated UAV coordinates relative to the radar (in the Cartesian coordinate system);
V_x, V_y and V_z are the velocity components of the UAV in the x, y and z axes, respectively.

Based on the flight logs stored, the time graphs of two indicators that will be used in the multisensory data fusion process are presented below: the probability of UAV detection (Figure 4) and the detection confidence level (Figure 5).

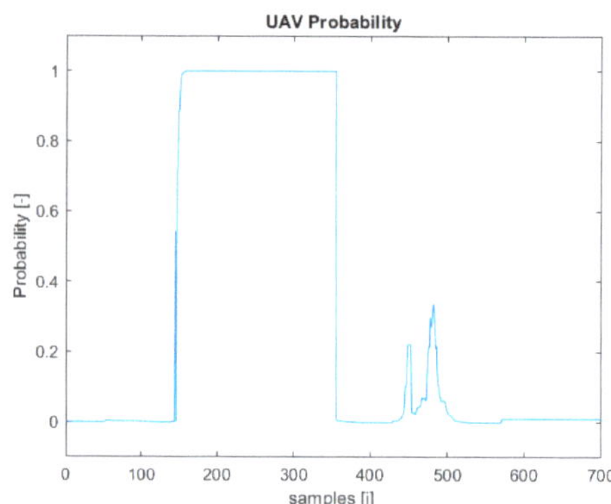

Figure 4. Probability of UAV detection plot.

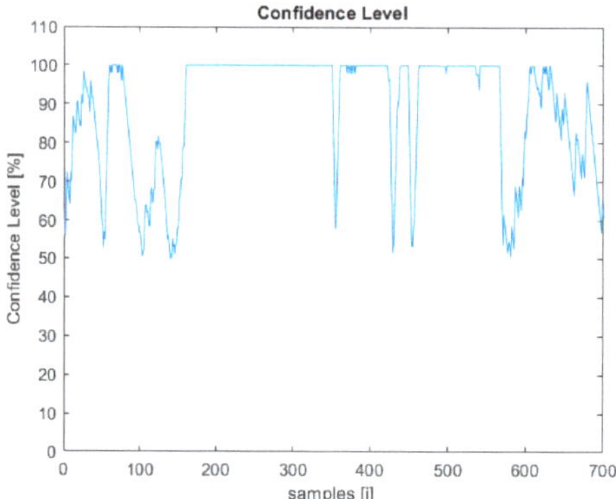

Figure 5. Radar detection confidence level plot.

Information from several indicators, such as the detection probability (Figure 4) and the confidence level (Figure 5), may indicate detections classified with high probability as the UAV.

6. Sensory Fusion Concept

The flight logs contain information acquired by three types of sensors. Each of them has a different operating principle, and thus is a source of broad-spectrum data that complement each other. Moreover, each aforementioned information source bears some uncertainty and inaccuracy; therefore, the fusion of the information provided by all these sources makes it possible to improve the certainty of detection. There are several known data fusion methods based on, e.g., the Kalman filter, complementary filter, weighted function or prediction methods. The most common method of multisource data fusion is based on the Kalman filter concept [28]. Such an approach fits well in the situations when the dynamics of individual data sources vary significantly, which is not the issue of the study presented. The performed in-depth data analysis focused our investigation on data fusion through their conditional complementarity depending on the current conditions, imposed by the type of the identified UAV and its equipment (e.g., the presence of the ADS-B system). Moreover, data selected to be fused are the result of multistage analysis and extraction. For this reason, the application of the Kalman filter-based approach, which was nevertheless considered, was ultimately abandoned at the early stage of the project development. The conditional complementary filtration, which is the second reasonable methodology for multi-source data fusion, was used instead. The data analysis carried out in the previous section allowed for the formulation of several observations:

- Each sensor works asynchronously in an event-based way;
- Data provided by the AeroScope and EchoGuard radar systems contain interference in the form of short duration pulses of high amplitude;
- The ADS-B system provides information about all the aircrafts in the airspace which are within the range of the ADS-B receiver, both manned and unmanned;
- None of these sensors allows a complete detection, recognition and identification procedure;
- The desired effect can be achieved by fusing data from several sensors on the basis of intelligent information complementarity.

Figure 6 illustrates the concept of the data fusion process. As can be seen, this process is performed sequentially, through three stages. There are six independent information sources at the process input: ADS-B and AEROSCOPE transponders as well as four ECHOGUARD radar antennas. Since the data coming from transponders have different features than those provided by radar, they are processed independently. At the first pre-processing stage, outliers are detected and removed from the data sets. In the next phase, the data are synchronized in relation to the system initialization moment, with the given sampling period.

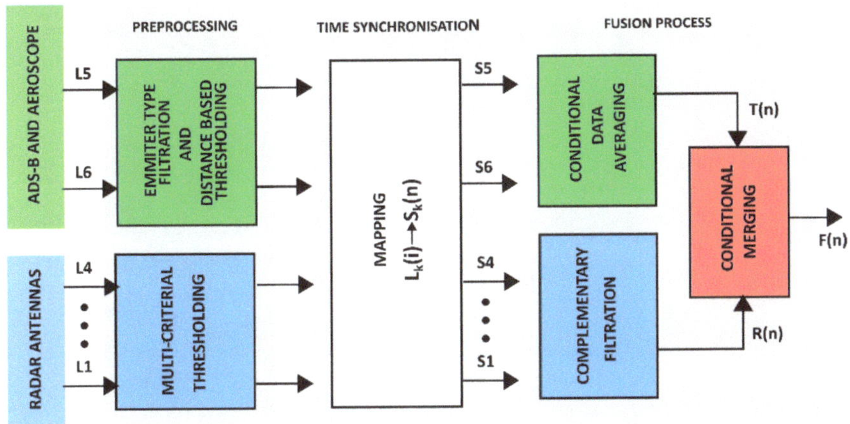

Figure 6. Diagram illustrating sensory fusion concept.

In the last stage synchronized data are fused in two steps. In the first one, independently, detections registered by transponders are merged using an averaging operation while detections recorded by the four radar antennas are fused with the complementary filter. After this, these two independent channels are combined by the conditional merging.

6.1. Data Pre-Processing

Since it is much easier to analyse and process distance-based indices in the Cartesian space, all data on the location of detected objects are converted to the common Cartesian coordinate system, fixed to the centre of the radar station and oriented as follows: North-East-Up. Transponder data are converted from geographical coordinate space (lon-lat-alt). Radar readings are measured in a Cartesian frame fixed to individual radar antennas, and these data also have to be converted into the common coordinates frame.

As was mentioned above, the first step of the procedure consists of removing bad samples which are not going to be used in the fusion process. Since the data packages emitted by transponders have a relatively long spatial range, the receivers which are part of the system described can record information coming from many, often very distant, flying objects. So, the first operation which has to be performed is removing these detections using the distance-based thresholding. The ADS-B sensor can receive data broadcast by large aircrafts which are out of the scope of the system described. Therefore, in the first stage of signal processing, AV emitter type filtration was used. This means that from among all the aircraft identified by the system, only the data related to unmanned aerial vehicles (UAVs) should be extracted. In this case, a dedicated conditional filtering was used in which the *'PingDetectionemitterType = 14'* parameter was used to extract the UAV. This parameter defines the category of the aircraft, and the value of '14' unambiguously determines the unmanned aerial vehicle.

Taking into account the radar, the distance-based thresholding is also the first reasonable criterion of removing outlying detections. However, the radar software algorithms mark each detection with two additional tags which are useful in further data analysis. The first one, which is named *UAV Probability*, ranges from 0 to 1 and describes the certainty that the object detected belongs to the UAV class. The second one is the *Confidence Level*, which takes the values from the range of [0–100] and reflects the confidence that this detection is not measurement noise. In the pre-processing, high values of these tags are used to narrow down the set of analysed data.

6.2. Data Synchronisation

The multi-sensory fusion method outlined so far is based on the assumption that all data provided by particular sensors of the system are synchronous. Such an assumption is necessary to analyse spatial relations between detected objects. This means that for each instance n of discrete time, data sets determining detections collected at the same time can be distinguished: $S_k(n)$, $k = 1, \ldots, K$, where n is the index of the given sensor. In real systems, such as the system described in this paper, both moments of detections and moments of recording them are event-related. This means that after classifying the given measurement (reading) as a detection of a UAV by the sensor algorithm, it is stored in the log-register with the time-stamp of the given sensor. The time-stamp is related to the particular sensor's clock. Therefore, the first stage of the data fusion process is data synchronisation. For each instance n of discrete time, the following mapping has to be made:

$$L_k(i) \rightarrow S_k(n) \tag{4}$$

where i denotes the detection index recorded in the log-register of the kth sensor, whereas n defines subsequent moments in time, $t_n = n\Delta t$, $n = 1, 2, \ldots$, discretised with the sampling period Δt. It is performed by labelling elements of original sets L_k with indices belonging to the given set $S_k(n)$. Each set contains detections acquired in the period of time $t \epsilon < (n-1)\Delta t, n\Delta t >$. In particular, these sets may be empty, which means there was not any detection in the given time interval. Each record of data provided by individual sensors is given a unique identification number. In each sample period n, multiple detections of the same ID can be registered. During the process of synchronising, these detections have to be merged into a single one. This operation can be performed by using various operators. In the case of this work, the averaging operation was applied.

6.3. ADS-B and AEROSCOPE Data Fusion

The idea of a complementary filter is well known from inertial measurement units IMU, the task of which is to estimate the orientation of the UAV based on measurements from independent sensors, characterised by the complementation of information in the frequency domain [29–34]. While analysing the sensory data acquired by the system, it was noticed that within some areas, the detections provided by multiple sensors complement each other. Detections of the ADS-B and AeroScope systems are very similar, provided that the observed UAV is a DJI platform. In cases where the observed area is violated by a UAV of another manufacturer, then the ADS-B system detections will complement the radar indications. Taking into account the above observations, a data fusion based on conditional complementarity was proposed (Figure 7).

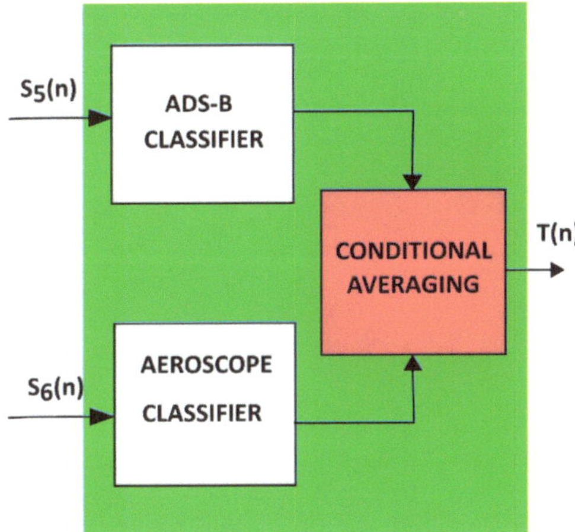

Figure 7. Transponders data fusion idea.

Considering the filter's activity in the assumed time interval, it checks first whether the ADS-B and AeroScope systems have detected the presence of the UAV. If both sensors registered detections, then these detections are averaged in the considered time interval. If the detection appeared only in the data doming from the ADS-B, it means that the detected UAV is not a DJI drone, and in this case, detections provided by the ADS-B are taken as the result of the fusion. However, when the ADS-B and AEROSCOPE systems do not show UAV activity, and the detections appear only in the radar readings, then most likely there is a UAV of a different manufacturer than DJI in the observed area, with neither an ADS-B nor an AEROSCOPE transponder installed. Therefore, the output data of this part of the system take the form of a conditional sum:

$$T(n) = S_5(n) \cup S_6(n) \cup \overline{S}_{5,6}(n) \tag{5}$$

where $\overline{S}_{5,6}(n)$ is the average of the elements of data sets S_5 and S_6.

6.4. Radar Data Fusion

Radar enables detection, localisation and motion parameter measurements of an object moving within its range. The detection range of a radar is dependent on its sensitivity and the spatial configuration of its antennas. Also, environmental conditions taken together with the aforementioned factors imply that the same object tracked by multiple antennas may be detected with different accuracy and certainty. For example, the certainty of an object detection moving at the limit of the sensor's range is usually lower than the object localised in the center of the detection area.

So, having information about detections coming from multiple radar antennas monitoring the same observation area, it is usually possible to improve detection accuracy and certainty. Figure 8 gives an interpretation of this case. There is a moving object within the range of two antennas of the radar station (S_1 and S_2; their detection areas are plotted with red and blue colors). The real localisation of the object in subsequent moments in time is marked with black circles, while its current location in the nth moment is marked with a blue circle. Detections of this object registered by antennas S_1 and S_2 are marked with the red circle and the blue star, respectively. The result of the fusion of information provided by these two sensors is plotted with the black circle.

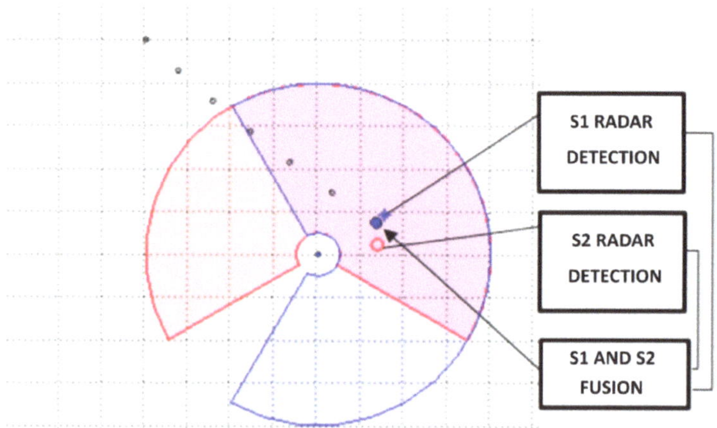

Figure 8. Geometrical interpretation of the case of multiple radar detection of the same object.

In the case of multiple detections, the problem is determining which of the detections represent the same object, then how to make their fusion. In this paper, we applied the distance-based clustering and complementary filtration with weighting factors calculated as functions of the certainty factor.

Let us assume that the given area is monitored by four radar antennas of different spatial configurations. Let us also assume that M unidentified objects are moving within this area. The set of detections captured by the kth antenna in the time n is denoted as

$$S_k(n) = \left\{ d_i^k \right\}, \; i = 1, 2, \ldots, M, \; k = 1 \ldots 4 \tag{6}$$

where

$$d_i^k = \left(P_i^k, c_i^k \right), \; c_i^k \in [0, 1] \tag{7}$$

The P_i^k denotes the ith detection of the kth antenna described in Cartesian spatial coordinates, while c_i^k is the detection certainty factor estimated by the internal algorithm of the radar system.

The first stage of the fusion procedure is identifying the given object among all detections captured by particular antennas. It is performed by determining the similarity (in the sense of the metric used) of elements belonging to the sets $S_1 \ldots S_4$. The similarity is related to the spatial proximity of the elements; therefore, the most convenient and intuitive metric is the Euclidean one, which was used in the described approach.

Two elements captured by different sensors at the same moment n in time are considered to be similar if the distance between them is smaller than some threshold value:

$$\left| d_i^k - d_j^l \right| < THR, \; k,l = 1,2,\ldots,4, \; i \in [1, M_k], \; j \in [1, M_j] \tag{8}$$

This threshold depends on many issues, such as the spatial resolution of the sensors, and is tuned experimentally. The result of the similarity checking process is the similarity matrix:

$$\mathbf{A_s} = \{a_{n,m}\}, \; n \in [0, N_s], \; m \in [1,4] \tag{9}$$

where m *is* equal to the number of sensors and n is the number of pairs of elements classified as similar, respectively. For example, the matrix row of elements [1 0 2 0] means that the first element of the set S_1 is similar to the second element of the set S_3, which in turn implies that antennas 1 and 3 detected the same physical object.

The next step of the data aggregation procedure is merging the detections of the same object captured by multiple radar antennas, indicated by the similarity matrix. Each row of this matrix indicates a pair of detections which are close enough in the sense of the threshold applied. In the approach presented, the complementary weighted average was used for merging the similar detections. The weighting factors are computed the way providing that the detection captured with a higher certainty factor also has a higher influence on the fusion result. Let us assume two detections classified as similar are given by

$$d_i^k = \left(P_i^k, c_i^k \right) \; \text{and} \; d_j^l = \left(P_j^l, c_j^l \right) \tag{10}$$

The fusion $\left(\widetilde{P}, \widetilde{c} \right)$ of these two elements is calculated as

$$\widetilde{P} = w_1 P_i^k + w_2 P_j^l \tag{11}$$

$$\widetilde{c} = \max\left(P_i^k, P_j^l \right) \tag{12}$$

The weighting factors are computed complementarily as the functions of certainty factors:

$$\begin{cases} w_1 = 0.5\left(1 + \dfrac{1}{1+e^{-\alpha(c_i^k - THR)}}\right), \; w_2 = 1 - w_1 & for \; c_i^k > c_j^l \\[4mm] w_1 = 0.5\left(1 + \dfrac{1}{1+e^{-\alpha(c_j^l - THR)}}\right), \; w_2 = 1 - w_1 & for \; c_i^k \leq c_j^l \end{cases} \tag{13}$$

After this stage of the fusion process, for the nth moment in time, the following vector of detections is obtained:

$$R(n) = \left[\widetilde{d}_1(n), \widetilde{d}_2(n), \ldots, \widetilde{d}_{M_n}(n) \right], \quad M_n \leq M \tag{14}$$

6.5. Final Fusion

At this stage of the process, detections coming from two separate information channels (transponders and radar) have to be merged into one. Let us denote the set containing fused readings coming from transponders ADS-B and AeroScope, recorded in the sampling period n as $T(n)$. On the other hand, there is a set of radar detections fused recorded in the same sampling period—$R(n)$. The fusion of these stets is performed using the following conditional scheme:

$$F(n) = \begin{cases} T(n) & if \; T(n) \neq \varnothing \cap R(n) = \varnothing \\ R(n) & if \; T(n) = \varnothing \cap R(n) \neq \varnothing \\ \widetilde{F}(n) & otherwise \end{cases} \tag{15}$$

The first two cases of this scheme are obvious and will not be commented upon. Let us take a closer look at the third one, describing the situation where both the transponders and the radar registered detections. In such a case, there is a need to distinguish two cases. First, when readings coming from the transponders and the radar refer to the same object, they have to be merged. Second, when detections registered by the radar and the transponder receivers are disjointed, both detections are stored in the files of the system records. Hence the following conditional procedure is performed:

$$\tilde{F}(n) = \begin{cases} T_i(n) \cup R_j(n) & if \ \left| T_i(n) - R_j(n) \right| > THR \\ T_i(n) & if \ \left| T_i(n) - R_j(n) \right| \leq THR \end{cases} \tag{16}$$

So, if the distance between detections $T_i(n)$ and $R_j(n)$ is greater than the assumed threshold THR, both detections are saved in the records. Otherwise, there is a high possibility that two detections refer to the same object. In this case, detection $T_i(n)$ coming from the transponder is kept, since it is more reliable.

6.6. Detection Identification and Tracking

At the last stage of the process of retrieving information upon violation of the observed airspace, identification of the detection and tracking the identified object must be provided. This is the most important information from the end user perspective, and performing these operations simultaneously is highly complex. Of course, all the preceding data processing is absolutely necessary to perform the last stage, and this must be highlighted. The identification of the objects detected by individual radar antennas is based on signatures given to them by the radar software algorithms. Similarly, data sent by transponders contain their own signatures identifying the objects. Further, while merging the detections, this information is lost. Therefore, it is very important to combine new incoming detections with those previously registered. In other words, the following similarity must be found:

$$F_i(n) \sim F_j(n-1), \ \ i = 1, 2, \ldots, \bar{\bar{F}}(n), \ \ j = 1, 2, \ldots, \bar{\bar{F}}(n-1), \tag{17}$$

where the time index n denotes the current detection while $n-1$ *is* the previously obtained one.

One of the possible solutions to the aforementioned problem is using the predictive approach. The method proposed consists of comparing the detection's prediction to the current detection, using the Euclidean metric:

$$F_i(n) \sim \hat{F}_j(n), \ \ i = 1, 2, \ldots, \bar{\bar{F}}(n), \ \ j = 1, 2, \ldots, \bar{\bar{F}}(n-1), \tag{18}$$

where $\hat{F}_j(n)$ denotes the detected objects's location prediction calculated for the current moment n. The prediction is the function of the past readings:

$$\hat{F}_j(n) = f\big(F_j(n), F_j(n-1), \ldots, F_j(n-H)\big), \tag{19}$$

where H is the number of past readings taken for calculating the prediction. Another option that allows the location prediction of the considered object to be obtained is using information on its velocity. Since the radar provides estimates of the detection's velocities measured in relation to three axes, it is easy to find the prediction of the object location using these data:

$$\hat{F}_j(n) = f\big(F_j(n-1), v_{x,j}(n-1), \ v_{y,j}(n-1), v_{z,j}(n-1)\big) \tag{20}$$

One more possibility of finding similarity between current and past detections is comparing them directly using the Euclidean metric.

In all the aforementioned options, if the distance between the current and previous detection are recognized as similar, the current detection is given to the identifier of the previous one. Otherwise, the current one is treated as a new detection and is given a new ID.

7. Results

The purpose of the task was to validate by simulation the algorithm developed for detecting single, as well as multiple, UAVs. As part of this task, the results generated by the algorithm implemented were verified by comparing them to the data obtained as a result of the planned experiment. Before commencing validation tests, measurement data were acquired by the sensory system during test flights. The tests covered:

- Single UAV flight—DJI Matrice 600;
- Simultaneous flight of two UAVs—DJI Matrice 600 and DJI Mavic 2.

In the case of using the DJI Matrice 600 platform, a full set of measurement data was provided, which includes ADS-B, AeroScope and radar (four sector antennas). In turn, using the DJI Mavic platform, AeroScope and radar data were provided.

7.1. Single UAV Detection

The first test consisted of performing an operator-controlled flight of the UAV of type DJI Matrice 600 within the monitored area of restricted airspace. As was mentioned before, this drone was equipped with both ADS-B and AREOSCOPE transponders. The radar antennas were mounted on the 8 m high mast. During the experiment, data coming from all sensors were saved. After the flight, the data acquired were post-processed. Time synchronisation with the sampling period equal to 1 [s] was forced. Figure 9 shows the result of the fusion procedure described in the previous section, presented using a 3D graph. Detections registered by the radar system, after filtering and merging data coming from four antennas, are marked with the blue crosses. In the case of radar data fusion, additional detections appeared, clearly visible on the 2D projection of the flight trajectory (Figure 10), which were the result of the assumed threshold values of complementary filtration. The threshold values used in the filtration process were: PUAV = 0.7 and CL = 80%.

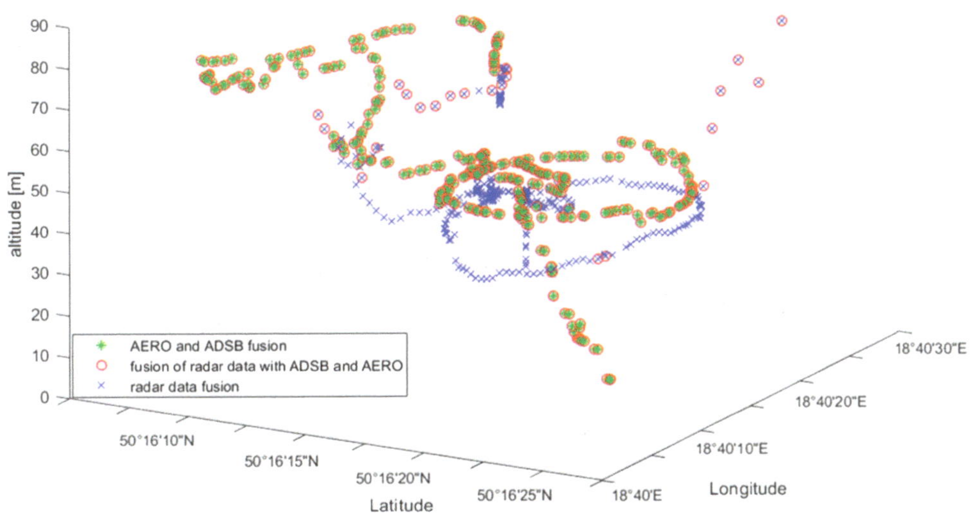

Figure 9. UAV flight path registered by various sensors after the information fusion.

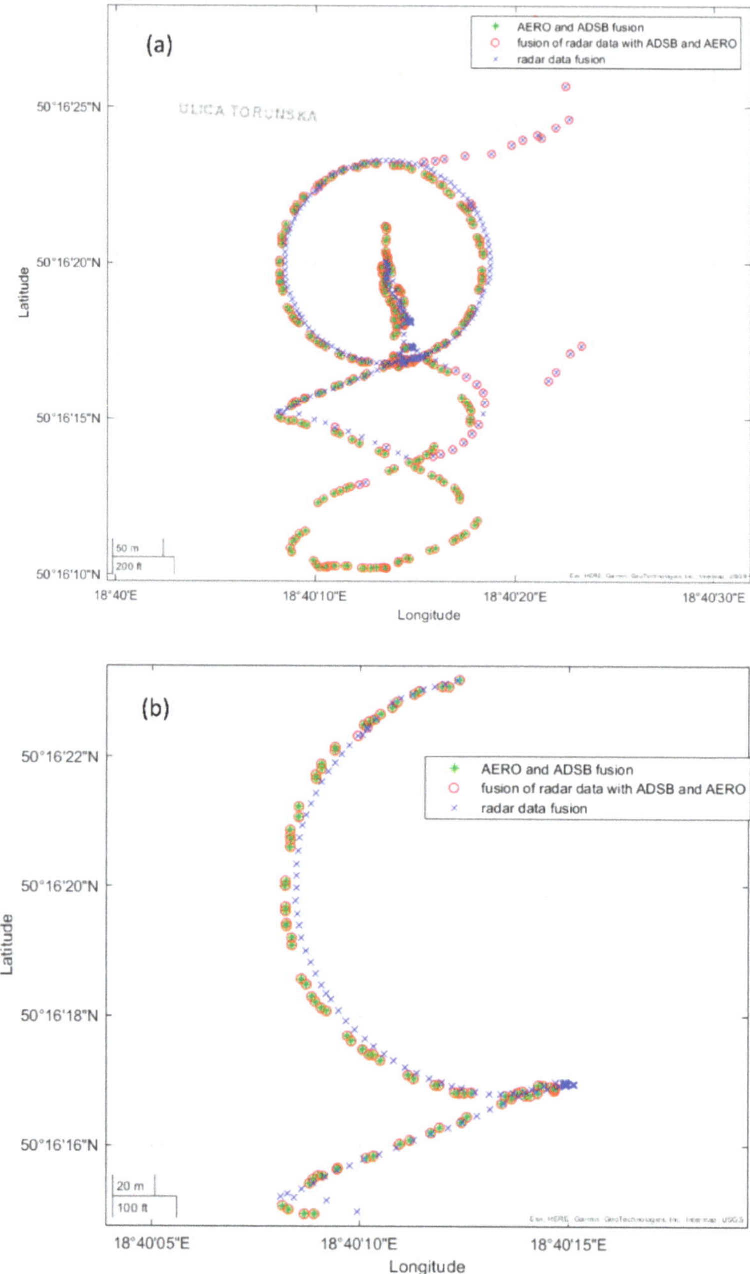

Figure 10. UAV flight path presented in 2D geographical coordinates (**a**) and its enlarged part (**b**).

On the other hand, the UAV flight path registered by transponder systems ADS-B and AREOSCOPE after the fusion process is plotted with a red circle. As one can notice by looking at this picture, there is a slight offset between the transponders' and radar's data regarding the flight altitude measurements. This phenomenon is illustrated more precisely in Figure 11. The reason for such a discrepancy in the presented results comes from the fact that the assembly offsets of the radar antennas were not measured precisely enough.

Finally, a complete fusion of the transponder and radar data are presented in Figure 9 with green stars. As was explained in the previous section, data coming from transponders, registered by a GPS system, are considered as more reliable than radar measurements. Therefore, in cases where there are both radar and transponder data in the given sampling period, the fusion results in neglecting the radar readings.

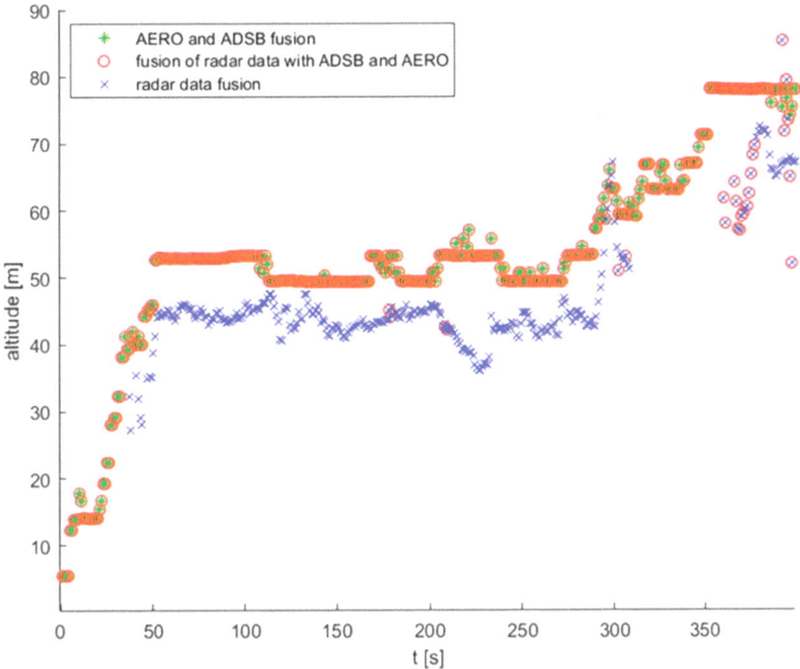

Figure 11. UAV flight altitude plot.

Figure 10 shows the 2D interpretation of the aforementioned fusion aspects: the entire flight path (a) and the enlarged part of the flight path for the time interval of 200–300 [s].

7.2. Multiple UAV Detection

System tests were performed comprehensively and followed the system integration stage. At this stage of the project, the end-to-end test was performed. The test scenario assumed the simultaneous flights of two drones—the Mavic 2 and Matrice 600 Pro. The flight altitude of the first one was about 50 [m], whereas the second drone was flying at the altitude of 30 [m]. The radar antennas assembly configuration was the same as in the experiment described in Section 7.1. Also, the notation used for marking individual stages of data processing is the same. Again, as in the previous scenario, data acquired from transponders were used as the reference to the radar measurements. This time, let us start the results analysis from observing the recorded altitude data, presented in Figure 12.

Let us consider the flight of the M600 drone. Analysing the AREOSCOPE data, we can see an almost flat, very precise record of the flight altitude at the level of 32 [m]. Taking into account the ADS-B readings, also mounted onboard the M600, the readings are not so accurate and oscillations of the amplitude of about 10 [m] can be observed. As for the radar detections of this object, a slight offset of about −5 [m] can be seen. Finally, after the fusion of data from these three sources, the M600 flight altitude path was extracted (black line). On the other hand, analysing the Mavic 2 drone flight, we can conclude that the data describing the altitude registered by radar and AEROSCOPE are more similar to each other.

As in the first case, the fusion procedure gave a quite reliable flight path identification (red line).

Figure 13 shows detected objects on the longitude–latitude plane. Readings from transponders are marked as explained by the picture legend. As mentioned before, they are considered as a reference to the data fusion results. We can observe a slight offset between the ADS-B and AEROSCOPE data identifying the M600 drone. Looking at the radar fusion results, it can be noticed that both drones were detected, identified and tracked.

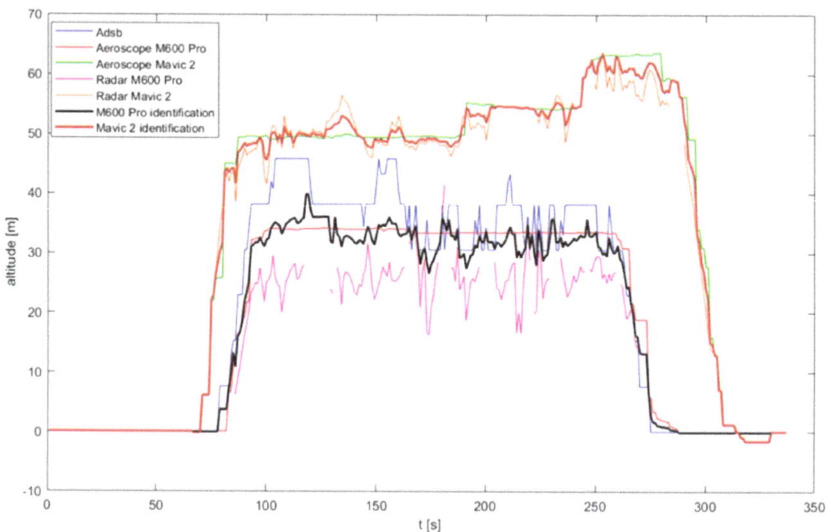

Figure 12. Multiple UAV flight altitudes–indirect detections and identifications.

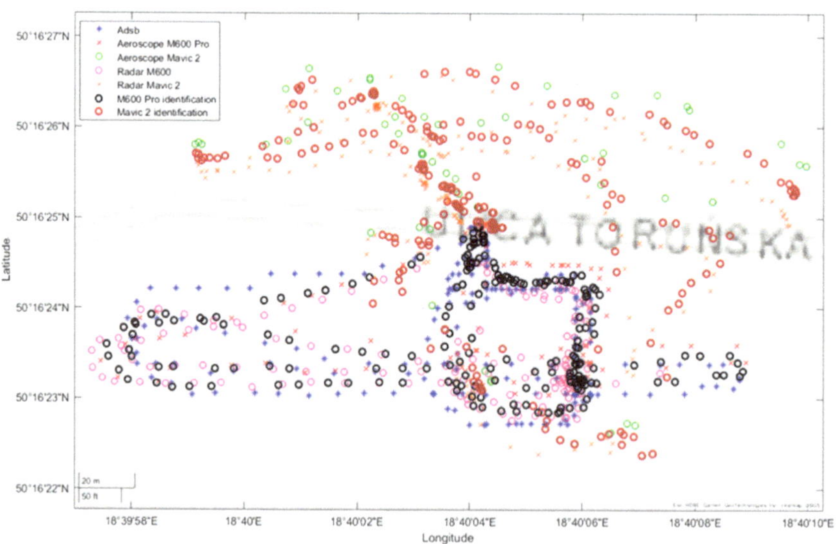

Figure 13. Flight trajectory in 2D space—indirect detections and identifications.

Finally, the fusion of all informational channels results in registering precise tracks of objects violating the observation area. This is shown even better in the last figure as a 3D plot (Figure 14). Since transponders' data contain information about the type of the UAV, these objects were identified as two drones: M600 and Mavic2. Of course, due to only having radar readings, it is not possible to perform high level identification. Nevertheless, as is shown in Figure 13, the radar was able to identify two separate tracks of UAVs flying within the monitored space.

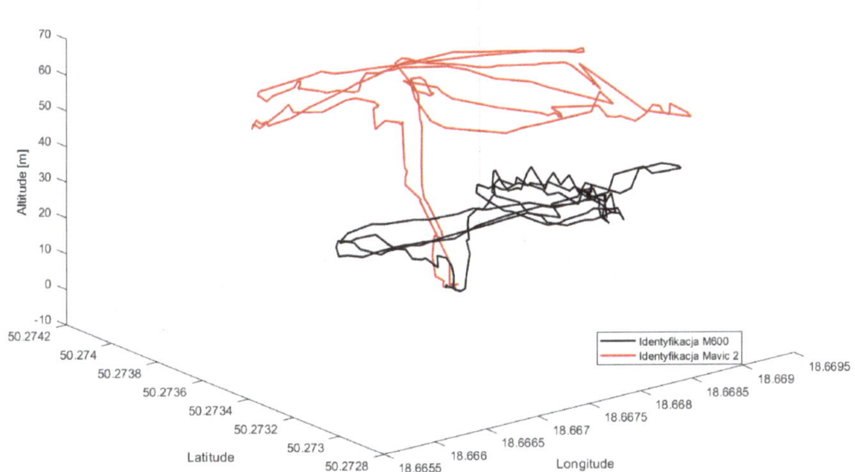

Figure 14. Flight trajectory in 3D space—final result of data fusion.

8. Conclusions

This paper concerned the problem of the detection of unmanned aerial vehicles violating restricted airspace. At the outset, the review of available market solutions in the field of anti-drone systems and existing detection technologies made the reader aware of the possibilities of counteracting unmanned aerial vehicles. The performed analysis also showed the shortcomings of the existing anti-drone systems and indicated new research directions. In the next stage, the data provided by the sensory system during field tests in real conditions was analysed. The algorithm of fusing data acquired by multiple sensors, enabling the UAV detection, identification and recognition in 3D space, was proposed. The developed detection system includes the following three subsystems: pre-processing, time synchronisation and data fusion, which is based on complementary conditional filtering. The efficiency of the proposed solution was tested on the basis of rapid prototyping in the MATLAB simulation environment using data from the real sensory system obtained during controlled UAV flights.

It must be concluded that if a UAV entering the range of the monitoring system is equipped with an ADS-B or AEROSCOPE transponder, there is no problem with detecting and identifying the aerial vehicle. Unfortunately, the most likely scenario is that the UAV violating the restricted airspace will intentionally not be equipped with such devices. Then, the system detection abilities are based on radar readings. The radar used in this project is very sensitive and is able to detect small objects—either miniature drones or other objects. This property of the sensor raises further problems with data interpretation. Usually, the radar captures many more objects than the ones being subject to the observation. In this paper, we proposed multistage filtration to reject the false detections. The efficiency of the proposed approach was proved by multiple tests. The next crucial problem addressed in this paper was the data fusion of detections registered by multiple radar antennas with overlapping fields of view. The complementary filter-based fusion, presented in Section 6.3, solved this problem in a satisfactory way. The next issue is the identification and tracking of the detected object using radar readings. In the fusion process, unique signatures of detections given to the objects by the radar software disappear and new identifications have to be given to the fused data. In the next phases of the system operation, this identification must be tracked to provide continuous observation of the identified object. This problem was solved in this project by using the predictive clustering method. In this case, satisfactory results were obtained as well. However, though the operational abilities of the monitoring system based on the presented methodology were in general satisfactory, further works on improving its efficiency are required. The most crucial issues that must be revisited are the radar data fusion in terms of overlapping observation areas and the detection tracking. Applying more sophisticated methods of providing a more reliable prediction of the tracked object seems to be a favourable starting point for the method's improvement.

Author Contributions: Conceptualization, J.D. and R.C.; methodology, R.C. and K.S.; software and validation, R.C. and K.S.; formal analysis R.C., J.D. and K.S.; investigation, R.C. and K.S.; resources R.C., J.D. and K.S.; data curation, R.C., J.D. and K.S.; writing—original draft preparation, J.D., R.C. and K.S.; writing—review and editing, J.D.; visualization, R.C. and K.S.; supervision, J.D.; project administration, J.D. All authors have read and agreed to the published version of the manuscript.

Funding: This research received no external funding.

Institutional Review Board Statement: Not applicable.

Informed Consent Statement: Not applicable.

Conflicts of Interest: The authors declare no conflict of interest.

Glossary

ICAO	aircraft type code
lat_i	latitude indicated by the ith sensor
$long_i$	longitude indicated by the ith sensor
h	altitude
y	heading
V	flight velocity
V_{xy}	horizontal velocity
V_z	vertical velocity
ET	category of the aircraft emitting the signal
d	distance of the UAV from the sensory system
DT	UAV type
Did	UAV identifier
P_{UAV}	probability of UAV detection
CL	confidence level
RCS	radar track RCS (radar cross-section)
az	estimated azimuth
el	estimated elevation
R	estimated distance between UAV and radar
x, y, z	estimated UAV coordinates relative to the radar (in Cartesian coordinate system)
V_x, V_y, V_z	velocity components of UAV in the x, y and z axes, respectively
$S_k(n)$	data sets determining detections collected in the same time
k	sensor index
n	subsequent moments in time $t_n = n\Delta t$, $n = 1, 2, \dots$
Δt	sampling period
$L_k(i)$	mapped detection set
i	detection index recorded in the log-register
$T(n)$	set of fused detections from ADS-B and AREOSCOPE
P_i^k	ith detection of kth antena
c_i^k	detection certainty factor
$\mathbf{A_s}$	detection similarity matrix
$R(n)$	set of fused radar detections
$F(n)$	fusion of $T(n)$ and $R(n)$ sets

References

1. Yang, J.; Thomas, A.G.; Singh, S.; Baldi, S.; Wang, X. A Semi-Physical Platform for Guidance and Formations of Fixed-Wing Unmanned Aerial Vehicles. *Sensors* **2020**, *20*, 1136. [CrossRef]
2. Yuan, X.; Xie, Z.; Tan, X. Computation Offloading in UAV-Enabled Edge Computing: A Stackelberg Game Approach. *Sensors* **2022**, *22*, 3854. [CrossRef]
3. Lykou, G.; Moustakas, D.; Gritzalis, D. Defending Airports from UAS: A Survey on Cyber-Attacks and Counter-Drone Sensing Technologies. *Sensors* **2020**, *20*, 3537. [CrossRef]
4. Wojtanowski, J.; Zygmunt, M.; Drozd, T.; Jakubaszek, M.; Życzkowski, M.; Muzal, M. Distinguishing Drones from Birds in a UAV Searching Laser Scanner Based on Echo Depolarization Measurement. *Sensors* **2021**, *21*, 5597. [CrossRef]
5. Flak, P. Drone Detection Sensor with Continuous 2.4 GHz ISM Band Coverage Based on Cost-Effective SDR Platform. *IEEE Access* **2021**, *9*, 114574–114586. [CrossRef]
6. Yang, S.; Qin, H.; Liang, X.; Gulliver, T.A. An Improved Unauthorized Unmanned Aerial Vehicle Detection Algorithm Using Radiofrequency-Based Statistical Fingerprint Analysis. *Sensors* **2019**, *19*, 274. [CrossRef] [PubMed]
7. Garcia, A.J.; Lee, J.M.; Kim, D.S. Anti-Drone System: A Visual-based Drone Detection using Neural Networks. In Proceedings of the 2020 International Conference on Information and Communication Technology Convergence (ICTC), Jeju, Korea, 21–23 October 2020. [CrossRef]
8. Martian, A.; Chiper, F.L.; Craciunescu, R.; Vladeanu, C.; Fratu, O.; Marghescu, I. RF Based UAV Detection and Defense Systems: Survey and a Novel Solution. In Proceedings of the International Black Sea Conference on Communications and Networking (BlackSeaCom), Bucharest, Romania, 24–28 May 2021. [CrossRef]
9. Sakharov, K.Y.; Sukhov, A.V.; Ugolev, V.L.; Gurevich, Y.M. Study of UWB Electromagnetic Pulse Impact on Commercial Unmanned Aerial Vehicle. In Proceedings of the 2018 International Symposium on Electromagnetic Compatibility (EMC EUROPE), Amsterdam, The Netherlands, 27–30 August 2018. [CrossRef]
10. Semkin, V.; Yin, M.; Hu, Y.; Mezzavilla, M.; Rangan, S. Drone Detection and Classification Based on Radar Cross Section Signatures. In Proceedings of the 2020 International Symposium on Antennas and Propagation (ISAP), Osaka, Japan, 25–28 January 2021. [CrossRef]
11. Molchanov, P.A.; Contarino, V.M. New distributed radar technology based on UAV or UGV application. Radar Sensor Technology XVII. In Proceedings of the Proceedings Volume 8714 Spie Defense, Security, and Sensing, Baltimore, MD, USA, 29 April–3 May 2013. [CrossRef]
12. Ferreira, R.; Gaspar, J.; Sebastao, P.; Souto, N. A Software Defined Radio Based Anti-UAV Mobile System with Jamming and Spoofing Capabilities. *Sensors* **2022**, *22*, 1487. [CrossRef] [PubMed]
13. Sazdić-Jotić, B.M.; Obradović, D.R.; Bujakovic, D.M.; Bondžulić, B.P. Feature Extraction for Drone Classification. In Proceedings of the 14th International Conference on Advanced Technologies, Systems and Services in Telecommunications (TELSIKS), Nis, Serbia, 23–25 October 2019. [CrossRef]
14. Par, S.; Kim, H.T.; Lee, S.; Joo, H.; Kim, H. Survey on Anti-Drone Systems: Components, Designs, and Challenges. *IEEE Access* **2021**, *9*, 42635–42659. [CrossRef]
15. Christnacher, F.; Hengy, S.; Laurenzis, M.; Matwyschuk, A.; Naz, P.; Schertzer, S.; Schmitt, G. Optical and acoustical UAV detection. In Proceedings of the Proceedings Volume 9988, Electro-Optical Remote Sensing X, 99880B (2016), Edinburgh, UK, 21 October 2016. [CrossRef]
16. Oh, H.M.; Lee, H.; Kim, M.Y. Comparing Convolutional Neural Network (CNN) models for machine learning-based drone and bird classification of anti-drone system. In Proceedings of the 19th International Conference on Control, Automation and Systems (ICCAS), Jeju, Korea, 15–18 October 2019.
17. Matuszewski, J.; Kraszewski, T. Evaluation of emitter location accuracy with the modified triangulation method by means of maximum likelihood estimators. *Metrol. Meas. Syst.* **2021**, *28*, 781–802. [CrossRef]
18. Farlik, J.; Kratky, M.; Casar, J.; Stary, V. Radar Cross Section and Detection of Small Unmanned Aerial Vehicles. In Proceedings of the 17th International Conference on Mechatronics—Mechatronika (ME), Prague, Czech Republic, 7–9 December 2016.
19. Abunada, A.H.; Osman, A.Y. Design and Implementation of a RF Based Anti-Drone System. In Proceedings of the 2020 IEEE International Conference on Informatics, IoT, and Enabling Technologies (ICIoT), Doha, Quatar, 2–5 February 2020.
20. Li, T.; Wen, B.; Tian, Y.; Li, Z. Numerical Simulation and Experimental Analysis of Small Drone Rotor Blade Polarimetry Based on RCS and Micro-Doppler Signature. *IEEE Antennas Wirel. Propag. Lett.* **2019**, *18*, 187–191. [CrossRef]
21. Ezuma, M.; Erden, F.; Anjinappa, C.K.; Ozdemir, O.; Guvenc, I. Micro-UAV detection and classification from RF fingerprints using machine learning techniques. In Proceedings of the 2019 IEEE Aerospace Conference, Big Sky, MT, USA, 2–9 March 2019. [CrossRef]
22. Shi, W.; Li, J.; Xu, W.; Zhou, H.; Zhang, N.; Zhang, S.; Shen, X. Multiple Drone-Cell Deployment Analyses and Optimization in Drone Assisted Radio Access Networks. *IEEE Access* **2018**, *6*, 12518–12529. [CrossRef]
23. Joseph, M. *Software Radio Architecture: Object-Oriented Approaches to Wireless Systems Engineering*; John Wiley & Sons Inc.: Hoboken, NJ, USA, 2000.
24. Tuttlebee, W. *Software Defined Radio: Origins, Drivers and International Perspectives*; John Wiley & Sons Inc.: Chichester, UK, 2002.
25. Zhang, W.; Ning, Y.; Sou, C. A Method Based on Multi-Sensor Data Fusion for UAV Safety Distance Diagnosis. *Electronics* **2019**, *8*, 1467. [CrossRef]

26. Dudczyk, J.; Kawalec, A. Fast-decision identification algorithm of emission source pattern in database. *Bull. Pol. Acad. Sci. Tech Sci.* **2015**, *63*, 385–389. [CrossRef]
27. Szafranski, G.; Czyba, R.; Janusz, W.; Blotnicki, W. Altitude estimation for the uav's applications based on sensors fusion algorithm. In Proceedings of the Unmanned Aircraft Systems (ICUAS), 2013 International Conference on, Atlanta, GA, USA, 28–31 May 2013; pp. 508–515.
28. Gibbs, B.P. *Advanced Kalman Filtering, Least-Squares and Modeling: A Practical Handbook*; John Wiley & Sons, Inc.: Hoboken, NJ, USA, 2011.
29. Błachuta, M.; Grygiel, R.; Czyba, R.; Szafrański, G. Attitude and heading reference system based on 3D complementary filter. In Proceedings of the 19th International Conference on Methods and Models in Automation and Robotics (MMAR), Miedzyzdroje, Poland, 2–5 September 2014; pp. 851–856, ISBN 978-1-4799-5082-9. [CrossRef]
30. Błachuta, M.; Czyba, R.; Janusz, W.; Szafrański, G. Data Fusion Algorithm for the Altitude and Vertical Speed Estimation of the VTOL Platform. *J. Intell. Robot. Syst.* **2014**, *74*, 413–420. [CrossRef]
31. Euston, M.; Coote, P.; Mahony, R.; Kim, J.; Hamel, T. A complementary filter for attitude estimation of a fixed-wing UAV. In Proceedings of the IEEE/RSJ International Conference on Intelligent Robots and Systems, Nice, France, 22–26 September 2008, pp. 340–345.
32. Tae, S.Y.; Sung, K.H.; Hyok, M.Y.; Sungsu, P. Gainscheduled complementary filter design for a MEMS based attitude and heading reference system. *Sensors* **2011**, *11*, 3816–3830. [CrossRef]
33. Raptis, I.A.; Valavanis, K.P. *Linear and Nonlinear Control of Small-Scale Unmanned Helicopters*; Springer: London, UK, 2011.
34. Mahony, R.; Hamel, T.; Pflimlin, J.M. Nonlinear Complementary Filters on the Special Orthogonal Group. *IEEE Trans. Autom. Control* **2008**, *53*, 1203–1218. [CrossRef]

Article

Pedestrian and Animal Recognition Using Doppler Radar Signature and Deep Learning

Danny Buchman [1,*], Michail Drozdov [2], Tomas Krilavičius [1], Rytis Maskeliūnas [1] and Robertas Damaševičius [1]

1   Department of Applied Informatics, Vytautas Magnus University, 44404 Kaunas, Lithuania; tomas.krilavicius@vdu.lt (T.K.); rytis.maskeliunas@vdu.lt (R.M.); robertas.damasevicius@vdu.lt (R.D.)
2   JVC Sonderus, 05200 Vilnius, Lithuania; michail.drozdov@gmail.com
*   Correspondence: danny.buchman@vdu.lt

Abstract: Pedestrian occurrences in images and videos must be accurately recognized in a number of applications that may improve the quality of human life. Radar can be used to identify pedestrians. When distinct portions of an object move in front of a radar, micro-Doppler signals are produced that may be utilized to identify the object. Using a deep-learning network and time–frequency analysis, we offer a method for classifying pedestrians and animals based on their micro-Doppler radar signature features. Based on these signatures, we employed a convolutional neural network (CNN) to recognize pedestrians and animals. The proposed approach was evaluated on the MAFAT Radar Challenge dataset. Encouraging results were obtained, with an AUC (Area Under Curve) value of 0.95 on the public test set and over 0.85 on the final (private) test set. The proposed DNN architecture, in contrast to more common shallow CNN architectures, is one of the first attempts to use such an approach in the domain of radar data. The use of the synthetic radar data, which greatly improved the final result, is the other novel aspect of our work.

Keywords: doppler radar; micro-Doppler signature; pedestrian recognition; animal recognition; deep learning

Citation: Buchman, D.; Drozdov, M.; Krilavičius, T.; Maskeliūnas, R.; Damaševičius, R. Pedestrian and Animal Recognition Using Doppler Radar Signature and Deep Learning. *Sensors* **2022**, *22*, 3456. https://doi.org/10.3390/s22093456

Academic Editors: Ram M. Narayanan, Janusz Dudczyk and Piotr Samczyński

Received: 16 February 2022
Accepted: 19 April 2022
Published: 1 May 2022

Publisher's Note: MDPI stays neutral with regard to jurisdictional claims in published maps and institutional affiliations.

1. Introduction

Artificial intelligence (AI), pattern recognition, machine learning (ML), and deep learning (DL) have recently gained popularity in a variety of domains of application, including autonomous driving [1], Internet-of-Things (IoT) [2], robots [3], smart mobility [4], etc. These applications collect data from their surroundings by sensing it; then they analyze the collected data, making choices and taking actions depending on the analysis [5]. Furthermore, computational intelligence methods and inference techniques based on deep learning are being intensively explored to improve the accuracy of computer vision systems.

One of the most significant challenges in computer vision is object tracking [6,7]. It offers a wide range of real-world applications, including robotics, medical imaging, traffic monitoring, autonomous vehicle tracking, surveillance [8], etc. Despite the obstacles of visual tracking, researchers are encouraged to develop quicker and better approaches, including resilience to strong occlusions, extreme size shift, discontinuities, precise localization, multi-object tracking, and failure recovery [9]. Despite achievements in resolving multiple obstacles under a variety of conditions, the basic issues remain complicated and difficult [10]. Because of its widespread applications in domains, such as gesture recognition [11], driver tracking [12], human action recognition [13], sports analysis [14], industrial work activity [15], monitoring the condition of industrial machinery [16], visual surveillance [17], and healthcare and rehabilitation [18], visual object tracking (VOT) is an active research issue in computer vision and machine learning. However, tracking is complicated by features such as partial or full occlusion, backdrop clutter, light change, deformation, and other environmental factors [19].

Pedestrian detection is a critical problem in many intelligent video surveillance systems because it offers critical information for semantic comprehension of video [20]. Accurate recognition of individual pedestrian occurrences in images and videos is critical in a variety of applications that might improve the quality of human existence. This pedestrian detection can be performed using radar [21–24].

Today, radar has multiple applications—from homeland security, through local radar for automotive [25], city surveillance [26], military [27] and even for healthcare [28] purposes. For example, a radar sensor can be capable of detecting the presence of a worker's activity and highlighting movements away from the workstation [29]. Target categorization is a critical radar activity in a wide range of security and military applications. Some applications, such as video surveillance [30], make use of electro-optical (EO) sensors. Radar offers substantial advantages over EO sensors in terms of resistance to harsh weather and poor illumination conditions, low cost, and durability, according to [31].

The micro-Doppler signature created by a subject reacting to an active emitter, for example, a radar, laser, or even acoustic emitters can be used to monitor the subject's minuscule micromotions, or even only sections of a subject. The micro-Doppler signatures are generated by the kinematic parameters of the object's motion and may be used to acquire the prominent elements of the object's motion and, in certain cases, to identify the object [32]. The target classification using the radar data traditionally uses one or several of the approaches highlighted in [33]:

1. Classification based on target radar cross-section (RCS) estimates.
2. Classification based on target RCS ratios.
3. Classification based on target RCS distributions.
4. Classification based on target modulation signatures.
5. Classification based on the target polarization scattering matrix.
6. Classification based on other scattering mechanisms.
7. Classification based on target kinematics.

Although many of these methods are more useful while dealing with man-made objects such as planes, [34], ships [35], drones [36], helicopters, [37] and other vehicles [38], there are some common problems with a classification of a walking person, an animal, a cyclist, or a group or its moving patterns [39,40]. The major problem while using any radar cross-section-based (RCS) approach is the calibration procedures of the radar. The result of RCS estimation is highly sensitive to the range of the target, the material, and the aspect angle.

For a long time, the most promising approaches to target classification were based on the decomposition of manually specified feature vectors using one or the other of several decomposition techniques [41–43]. The most widely used decomposition methods are principal component analysis (PCA) [42,43] and singular value decomposition (SVD) [41]. However, recently, deep-learning methods such as deep convolutional neural networks (DCNNs) have been adopted for radar-based target detection and recognition tasks [44–46]. DCNNs have been used to process several forms of millimeter-wave radar data [47] as well as light detection and ranging (LiDAR) data [48].

The MAFAT radar challenge [49] is a perfect opportunity to test different approaches to target classification without investing a huge effort in data acquisition, annotation, and other required steps before the data can be used for the analysis. The contributions of this study are as follows: a novel custom deep-learning architecture for solving the animal and pedestrian recognition problem, which has not been used before. To the best of our knowledge, the suggested deep neural network (DNN) architecture is one of the first attempts to employ such a method in the domain of radar data as opposed to more usual shallow convolutional neural network (CNN) designs. The other innovative component of our study is the use of synthetic radar data, which enhanced the final outcome considerably. Our competition technique might be utilized as a foundation for future implementations of radar classification based on CNN.

The paper is arranged as explained further. Section 2 analyzes state-of-the-art work. The problem of the MAFAT radar challenge, the data pipeline, and the proposed deep neural architecture are described in Section 3. The dataset used in this study and the experimental results are presented and analyzed in Section 4. Finally, the results of this study are discussed in Section 5.

2. Related Work

Some examples of using neural networks as the main classification module can be found as early as 1996 [50]. More recent related results include [51,52]. The common trend in this research is that very small data samples are used (tens or hundreds) and networks are very shallow (3–4 layers).

Gadde et al. [53] proposed strategies for analyzing radar data and using them to detect geriatric falls. The disparity in radar signal returns and the Doppler shift are caused by human motor activity. Because the signals were not stationary, therefore, the time–frequency analysis was critical in detecting movement, such as a fall. The article used real fall data to demonstrate the efficacy of preexisting models. The initial fall data also aided in revealing some of the difficulties encountered by technology improvements for fall detection.

Ma et al. [54] proposed MHT-PCD-Speed, a revolutionary model-free detection-based tracking technique for identifying and following moving objects in Doppler LiDAR scans. According to the findings, using Doppler radar images can improve tracking reliability and raise the precision of dynamic state estimates.

In [55] Han et al. were using shallow (two and six layers) convolutional neural networks (CNNs) to classify objects based on their radar returns. Authors performed their analysis using around 4000 samples of signals reflected from unmanned aerial vehicles (UAVs), people, cars, and other objects and achieved a modest total classification accuracy of around 0.48 using an augmented version of their dataset.

In yet another example [56] of using CNN architecture for the radar target classification, Stadelmayer et al. proposed constrained CCNs. Parameters of convolutional filters used in the first layer of such networks were learned during the process of training. Reported accuracy of the classification of different human activities (walking, idle, arm movements, etc.) in a controlled experimental environment were above 0.99 which is above other state-of-the-art methods mentioned in the publication.

Wan et al. in [57] solved the plane classification and outlier rejection problems using high-resolution radar (HRR) data and CNN architecture consisting of the classification part and the decoder part (for the target rejection). The authors were able to show the classification accuracy dependencies on the network architecture, the amount of training data, and hyper-parameters, but in all the cases it was well above 0.9. The number of samples used during the training was of the order of 100,000 which is much more than in other discussed publications.

Dadon et al. [58] presented a deep-learning-based technique for classifying ground-moving radar objects. The proposed technique learns the micro-Doppler signatures of radar objects in the 2D radar echoes domain. This study demonstrated that a CNN model can do well in classification. It also demonstrated that efficient data augmentation and regularization increase classification performance and decrease overfitting.

Tiwari et al. [59] developed a unique concatenated CNN model that takes the geolocation type and the radar signal data as input and conducts a binary classification to identify animals and persons. The suggested model has an AUC of 99%.

A common step in the data processing in the aforementioned publications is a representation of the radar data as a spectrogram over a slow time which we also use in our approach.

3. Methods

3.1. Problem Definition

This section describes the radar target classification problem. Moving targets lit by a radar system exhibit frequency modulation as a result of the time-varying delay between the target and the sensor. As follows from the Doppler effect, the major bulk translation of the object toward or away from the sensor causes a Doppler shift of the echo. The target velocities are estimated by:

$$f_d \approx 2v\frac{f_t}{c} \tag{1}$$

where f_d is Doppler frequency, f_t is the carrier frequency, v is target radial velocity, and c is the speed of light.

The target, on the other hand, may have sections with extra motions. These can add frequency modulations around the main Doppler shift, i.e., micro-Doppler modulations Chen [60,61] simulated radar micro-Doppler signatures for a variety of objects, including revolving cylinders, vibrating scatterers, and people targets. The scientists also demonstrated that time–frequency analysis of the received signal is a viable method for extracting the micro-Doppler signature, yielding additional information on the target that can be utilized for classification and recognition. Micro-Doppler may be thought of as a unique signature of the target that gives extra information about the target in addition to current methods for target recognition.

3.2. MAFAT Radar Challenge

The goal of the MAFAT radar challenge participants is to classify segments of human or animal radar tracks using an I/Q signal matrix as input (Figure 1). The proposed task is a binary classification task; tracked objects are people or animals. The data is real data collected from different geographical locations, with different time, sensors and quality (i.e., signal to noise ratio (SNR)).

The competition is divided into two parts with different conditions: a public phase with an unlimited number of applications, evaluated on a subset of the public test suite; a private stage where teams are limited to two entries, and where models are evaluated on completely new and unseen data.

The goal of the competition is to classify radar segment data as humans or animals, using ROC AUC as the metric.

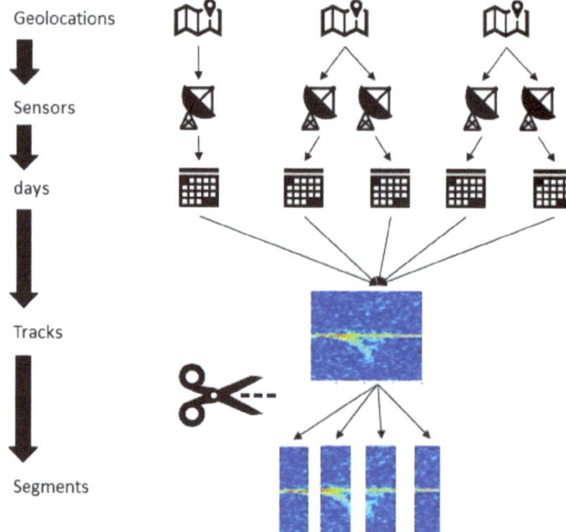

Figure 1. Explanation of MAFAT data.

3.2.1. Data

The dataset consists of signals recorded by ground Doppler-pulse radar. Each radar "stares" at a fixed, wide area of interest. Whenever an animal or a human moves within the radar's covered area, it is detected and tracked. The dataset contains records of those tracks. The tracks in the dataset are split into 32 time-unit segments. Each record in the dataset represents a single segment. The dataset is split to training and test sets; the training set contains the actual labels (humans or animals).

A segment consists of a matrix with I/Q values and metadata. The matrix of each segment has a size of 32×128. The X-axis represents the pulse transmission time, also known as "slow-time". The Y-axis represents the reception time of signals with respect to pulse transmission time divided into 128 equal sized bins, also known as "fast-time". The Y-axis is usually referred to as "range" or "velocity" as wave propagation depends on the speed of light. For example, for pulse repetition interval (PRI) of 128 ms, each Y-axis is a bin of 1 ms. For pulse sent in t(n) and a signal received in t(n+m), where 0 < m <= 128, the signal is set in the "m" bin of pulse n (the numbers are not the real numbers and are given only for the sake of the example).

The radar's raw, original received signal is a wave defined by amplitude, frequency, and phase. Frequency and phase are treated as a single-phase parameter. Amplitude and phase are represented in polar coordinates relative to the transmitted burst/wave. Polar coordinate calculations require frequent sine operations, making calculations time-consuming. Therefore, upon reception, the raw data are converted to Cartesian coordinates, i.e., I/Q values. The values in the matrix are complex numbers: I represents the real part, and Q represents the imaginary part.

The I/Q matrices that are supplied to participants have been standardized, but they have not been transformed or processed in any other way. Therefore, the data represent the raw signal. Different preprocessing and transformation methods, such as Fourier transform, can and should be used to model the data and extract meaningful features. For more information, see "Signal Processing" methods or view the links at the bottom for more information.

The metadata of a segment includes track id, location id, location type, day index, sensor id, and the SNR level. The segments were collected from several different geographic locations, and a unique id was given per location. Each location consists of one or more sensors; a sensor belongs to a single location. A unique id was given per sensor. Each sensor has been used in one or more days, and each day is represented by an index. A single track appears in a single location, sensor, and day. The segments were taken from longer tracks, and each track was given a unique id.

The task of classifying signals to humans and animals is difficult, and it is more challenging in short segments and low SNR signals. One way to view the data is to visualize the signals as a spectrogram. A spectrogram is depicted as a heat map with intensity shown by a color palette.

To generate a spectogram, the I/Q matrix was transformed and processed using Hann windowing and FFT (fast Fourier transform) and calculating the median. We then set as the minimum value of the I/Q matrix and at the end pseudo-coloring.

The images shown in Figure 2 are spectrograms of low and high SNR segments of animals and humans. The white dots are the Doppler burst vector which mark the target's center-of-mass.

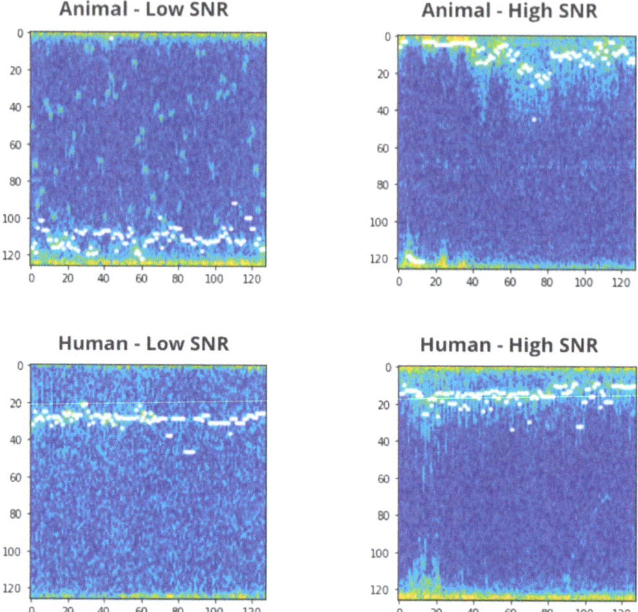

Figure 2. Examples of spectograms.

3.2.2. Generalization Considerations

Adjacent segments that can be combined to a whole track can be found in the training and auxiliary datasets but not in the test set. The participants' goal is to classify every tracked object correctly based on a single segment and not to use the correlation that exists between multiple segments in a track for the classification task. Therefore, most of the records in the test set are single segments that were randomly selected from a full track. In cases where the track was long enough, more than one segment of the same track may be in the test set. Note that they will not be sequential.

The classification should be performed on a single segment level, i.e., the trained models should receive a single segment as input and predict the class of this segment as an output. The class of every segment should be inferred separately based on the features that are extracted only from this specific segment, regardless of any other segment in the test set. The prediction should also be stable, given that the same segment and the same output are expected.

Generalizing to new, unseen, geographic locations such as positioning a radar in a new location changes many things. The terrain, the weather, the objects in the location, and reflections—all these factors may vary from one location to another. The ability to classify a tracked object correctly should be impervious to the changes involved in positioning a radar in new locations. The trained models will be challenged to classify humans or animals on radar tracks that were captured in new location sites, unrepresented in the training set.

The training and test sets contain the following:

- A total of 1510 tracks in the training set;
- A total of 106 segments in the public test set and 6656 segments in the training set;
- In total, there are 566 high SNR tracks and 1144 low SNR tracks in the training set; *200 tracks are high SNR in one part and low SNR in the other;
- In total, there are 2465 high SNR segments and 4191 low SNR segments in the training set;

- Segments are taken from multiple locations. A location is not guaranteed to be a single dataset, but since the goal is to train models that can generalize well to new, unseen, locations—several locations are in the training or the test datasets only;
- It should be mentioned that the data in the training set and in the test set do not necessarily come from the same distribution. Participants are encouraged to split the training set into training and validation sets (via cross-validation or other methods) in such way that the validation set will resemble the test set.

3.3. Data Pipeline

The goal of the developed classification module is to classify the already tracked object from the Doppler velocity graphs of this target. The classification module should be able to distinguish between the person and animal while using just Doppler velocities graphs.

Because of their Doppler resemblance, it is difficult to distinguish between people and animals. However, because they are nonrigid entities, changing movements of their sections cause extra modulations in the radar echoes [60]. These micro-Doppler modulations were used for radar target categorization [62]. The categorization of objects using feature extraction by CNNs has received a great deal of attention in the literature [63,64]. It was demonstrated that CNN trained on visual data may exceed human classification capabilities when subjected to visual distortions [63,65–67].

Some additional complications that are not present in other datasets collected to test a specific method include the low signal-to-noise ratio, targets that have unexpected/uncontrolled behavior, different sensors used, and other measurement conditions which should not be used while classifying.

3.4. Neural Network Design

We investigated the impact of the CNN architecture and data augmentation on the radar target classification performance. Efficient CNN training requires a large, diverse, and well-balanced dataset [68]. The dataset in [49] is small and highly imbalanced.

We studied the effect of the CNN architecture and data augmentation on the classification performance of radar targets. Efficient CNN training necessitates a large, diversified, and well-balanced training dataset, according to [68]. The MAFAT data set [49] is severely unbalanced and tiny. As a result, simple DL algorithms are prone to overfitting. We demonstrated how the proper configuration of well-known regularization approaches may enhance model performance under the ROC-AUC criterion. [69]. We reviewed different CNNs and achieved height performance results in competition by mixing and training different types of CNNs and modifying network layers, selection and training methods, and data balancing techniques to prevent overfitting

Our high-level approach to the competition could be summarized as applying image classification techniques to the preprocessed radar raw data (Figure 3). Some preliminary testing has shown that most simple methods traditionally used for the classification such as logistic regression, linear regression, random forest, and decision tree did not reach the result of the baseline model. The development was focused on data selection, the CNN architecture, and various methods of addressing the overfitting and data leakage which were obvious from the difference between the validation ROC–AUC and the one of the public test set.

We applied 82 times to MAFAT [49] for comparing our tests results with the public test set, as it was not available for training or testing, and we only received information about ACU of the applied model. In the final solution, an ensemble of two models was used to improve the accuracy of classification. One architecture is a deep neural network inspired by the ResNet [70]. The complete architecture description is provided in Figures 4–6.

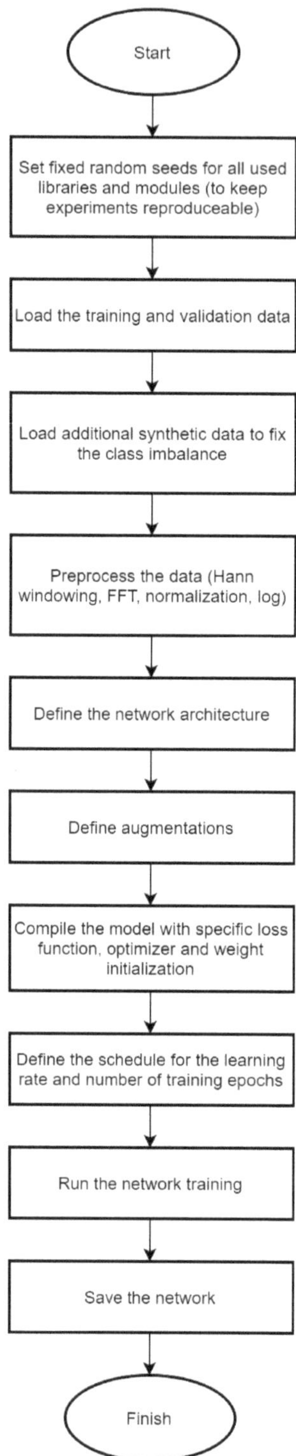

Figure 3. Basic process structure of the radar data classification model training.

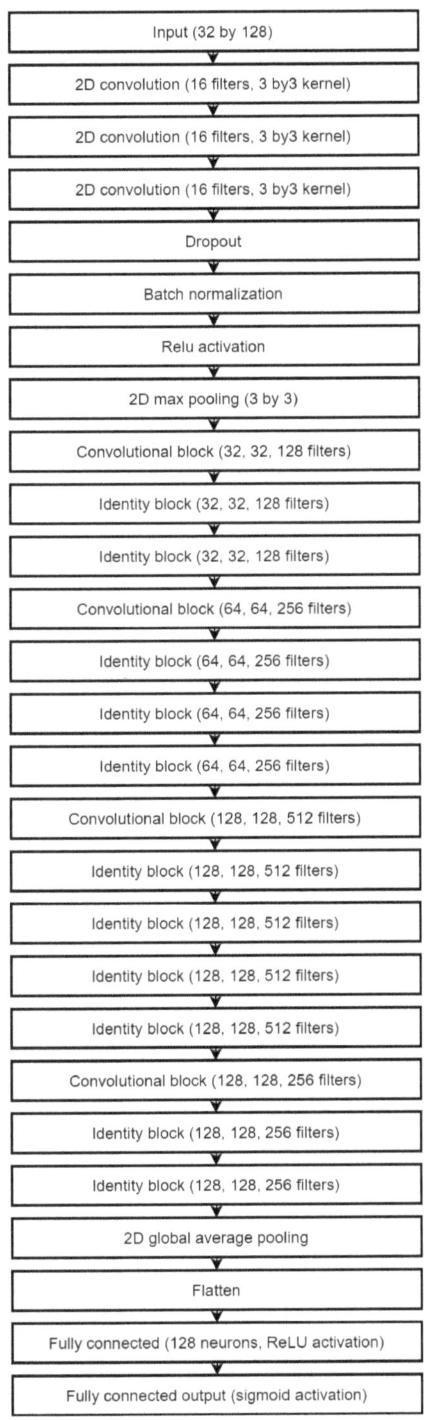

Figure 4. The architecture of the main classification model.

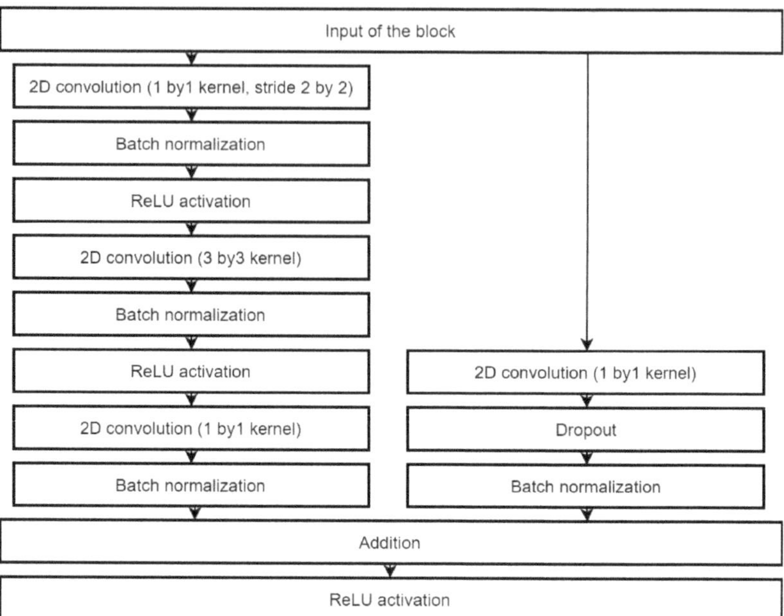

Figure 5. The convolution residual block used in the main classification model.

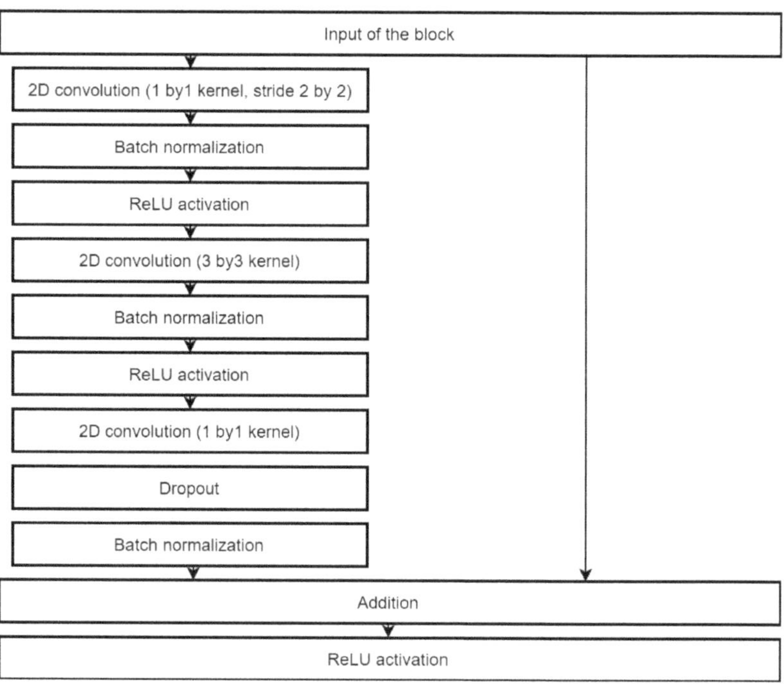

Figure 6. The convolution identity block used in the main classification model.

A shallower network was used as a secondary classification model and improved the total accuracy and especially the accuracy of true negatives (animal class). This architecture is shown in Figure 7. All convolution layers use ReLU activations. Both networks split into convolution stages which contain two to four convolution layers followed by the batch normalization and max pooling layers. Inside the stage, convolutions use the "same" padding method, which keeps the original size of inputs. Each stage increases the number of filters used but reduces each dimension two times. The weight regularization is used extensively in the secondary network: the L2 norm with weight decay of 0.001 is used in layers before each max pooling and the L1 norm in the final convolutional layer. The L2 norm with the same decay value is used in the fully connected layer of the main network.

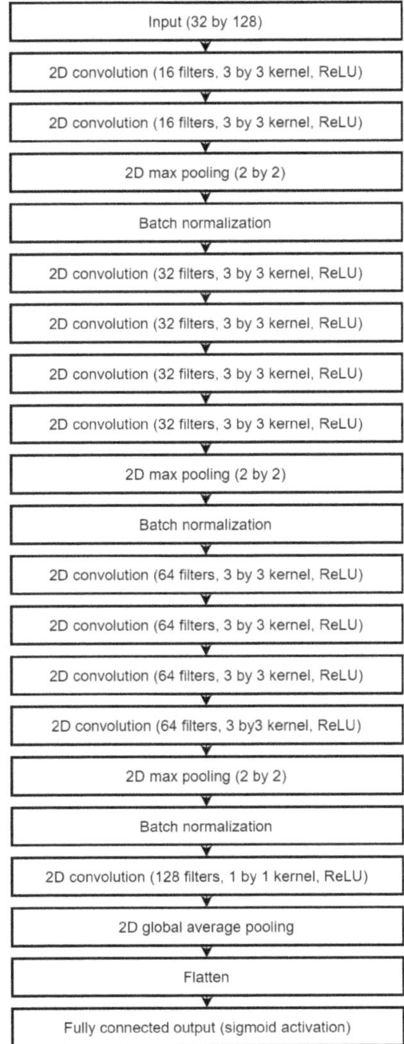

Figure 7. The architecture of the secondary classification model.

4. Experiments and Results

4.1. Data Set

Data were obtained from different locations and using different radar models during different times of the day. The raw data were I/Q signals grouped into matrices of 32 by 128. The first dimension is the pulse transmission time or the slow time, while the second dimension represents the reception time with respect to the single transmission time or the fast time. The result of the FFT along the second dimension would produce the Doppler velocity graph. No description was given to establish the exact ranges for each of the dimensions, but it was assured that the entire dataset was normalized.

Each 32 × 128 matrix is called a segment. The complete object-tracking event is called a track, and it could take a much longer time than the one during which 32 pulses are transmitted. As a consequence, a single track can have one or more segments that are more or less informative for the task of classification. Although the training dataset contained all the data necessary to construct complete tracks from separate segments, the classification module should classify the object using a single segment without using remaining segments of the same track.

The high imbalance of the data with regard to class of the object was observed. There were many more segments containing animal data than those containing person data (ratio of approximately 5 to 1). Some measurement locations were heavily underrepresented—out of seven different locations, the data were collected in more than half of samples came from the third location. Understanding this, the organizers provided so-called auxiliary experiment data—the data collected at the same locations while performing controlled experiments. In these experiments, much more data of people tracking were obtained. It was not clear at the start, however, if the synthetic data would provide any benefit to the solution.

Finally, some additional data (such as background data for different locations and data with synthetically added noise) were provided.

4.2. Data Presprocessing

Some of the most common preprocessing steps, such as loading the data, splitting into training and validation, the FFT, and converting to the logarithmic scale, were provided by the MAFAT radar challenge organizers.

The preprocessing stage was kept from the baseline implementation although different options (such as working with the original I/Q data) were possible. A first Hann windowing function was used to suppress sidelobes resulting after the FFT characteristics. Then, FFT followed and an absolute of the resulting spectrum was taken. Lastly, a logarithm of each value was calculated and result was normalized.

4.3. Image Augmentation

Image augmentations is one of the options to reduce overfitting. In the final solution, a very conservative set of augmentations was used:

1. Cyclic width shift of 0.25;
2. Height shift of 0.05;
3. Vertical and horizontal flips of the data.

These augmentations proved to work well while evaluating on the public test set, but the reduced performance on the private test set showed that more image augmentation could be used to reduce the overfitting. The strategy of creating new segments from joined tracks by shifting the sampling of segments from the track in the slow time dimension was considered and tested but did not show meaningful improvement.

At the data exploration step it was quickly observed that adding positive samples from the auxiliary synthetic data not only made the training set more balanced but greatly improved the public test score. Aside from the main training data additional chunk of 7000 segments was loaded from this set in the final solution. Approximately 15,800 total segments were, used for the training 8400 of which were of *person* class.

4.4. Model Hyperparameters

The model was compiled with an Adam [71] optimizer which is currently the default choice for the training of convolutional neural networks. Binary cross entropy loss function was used while training.

The learning rate was scheduled to increase and drop cyclically (the cyclic learning [72]) which allows it to avoid the weights state staying in a local minima of the multidimensional loss. This proved to be one of the main sources of the score increase. The learning rate started at 0.003 then dropped to 0.00005 at the end of each cycle. CLR implementation for Keras was taken from [73]. The learning process consisting of two to five cycles was tested with the duration of a single cycle between 20 and 45 epochs. A typical learning rate schedule produced by the library is shown in Figure 8. Some typical examples of various metrics obtained in the process of K-fold validation are shown in Figure 9a–c.

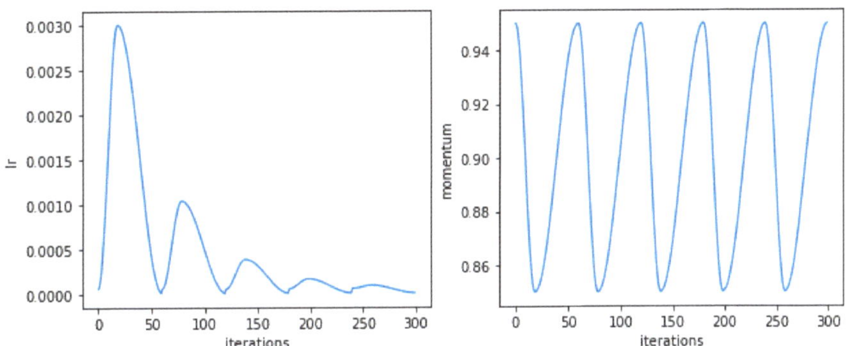

Figure 8. Test run of the learning rate schedule with five cycles produced by a software tool available from [73].

4.5. Neural Network Training

The splitting into training and validation sets proved to be one of the biggest challenges. The baseline splitting was defined as using three first segments of tracks measured in locations different than those of the training set. The validation set would contain similar numbers of positive and negative examples. After few submissions, it became clear that this validation set does not correlate to the public test—improving ROC AUC from 0.94 to 0.98 would not improve the public test score. To minimize overfitting for the public test, we had to switch to the K-fold validation at some stage of the method development. Special care was taken to avoid the obvious source of the data leakage while evaluating—no two segments of the same track could be used in different folds (Table 1).

Table 1. Experimental results with different K-folds.

K-Fold Experiments Results			
K-Fold	**7**	**5**	**10**
Accuracy	94.035 (±0.5777)	89.708 (±0.5129)	91.855 (±3.0449)
ROC AUC	98.347 (±0.2502)	97.181 (±0.1869)	98.510 (±0.6938)
Loss	0.1776	0.3074	0.2828

The important aspect of the training procedure is that after each cycle, network weights are saved. In the competition, this allowed us to improve the final result by ensembling the same network architecture using different sets (obtained at the end of each cycle) of weights. Although this is impractical while deploying any real-world system, the knowledge distillation could have been implemented to reduce the memory footprint and the processing cost while keeping relatively good performing weights.

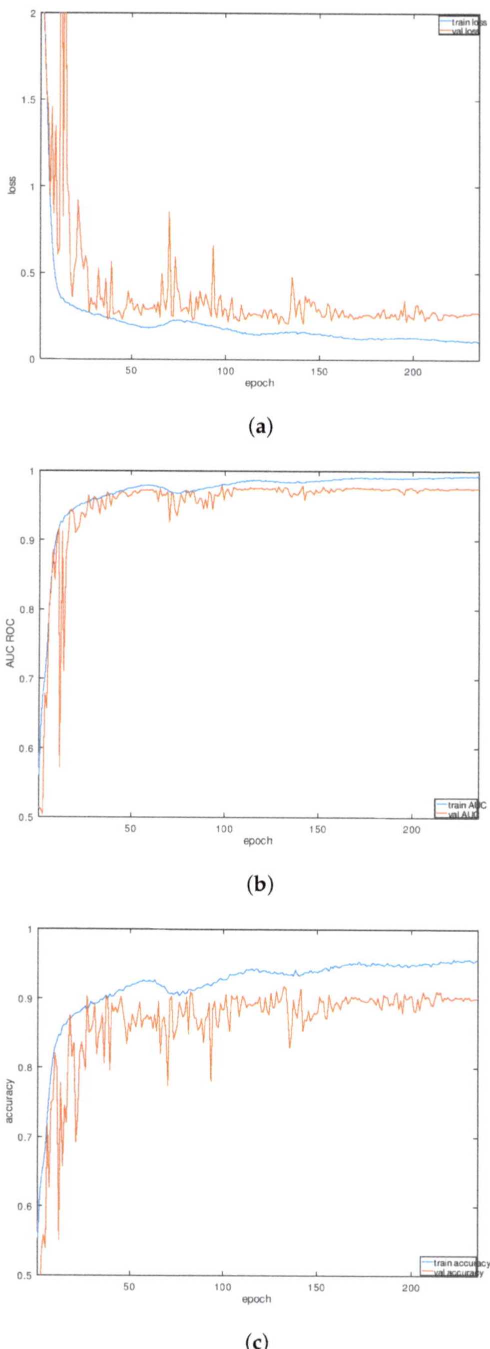

(a)

(b)

(c)

Figure 9. Training performance: (**a**) loss; (**b**) ROC AUC; (**c**) accuracy.

5. Discussion and Conclusions

This study presented a CNN-based technique for exploiting micro-Doppler signals to classify people and animals using radar. Encouraging results were achieved during the competition—AUC of the ROC around 0.95 on the public test set and above 0.85 on the final (private) test set. Problems of an unbalanced dataset and out-of-sample testing data were the main hurdles while optimizing our processing pipeline. Final results demonstrated a missed opportunity of improving the generalization ability of our networks—an obvious example of overfitting to the public test set was observed. On the other hand, all techniques we employed were a focused effort in this regard, and only the lack of the clear improvement criteria stopped us from achieving even better results.

The proposed DNN architecture, in contrast to more common shallow CNN architectures, is one of the first attempts to use such an approach in the domain of radar data. The usage of the synthetic radar data, which greatly improved the final result, is the other novel aspect of our work. Our solution in the competition could be used as a building block to future implementation of radar classification based on CNN.

During evaluation, we noticed that different network structures perform differently on animal and human recognition, and that is the main reason to use two networks for recognition of humans and animals. Combining the results helps to prevent fouls defections, where there is no target in image, but the network still recognizes one.

Future work is required to optimize the implementation to be able to run calcification in real-time directly since it now requires high resource usage. Exploring competitors, work [58,74] will also improve the solution.

Alternatively, the knowledge distillation technique would be a promising way to reduce the amount of processing and improve the decision latency. Another possible way of improvement is based on the fact that each foster network predicts another object, and if we add more information (data), it will be possible to perform transfer learning and strengthen the accuracy of the model to provide the an option to build an independent system that works in parallel (using Kubernetes).

Author Contributions: Conceptualization, T.K.; Formal analysis, D.B., M.D., R.M. and R.D.; Funding acquisition, T.K.; Investigation, D.B.; Methodology, R.D.; Software, D.B.; Supervision, T.K.; Validation, D.B., M.D., T.K. and R.M.; Writing—original draft, D.B., M.D. and T.K.; Writing—review & editing, R.M. and R.D. All authors have contributed equally to this manuscript. All authors have read and agreed to the published version of the manuscript.

Funding: This research was funded by "Development of doctoral studies", grant No. 09.3.3-ESFA-V-711-01-0001.

Institutional Review Board Statement: Not applicable.

Informed Consent Statement: Not applicable.

Data Availability Statement: The MAFAT Challenge dataset is available at https://competitions.codalab.org/competitions/25389#learn_the_details-data (accessed on 16 January 2022).

Conflicts of Interest: The authors declare no conflict of interest.

References

1. Guo, Z.; Huang, Y.; Hu, X.; Wei, H.; Zhao, B. A survey on deep learningbased approaches for scene understanding in autonomous driving. *Electronics* **2021**, *10*, 471. [CrossRef]
2. Mahdavinejad, M.S.; Rezvan, M.; Barekatain, M.; Adibi, P.; Barnaghi, P.; Sheth, A.P. Machine learning for internet of things data analysis: A survey. *Digit. Commun. Netw.* **2018**, *4*, 161–175. [CrossRef]
3. Hua, J.; Zeng, L.; Li, G.; Ju, Z. Learning for a robot: Deep reinforcement learning, imitation learning, transfer learning. *Sensors* **2021**, *21*, 1278. [CrossRef] [PubMed]
4. Mauri, A.; Khemmar, R.; Decoux, B.; Ragot, N.; Rossi, R.; Trabelsi, R.; Boutteau, R.; Ertaud, J.; Savatier, X. Deep learning for real-time 3D multi-object detection, localisation, and tracking: Application to smart mobility. *Sensors* **2020**, *20*, 532. [CrossRef]
5. Cheng, G.; Han, J.; Lu, X. Remote Sensing Image Scene Classification: Benchmark and State of the Art. *Proc. IEEE* **2017**, *105*, 1865–1883. [CrossRef]
6. Luo, W.; Xing, J.; Milan, A.; Zhang, X.; Liu, W.; Kim, T. Multiple object tracking: A literature review. *Artif. Intell.* **2021**, *293*, 103448. [CrossRef]
7. Masood, H.; Zafar, A.; Ali, M.U.; Hussain, T.; Khan, M.A.; Tariq, U.; Damaševičius, R. Tracking of a Fixed-Shape Moving Object Based on the Gradient Descent Method. *Sensors* **2022**, *22*, 1098. [CrossRef]

8. Ge, H.; Zhu, Z.; Lou, K.; Wei, W.; Liu, R.; Damaševičius, R.; Woźniak, M. Classification of infrared objects in manifold space using kullback-leibler divergence of gaussian distributions of image points. *Symmetry* **2020**, *12*, 434. [CrossRef]
9. Zhou, B.; Duan, X.; Ye, D.; Wei, W.; Woźniak, M.; Połap, D.; Damaševičius, R. Multi-level features extraction for discontinuous target tracking in remote sensing image monitoring. *Sensors* **2019**, *19*, 4855. [CrossRef]
10. Kalake, L.; Wan, W.; Hou, L. Analysis Based on Recent Deep Learning Approaches Applied in Real-Time Multi-Object Tracking: A Review. *IEEE Access* **2021**, *9*, 32650–32671. [CrossRef]
11. Mujahid, A.; Awan, M.J.; Yasin, A.; Mohammed, M.A.; Damaševičius, R.; Maskeliūnas, R.; Abdulkareem, K.H. Real-time hand gesture recognition based on deep learning YOLOv3 model. *Appl. Sci.* **2021**, *11*, 4164. [CrossRef]
12. Ali, S.F.; Aslam, A.S.; Awan, M.J.; Yasin, A.; Damaševičius, R. Pose estimation of driver's head panning based on interpolation and motion vectors under a boosting framework. *Appl. Sci.* **2021**, *11*, 11600. [CrossRef]
13. Kiran, S.; Khan, M.A.; Javed, M.Y.; Alhaisoni, M.; Tariq, U.; Nam, Y.; Damaševĭcius, R.; Sharif, M. Multi-Layered Deep Learning Features Fusion for Human Action Recognition. *Comput. Mater. Contin.* **2021**, *69*, 4061–4075. [CrossRef]
14. Žemgulys, J.; Raudonis, V.; Maskeliūnas, R.; Damaševičius, R. Recognition of basketball referee signals from real-time videos. *J. Ambient. Intell. Humaniz. Comput.* **2020**, *11*, 979–991. [CrossRef]
15. Patalas-maliszewska, J.; Halikowski, D.; Damaševičius, R. An automated recognition of work activity in industrial manufacturing using convolutional neural networks. *Electronics* **2021**, *10*, 2946. [CrossRef]
16. Huang, Q.; Pan, C.; Liu, H. A Multi-sensor Fusion Algorithm for Monitoring the Health Condition of Conveyor Belt in Process Industry. In Proceedings of the 2021 3rd International Conference on Industrial Artificial Intelligence (IAI), Shenyang, China, 8–11 November 2021; IEEE: Red Hook, NY, USA, 2021. [CrossRef]
17. Bai, Z.; Li, Y.; Chen, X.; Yi, T.; Wei, W.; Wozniak, M.; Damasevicius, R. Real-time video stitching for mine surveillance using a hybrid image registration method. *Electronics* **2020**, *9*, 1336. [CrossRef]
18. Ryselis, K.; Petkus, T.; Blažauskas, T.; Maskeliūnas, R.; Damaševičius, R. Multiple Kinect based system to monitor and analyze key performance indicators of physical training. *Hum.-Centric Comput. Inf. Sci.* **2020**, *10*, 4733. [CrossRef]
19. Mondal, A. Occluded object tracking using object-background prototypes and particle filter. *Appl Intell* **2021**, *51*, 5259–5279. [CrossRef]
20. Heikkilä, J.; Silvén, O. A real-time system for monitoring of cyclists and pedestrians. *Image Vis. Comput.* **2004**, *22*, 563–570. [CrossRef]
21. Peng, X.; Shan, J. Detection and tracking of pedestrians using doppler lidar. *Remote Sens.* **2021**, *13*, 2952. [CrossRef]
22. Held, P.; Steinhauser, D.; Koch, A.; Brandmeier, T.; Schwarz, U.T. A Novel Approach for Model-Based Pedestrian Tracking Using Automotive Radar. *IEEE Trans. Intell. Transp. Syst.* **2021**, 1–14. [CrossRef]
23. Severino, J.V.B.; Zimmer, A.; Brandmeier, T.; Freire, R.Z. Pedestrian recognition using micro Doppler effects of radar signals based on machine learning and multi-objective optimization. *Expert Syst. Appl.* **2019**, *136*, 304–315. [CrossRef]
24. Ninos, A.; Hasch, J.; Heizmann, M.; Zwick, T. Radar-Based Robust People Tracking and Consumer Applications. *IEEE Sens. J.* **2022**, *22*, 3726–3735. [CrossRef]
25. Gao, X.; Xing, G.; Roy, S.; Liu, H. RAMP-CNN: A Novel Neural Network for Enhanced Automotive Radar Object Recognition. *IEEE Sens. J.* **2021**, *21*, 5119–5132. [CrossRef]
26. Wang, Z.; Miao, X.; Huang, Z.; Luo, H. Research of target detection and classification techniques using millimeter-wave radar and vision sensors. *Remote Sens.* **2021**, *13*, 1064. [CrossRef]
27. Dudczyk, J. A method of feature selection in the aspect of specific identification of radar signals. *Bull. Pol. Acad. Sci. Tech. Sci.* **2017**, *65*, 113–119. [CrossRef]
28. Pisa, S.; Pittella, E.; Piuzzi, E. A survey of radar systems for medical applications. *IEEE Aerosp. Electron. Syst. Mag.* **2016**, *31*, 64–81. [CrossRef]
29. Cardillo, E.; Caddemi, A. Feasibility Study to Preserve the Health of an Industry 4.0 Worker: A Radar System for Monitoring the Sitting-Time. In Proceedings of the 2019 II Workshop on Metrology for Industry 4.0 and IoT (MetroInd4.0&IoT), Naples, Italy, 4–6 June 2019. [CrossRef]
30. Prasad, D.K.; Rajan, D.; Rachmawati, L.; Rajabally, E.; Quek, C. Video processing from electro-optical sensors for object detection and tracking in a maritime environment: A survey. *IEEE Trans. Intell. Transp. Syst.* **2017**, *18*, 1993–2016. [CrossRef]
31. Mishra, A.; Li, C. A review: Recent progress in the design and development of nonlinear radars. *Remote Sens.* **2021**, *13*, 4982. [CrossRef]
32. Tahmoush, D. Review of micro-Doppler signatures. *IET Radar Sonar Navig.* **2015**, *9*, 1140–1146. [CrossRef]
33. Anderson, S. Target Classification, Recognition and Identification with HF Radar. In Proceedings of the NATO Research and Technology Agency, Sensors and Electronics Technology Panel Symposium SET–080/RSY17/RFT: Target Identification and Recognition Using RF Systems, Oslo, Norway, 11–13 October 2004; p. 18.
34. Perl, E. Review of airport surface movement radar technology. *IEEE Aerosp. Electron. Syst. Mag.* **2006**, *21*, 24–27. [CrossRef]
35. Le Caillec, J.; Gorski, T.; Sicot, G.; Kawalec, A. Theoretical Performance of Space-Time Adaptive Processing for Ship Detection by High-Frequency Surface Wave Radars. *IEEE J. Ocean. Eng.* **2018**, *43*, 238–257. [CrossRef]
36. Coluccia, A.; Parisi, G.; Fascista, A. Detection and classification of multirotor drones in radar sensor networks: A review. *Sensors* **2020**, *20*, 4172. [CrossRef] [PubMed]

7. Baczyk, M.K.; Samczyński, P.; Kulpa, K.; Misiurewicz, J. Micro-Doppler signatures of helicopters in multistatic passive radars. *IET Radar Sonar Navig.* **2015**, *9*, 1276–1283. [CrossRef]
8. Zhou, T.; Yang, M.; Jiang, K.; Wong, H.; Yang, D. Mmw radar-based technologies in autonomous driving: A review. *Sensors* **2020**, *20*, 7283. [CrossRef]
9. Amiri, R.; Shahzadi, A. Micro-Doppler based target classification in ground surveillance radar systems. *Digit. Signal Process. Rev. J.* **2020**, *101*, 102702. [CrossRef]
10. Palffy, A.; Dong, J.; Kooij, J.F.P.; Gavrila, D.M. CNN Based Road User Detection Using the 3D Radar Cube. *IEEE Robot. Autom. Lett.* **2020**, *5*, 1263–1270. [CrossRef]
11. Fioranelli, F.; Ritchie, M.; Griffiths, H. Classification of Unarmed/Armed Personnel Using the NetRAD Multistatic Radar for Micro-Doppler and Singular Value Decomposition Features. *IEEE Geosci. Remote Sens. Lett.* **2015**, *12*, 1933–1937. [CrossRef]
12. Secmen, M. Radar target classification method with high accuracy and decision speed performance using MUSIC spectrum vectors and PCA projection. *Radio Sci.* **2011**, *46*, 1–9. [CrossRef]
13. Zabalza, J.; Clemente, C.; Di Caterina, G.; Ren, J.; Soraghan, J.J.; Marshall, S. Robust PCA micro-doppler classification using SVM on embedded systems. *IEEE Trans. Aerosp. Electron. Syst.* **2014**, *50*, 2304–2310. [CrossRef]
14. Jiang, W.; Ren, Y.; Liu, Y.; Leng, J. Artificial Neural Networks and Deep Learning Techniques Applied to Radar Target Detection: A Review. *Electronics* **2022**, *11*, 156. [CrossRef]
15. Van Eeden, W.D.; De Villiers, J.P.; Berndt, R.J.; Nel, W.A.; Blasch, E. Micro-Doppler radar classification of humans and animals in an operational environment. *Expert Syst. Appl.* **2018**, *102*, 1–11. [CrossRef]
16. Hou, F.; Lei, W.; Li, S.; Xi, J. Deep Learning-Based Subsurface Target Detection from GPR Scans. *IEEE Sens. J.* **2021**, *21*, 8161–8171. [CrossRef]
17. Abdu, F.J.; Zhang, Y.; Fu, M.; Li, Y.; Deng, Z. Application of deep learning on millimeter-wave radar signals: A review. *Sensors* **2021**, *21*, 1951. [CrossRef]
18. Wu, Y.; Wang, Y.; Zhang, S.; Ogai, H. Deep 3D Object Detection Networks Using LiDAR Data: A Review. *IEEE Sens. J.* **2021**, *21*, 1152–1171. [CrossRef]
19. MAFAT Radar Challenge Homepage. Available online: https://competitions.codalab.org/competitions/25389#learn_the_details-overview (accessed on 16 January 2022).
50. Jianjun, H.; Jingxiong, H.; Xie, W. Target Classification by Conventional Radar. In Proceedings of the International Radar Conference, Beijing, China, 8–10 October 1996; pp. 204–207. [CrossRef]
51. Ibrahim, N.K.; Abdullah, R.S.A.R.; Saripan, M.I. Artificial Neural Network Approach in Radar Target Classification. *J. Comput. Sci.* **2009**, *5*, 23. [CrossRef]
52. Ardon, G.; Simko, O.; Novoselsky, A. Aerial Radar Target Classification using Artificial Neural Networks. In Proceedings of the ICPRAM, Valletta, Malta, 22–24 February 2020; pp. 136–141. [CrossRef]
53. Gadde, A.; Amin, M.G.; Zhang, Y.D.; Ahmad, F. Fall detection and classifications based on time-scale radar signal characteristics. In Proceedings of the SPIE—The International Society for Optical Engineering, Baltimore, MD, USA, 29 May 2014; Volume 9077.
54. Ma, Y.; Anderson, J.; Crouch, S.; Shan, J. Moving object detection and tracking with doppler LiDAR. *Remote Sens.* **2019**, *11*, 1154. [CrossRef]
55. Han, H.; Kim, J.; Park, J.; Lee, Y.; Jo, H.; Park, Y.; Matson, E.; Park, S. Object classification on raw radar data using convolutional neural networks. In Proceedings of the 2019 IEEE Sensors Applications Symposium (SAS), Sophia Antipolis, Valbonne, France, 11–13 March 2019; pp. 1–6. [CrossRef]
56. Stadelmayer, T.; Santra, A.; Weigel, R.; Lurz, F. Data-Driven Radar Processing Using a Parametric Convolutional Neural Network for Human Activity Classification. *IEEE Sens. J.* **2021**, *21*, 19529–19540. [CrossRef]
57. Wan, J.; Chen, B.; Xu, B.; Liu, H.; Jin, L. Convolutional neural networks for radar HRRP target recognition and rejection. *EURASIP J. Adv. Signal Process.* **2019**, *2019*, 4962. [CrossRef]
58. Dadon, Y.D.; Yamin, S.; Feintuch, S.; Permuter, H.H.; Bilik, I.; Taberkian, J. Moving Target Classification Based on micro-Doppler Signatures Via Deep Learning. In Proceedings of the IEEE National Radar Conference—Proceedings, Atlanta, GA, USA, 8–14 May 2021; Volume 2021.
59. Tiwari, A.; Goomer, R.; Yenneti, S.S.S.; Mehta, S.; Mishra, V. Classification of Humans and Animals from Radar Signals using Multi-Input Mixed Data Model. In Proceedings of the 2021 International Conference on Computer Communication and Informatics, ICCCI 2021, Rhodes, Greece, 27–29 January 2021.
60. Chen, V.C.; Ling, H. *Time-Frequency Transforms for Radar Imaging and Signal Analysis*; Artech House: London, UK, 2001.
61. Chen, V.C. *The Micro-Doppler Effect in Radar*; Artech House: London, UK, 2011.
62. Bilik, I.; Tabrikian, J.; Cohen, A. Gmm-based target classification for ground surveillance doppler radar. *IEEE Trans. Aerosp. Electron. Syst.* **2006**, *42*, 267–278. [CrossRef]
63. Krizhevsky, A.E.A. Imagenet classification with deep convolutional neural networks. *Commun. ACM* **2017**, *60*, 84–90. [CrossRef]
64. Goodfellow, I.; Bengio, Y.; Courville, A. *Deep Learning*; MIT Press: Cambridge, MA, USA, 2016. Available online: http://www.deeplearningbook.org (accessed on 16 January 2022).
65. Lee, C.-Y.; Gallagher, P.W.; Tu, Z. Generalizing Pooling Functions in Convolutional Neural Networks: Mixed, Gated, and Tree. 2015. Available online: https://ieeexplore.ieee.org/document/7927440 (accessed on 16 January 2022). [CrossRef]

66. Dodge, S.F.; Karam, L.J. A study and comparison of human and deep learning recognition performance under visual distortions. In Proceedings of the 2017 26th International Conference on Computer Communication and Networks (ICCCN), Vancouver, BC, Canada, 31 July–3 August 2017; pp. 1–7. [CrossRef]
67. He, K.E.A. Deep residual learning for image recognition. In Proceedings of the IEEE Conference on Computer Vision and Pattern Recognition 2016, Las Vegas, NV, USA, 26 June–1 July 2016 ; pp. 770–778.
68. Batista, G.E.; Prati, R.C.; Monard, M.C. A study of the behavior of several methods for balancing machine learning training data. *SIGKDD Explor. Newsl.* **2004**, *6*, 20–29. [CrossRef]
69. Bradley, A.P. The use of the area under the roc curve in the evaluation of machine learning algorithms. *Pattern Recog.* **1997**, *30*, 1145–1159. [CrossRef]
70. He, K.; Zhang, X.; Ren, S.; Sun, J. Deep Residual Learning for Image Recognition. *arXiv* **2015**, arXiv:1512.03385.
71. Kingma, D.P.; Ba, J. Adam: A Method for Stochastic Optimization. 2017. Available online: http://xxx.lanl.gov/abs/1412.6980 (accessed on 16 February 2022).
72. Smith, L. Cyclical Learning Rates for Training Neural Networks. In Proceedings of the 2017 IEEE Winter Conference on Applications of Computer Vision (WACV), Santa Rosa, CA, USA, 24–31 March 2017; pp. 464–472. [CrossRef]
73. Kultavewuti, P. One Cycle & Cyclic Learning Rate for Keras. Available online: https://github.com/psklight/keras_one_cycle_clr (accessed on 16 January 2022).
74. Axon Pulse. MAFAT Radar Challenge: Solution by Axon Pulse. Available online: https://medium.com/axon-pulse/mafat-radar-challenge-solution-by-axon-pulse-a4f082e62b3e (accessed on 16 January 2022).

Article

Multi-Criteria Decision Making to Detect Multiple Moving Targets in Radar Using Digital Codes

Majid Alotaibi

Department of Computer Engineering, College of Computer and Information Systems, Umm Al-Qura University, Makkah 24382, Saudi Arabia; mmgethami@uqu.edu.sa

Abstract: Technological advancement in battlefield and surveillance applications switch the radar investigators to put more effort into it, numerous theories and models have been proposed to improve the process of target detection in Doppler tolerant radar. However, still, more effort is needed towards the minimization of the noise below the radar threshold limit to accurately detect the target. In this paper, a digital coding technique is being discussed to mitigate the noise and to create clear windows for desired target detection. Moreover, multi-criteria of digital code combinations are developed using discrete mathematics and all designed codes have been tested to investigate various target detection properties such as the auto-correlation, cross-correlation properties, and ambiguity function using mat-lab to optimize and enhance the static and moving target in presence of the Doppler in a multi-target environment.

Keywords: Doppler radar; target detection; digital codes; windows; mat-lab

Citation: Alotaibi, M. Multi-Criteria Decision Making to Detect Multiple Moving Targets in Radar Using Digital Codes. *Sensors* **2022**, *22*, 3176. https://doi.org/10.3390/s22093176

Academic Editors: Piotr Samczyński and Janusz Dudczyk

Received: 30 January 2022
Accepted: 18 March 2022
Published: 21 April 2022

Publisher's Note: MDPI stays neutral with regard to jurisdictional claims in published maps and institutional affiliations.

1. Introduction

To keep an eye on objects (static or moving), a radar system is an alone equipment to investigate the characteristics of the object for position and velocity. The radar system broadcasts electromagnetic waves in the direction of the target and takes the wave echo by the target to observe the different parameters such as range, the velocity of the desired target. The present-day radar system exploits many aerial fundamentals to transmit and receive the echoes to increase the probability of target detection. In the addition phase, array radars broadcast entirely consistent waveforms (probably scaled with the help of a complex constant). These waveforms are produced from their 'M' dissimilar broadcast antenna fundamentals to form a powerful transmitted signal in the favored path. Formation of the beam is performed merely by the receiving antenna array to guess the parameters of the radar to estimate the current position of the target. Therefore the broadcast degrees of freedom are restricted to one and the receiver's degrees of freedom are more than one (simply we can say 'n'). But multiple input and multiple output radars transmit unstable signals, these signal waveforms are obtained from their dissimilar broadcasting aerial fundamentals and use combined processing of the acknowledged signals from the dissimilar receiver array fundamentals. Though phase array radars make use of the only spatial variety, multiple input and multiple output radars utilize together spatial and signal waveform variety to get better results of the performance of the system. MIMO radars can be extensively spaced or collocated antennas [1,2]. Whereas the former design enhances the selection of the target detection to get better results. This pattern of target detection improves the resolution of the target and intervention refusal ability. However, the signal waveforms for MIMO radars affect the range. Whereas the signal waveform for single input single output (SISO) radar is considered for desired Doppler and delay resolution properties. The signal waveforms selected for MIMO radars be supposed to have desirable uncertainty in the delay. The resolution characteristics of the broadcast signal waveforms are considered with the help of the uncertainty function [3]. The uncertainty function of a broadcast signal waveform presents the matched filter output in the existence of Doppler and delay variance. If the delay (90 ms) and Doppler resolution are high (80–100 kHz), the uncertainty function is supposed to resemble a "thumbtack". The idea of the uncertainty function is to complete the MIMO system proposed by Antonio and Robey [4]. Chun and Vaidyanathan [5] have proved the characteristics of multiple input uncertainties function and further extended the MIMO uncertainty function for frequency hopping signal waveforms. The Phase coded pulse waveforms are frequently used for radar applications because they are having a high bandwidth-time product (500–100) Numerous phase-coded signal waveforms are presented formerly [6–9] having better cross-correlation and autocorrelation properties. Such signal waveforms are intended for high-quality Doppler and delay resolution properties. On the other hand, manipulating the phase-coded signal waveforms by optimizing the MIMO uncertainty function modifies the Doppler, delay and also improves the spatial resolution characteristics.

The existing approaches in radar employ optimization algorithms with a huge number of iterations which may increase the delay, to obtain the codes with good auto or cross-correlation and mitigate the noise in the detection of static targets. Doppler tolerant digital codes are also implemented by the state of the art researchers to detect the moving targets. But these codes may not have good auto and cross-correlation properties. In this paper, an approach is presented in which various multi-criteria of digital codes have been generated with optimal auto and cross-correlation properties to mitigate the noise and the same code is tested for Doppler tolerance to detect static and multi moving targets by creating a huge number of clear windows within the threshold limit of radar detection. Each of the codes performs better for all three parameters; it is the optimization of all the three functions to efficiently detect the desired object. The binary sequences with different bit lengths in this approach are generated by using discrete mathematical operations to the digital sequence.

The rest of the paper is organized as follows. In Section 2 we present the related work, Section 3 presents our proposed approach, in Section 4 discussion about the proposed approach is presented and the paper is concluded in Section 5.

2. Related Work

Lewis and Kretschmer [10,11] presented two different approaches in which they recommended extra codes called P3 and P4 codes and are created from LFMWT. Moreover, these codes are used to improve the Doppler tolerance, in particular when compared to P1 and P2 codes. In addition peak side lobes of P3 and P4 codes are enhanced further to overcome the gaps that existed in polyphase codes. But the performance of these codes corrupts a smaller amount with an increase in Doppler frequency. Lewis [12] proposed a method based on windowing mode to diminish the range versus time effects of side lobes in polyphase codes considerably. However, this technique simply decreases the peak sidelobe and an ingredient amount irrespective of the successful pulse compression ratio. Kretschmer and Welch [13] projected a method in which they demonstrated that the autocorrelation function of polyphase codes is having undesired range side lobes. Hence cannot be suitable to detect multi-target detection. This approach also discussed the effect of Sidelobe reduction with the help of amplitude weighting function (AWF) of polyphase codes in the receiver filter. Even though weighed windows whenever used in source and destination (i.e., Transmitter and Receiver), provides improved results. In this approach author also proved that the weight on the sender (Receiver) is more ideal as the weight on destination (transmit) results in power loss because the existing source (transmitter) power cannot be entirely utilized. Luszczyk and Mucha [14] proposed a model to reduce the effect of the range side lobe. In this approach, they used the Kaiser-Bessel weighing function of the p4 pulse compression signal waveform to reduce the range side lobes of P4. However, whenever a high Doppler occurs, the performance of this approach shows a poor response. Ajit Kumar Sahoo and Ganapati Panda [15] anticipated a firmness windowing method to minimize the consequences of side lobes in Doppler tolerant radars. However, the presented technique experiences a delay and cannot produce bigger windows or enhance the count of windows to sense numerous moving objects correctly. Sharma and Rajeswari [16] proposed a model to represent optimization for multiple input multiple output radar ambiguities. However due to huge mathematical complexity delay increases, also this approach has not had a large window so the probability of missing the target is more. Reddy and Anuradha [17] presented an approach, to improve the Signal to noise ratio of Mesosphere- Stratosphere-Troposphere (MST) with the help of Kaiser hamming and cosh hamming window function. Also, this approach increases the energy of the main lobe to amplify the merit factor; however, this approach fails to remove the effect of side lobes. Because the Doppler change continuously. Therefore cannot be suitable to detect multiple moving targets. Lewis and Kretschmer [18] proposed an approach in which they proved that one can use polyphase codes instead of using bi-phase codes, as the polyphase codes are also referred to compress the pulse of the given signal waveform to attain enhanced PSR and can be used to avoid the security problem. This method is very accurate to cause P1 and P2 polyphase codes. These polyphase codes are generated from the step estimation by the use of a modulation technique called linear frequency modulated waveform technique (LFMWT). Such polyphase codes have an additional tolerance to the restriction of the receiver side bandwidth.

Sindhura et al. [19] proposed a model, based on a wavelet, in this paper the authors compare the signal to noise ratio enhancement for Lower Atmospheric Signals. This approach is used to calculate the wind performance in the atmospheric boundary layer (ABL) and minor troposphere. SakhaWat et al. [20] presented an approach in which the author gets the information of GPS signals and can be used to present a novel use to remote-sensing since they are capable of providing valuable information concerning the reflecting face. However, this technique puts much attention towards image arrangement that too of fixed (static) targets only, hence cannot have an optimal use in moving target detection. Syed and Venay [21] projected a technique in which the main focus was to improve signal loss. Here they use p1, p3 codes, and hyperbolic frequency modulation (HFM) polyphase codes are used to increase the signal-to-noise ratio of the acknowledged signal. However, this approach has an adverse effect of delay, therefore not suitable to detect the multiple moving targets. Singh et al. [22,23] projected two different coding techniques to minimize the noise. No doubt the authors improved the count of the windows which can enhance the probability of target detection. However, these methods are limited to finding stationary and slow-moving objects only, since the period of the designed code vector is not as much of which minimizes the merit factor (MF) of the received echo signal and has noise in the region of the zero Doppler. In [24,25] two coding techniques have been presented to enhance the target detection in Doppler tolerant radar. Though, both techniques exploit the mathematical involvedness and maximize the delay parameter, which decreases the probability of target detection and could not be the optimum method to find -moving targets. Alotaibi [26] presented a technique of target detection using linear block coding, in which the author generates the code using well-known (6,3) block code and then inserts the odd and even parity to change the present position of the code word to achieve the result. However the presented approach uses more number of gates to complement the existing code to increase the length of the codeword which increases the hardware and delay therefore, may not be optimum to detect multi-moving targets in Doppler tolerant radar. The digital code techniques [24–26] are only concentrating on the Doppler but not on the auto and cross-correlation properties of the sequences which are considered to be important to detect the desired target in a multi-target environment. In [27] the authors designed the digital sequences with optimal properties to detect the static and moving targets. But the side noise in this method is more when compared to the presented approach (refer to Section 4 as comparative analysis of the existing approach and the proposed approach is depicted in the same section).

In this paper, an approach is being presented to optimize the interference to detect static and moving objects. The presented approach helps to detect multiple moving targets efficiently as the sequences obtained are tested for auto and cross-correlation along with the Doppler tolerant frequencies.

3. Proposed Approach

Initially, a concatenation series of decimal numbers having the equal number of zeros and ones when represented in hex code (i.e., from 0 to 15) are considered. This series is termed binary hex equal-weighted codes (BHEWC) and can be represented as

$$I_{CW} = \prod_{p=1}^{4} 3p, \ 1 \leq p \leq 4 \tag{1}$$

where 'I_{CW}' is concatenation series of BHEWC.

The concatenation series then will be 3, 6, 9, and 12; the same series in binary hex value can be represented as '0011011010011100'. The MSB bit of the initially designed code is zero and it doesn't satisfy the cyclic division rule (as the cyclic division process is used to generate the codes) so the hex value of the first decimal number is given one bit left shift (i.e., 0011 = 1001), and thus the concatenation series can now be represented as 1001011010011100. The remaining decimal numbers having equal weight (i.e., missed numbers in Equation (1)) can be formulated as

$$M_{B1} = \frac{I_{TS} + L_{TS}}{I_{TS}} \tag{2}$$

$$M_{B2} = 2\,M_{B1} \tag{3}$$

'I_{TS}' is the first decimal value of 'I_{CW}' and
'L_{TS}' is the last decimal value of 'I_{CW}'

Now from Equations (2) and (3) $M_{B1} = \frac{3+12}{3} = 5$ (Binary hex value 0101) and $M_{B2} = 2 \times 5 = 10$ (Binary hex value 1010) are the missing decimal numbers of 'I_{CW}' Later cyclic redundancy method is used to improve the target detection and minimize the noise near-zero Doppler, in which the generator codeword can be represented by 'G_{CW}' and the Radar codeword generation using cyclic process can be characterized as

$$(R_{CW})_{CP} = \frac{I_{CWp_i}}{G_{CW}}, \ i = 1, \ 2 \tag{4}$$

where 'I_{CWp_i}' is the code after appending '$p_1(q)$' or '$p_2(q)$' zeros

$$p_1(q) = p_2(q) = -1 \tag{5}$$

where r is the hex code length of M_{B1} or M_{B2} (as they are of the same length).

Therefore, the length is 4 bits for both the cases, thus $4 - 1 = 3$ zeros need to append to 'I_{CW}' to get I_{CWi} codes that are employed in the cyclic process to get the initial radar codes to test the target. In this approach, modulo 2 additions are employed to generate the codeword. However, the code sequence that is exploited to generate the initial codeword to detect the targets in Doppler tolerant radar can be represented as

$$R_{1CW} = r_{cp1} \oplus I_{CWp_1} \tag{6}$$

$$R_{2CW} = r_{cp2} \oplus I_{CWp_2} \tag{7}$$

where 'r_{cp1}', 'r_{cp2}' are the remainders when 'M_{B1}' and 'M_{B2}' are the divisors of Equation (4) respectively, However as the MSB (most significant bit) of M_{B1} is '0', therefore to satisfy the cyclic process one bit left shift has given to 'M_{B1}'. Hence the generated divisor can be represented as '1100' and the two initial code words that are used to generate the final radar target detection code words can be represented in Tables 1 and 2 respectively.

Table 1. Binary representation of R_{1CW}.

R_{1CW}	b_1	b_2	b_3																b_n
	1	0	0	1	0	1	1	0	1	0	0	1	1	1	0	1	1	0	0

$b_1, b_2, b_3, \ldots \ldots, b_n$ bit positions of code R_{1CW}.

Table 2. Binary representation of R_{2CW}.

R_{2CW}	c_1	c_2	c_3																c_n
	1	0	0	1	0	1	1	0	1	0	0	1	1	1	0	1	0	1	0

$c_1, c_2, c_3, \ldots \ldots, c_n$ bit positions of code R_{2CW}.

The final codeword matrix is formulated and can be represented in Table 3 whose first row is 'R_{1CW}' and first column is 'R_{2CW}'

Table 3. Final codewords using matrices.

r_{11}	r_{12}	-	-	-	-	-	-	-	-	-	-	-	-	-	-	-	-	r_{1n}
	-																	-
	-																	-
	-																	-
r_{n1}	r_{n2}	-	-	-	-	-	-	-	-	-	-	-	-	-	-	-	-	r_{nn}

The remaining elements of the matrices are formulated by modulo two operations of row and Column and can be represented as

$$r_{11}= b_1, \; r_{12}= b_2, \ldots \ldots r_{1n}= b_n \tag{8}$$

$$r_{21}= c_2, \; r_{31}= c_3, \ldots \ldots r_{n1}= c_n \tag{9}$$

$$r_{12}= b_2 \oplus c_2, \; r_{n2}= b_2 \oplus c_n, \; r_{nn}= b_n \oplus c_n \tag{10}$$

The matrix formed is represented in Figure 1. From the figure, 19×19 matrix without considering the initial codes R_{1CW} (first row) and R_{2CW} (first column) is obtained.

```
  1 0 0 1 0 1 1 0 1 0 0 1 1 1 0 1 1 0 0
  ---------------------------------------
1 | 0 1 1 0 1 0 0 1 0 1 1 0 0 0 1 0 0 1 1
0 | 1 0 0 1 0 1 1 0 1 0 0 1 1 1 0 1 1 0 0
0 | 1 0 0 1 0 1 1 0 1 0 0 1 1 1 0 1 1 0 0
1 | 0 1 1 0 1 0 0 1 0 1 1 0 0 0 1 0 0 1 1
0 | 1 0 0 1 0 1 1 0 1 0 0 1 1 1 0 1 1 0 0
1 | 0 1 1 0 1 0 0 1 0 1 1 0 0 0 1 0 0 1 1
1 | 0 1 1 0 1 0 0 1 0 1 1 0 0 0 1 0 0 1 1
0 | 1 0 0 1 0 1 1 0 1 0 0 1 1 1 0 1 1 0 0
1 | 0 1 1 0 1 0 0 1 0 1 1 0 0 0 1 0 0 1 1
0 | 1 0 0 1 0 1 1 0 1 0 0 1 1 1 0 1 1 0 0
0 | 1 0 0 1 0 1 1 0 1 0 0 1 1 1 0 1 1 0 0
1 | 0 1 1 0 1 0 0 1 0 1 1 0 0 0 1 0 0 1 1
1 | 0 1 1 0 1 0 0 1 0 1 1 0 0 0 1 0 0 1 1
1 | 0 1 1 0 1 0 0 1 0 1 1 0 0 0 1 0 0 1 1
0 | 1 0 0 1 0 1 1 0 1 0 0 1 1 1 0 1 1 0 0
1 | 0 1 1 0 1 0 0 1 0 1 1 0 0 0 1 0 0 1 1
0 | 1 0 0 1 0 1 1 0 1 0 0 1 1 1 0 1 1 0 0
1 | 0 1 1 0 1 0 0 1 0 1 1 0 0 0 1 0 0 1 1
0 | 1 0 0 1 0 1 1 0 1 0 0 1 1 1 0 1 1 0 0
```

Figure 1. Binary code matrix.

To design the codes which are Doppler tolerant and with optimal autocorrelation and cross-correlation properties can be generated by performing various mathematical operations on the matrix such as

a. Concatenation of rows,
b. complement operation is performed on even terms,
c. if continuous series of three 1's and 0's occur in the sequence then the middle term is complemented and

The bits which are at the position of the sum of the power of 2 are complemented (i.e., multi-criteria creation of digital codes to test the target).

Note that one need not perform all the operations to obtain the sequence with good properties. The sequence obtained after performing each operation is tested and the sequence with the best properties (maybe after performing one operation, two operations, all the operations, or no operation) is selected. Figure 2 shows the flow diagram of the proposed approach. The codes of different lengths (here 19-bit, 95-bit, and 190-bit codes) are generated from the above operations. The ambiguity, auto-correlation, and cross-correlation which are the optimal characteristic properties of static and moving target detection in a multi-target environment are observed using the mat-lab tool. For testing the sequence for auto and cross-correlation properties the 0's of the sequence is replaced by -1 to obtain the actual length of the sequence and for the ambiguity function, this change is not necessary. The simulation parameters are shown in Table 4.

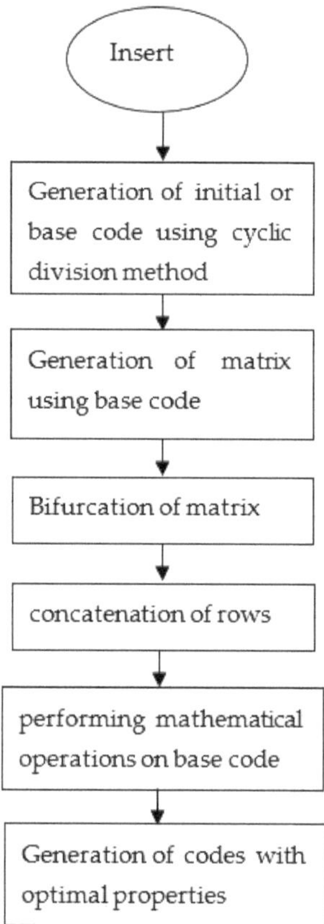

Figure 2. Flow diagram of the proposed approach.

Table 4. Simulation Parameters.

Transmitter	
Power of pulse	470 kW
Frequency	2800 MHz to 3100 MHz
RF duty cycle	Max 0.004
Width of the pulse	1.55 µs and 4.50 µs
Receiver	
Intermediate frequency	55.5 MHz
Range (dynamic)	95 dB
noise power of the system	−112 dBm
Target (in σ *meter square*)	
Man	0.14 to 1.05
Aircraft C-54	10 to 1000
A free electron	8×10^{-30}
Birds	10^{-3}

The 19-bit code R_{1CW} and R_{2CW} are directly tested (without performing any mathematical operation) for the above properties and R_{2CW} shows better response, unlike R_{1CW}. Figures 3–5 depict the auto-correlation, cross-correlation, and Doppler tolerance of the code R_{2CW}.

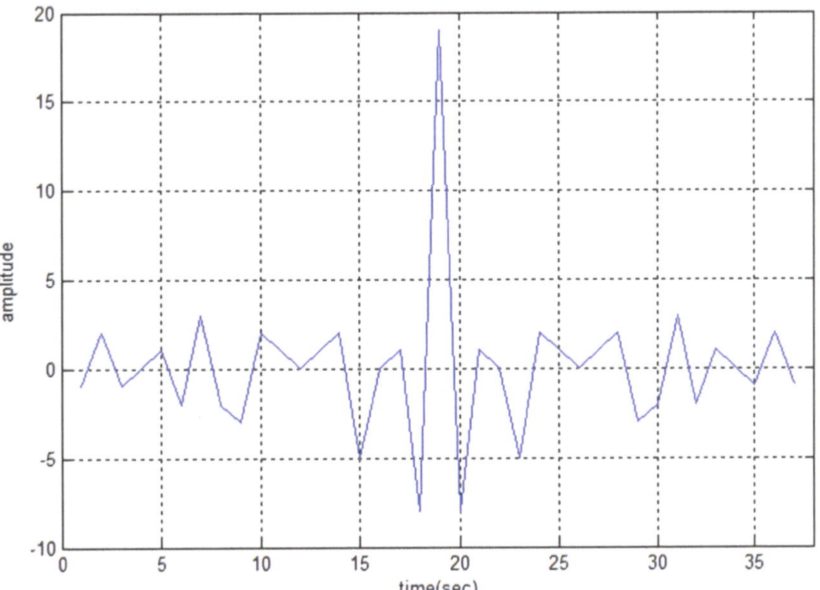

Figure 3. Auto-Correlation property of R_{2CW}.

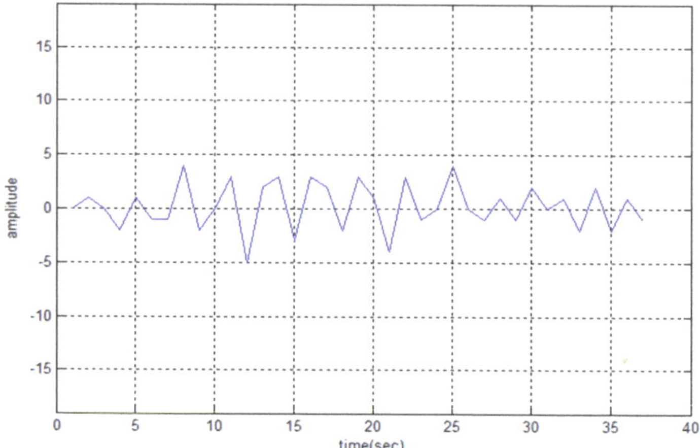

Figure 4. Cross-Correlation property of R_{2CW}.

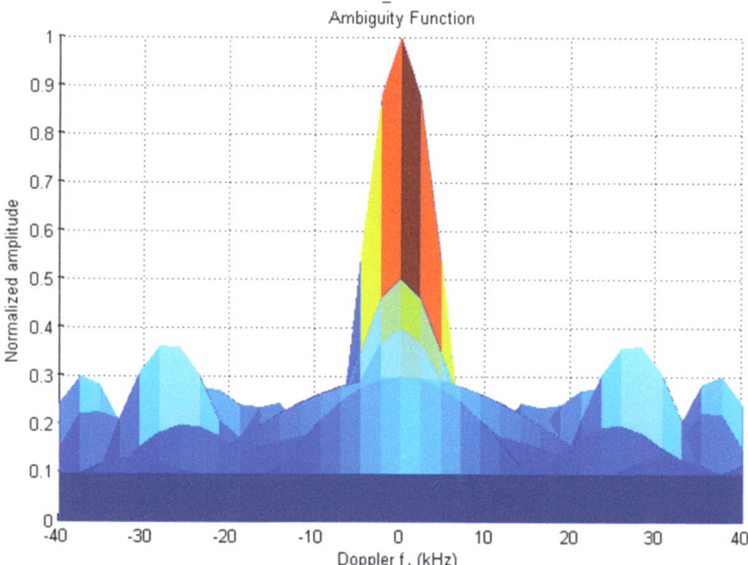

Figure 5. Ambiguity-Function of R_{2CW}.

From the graph, the side noise amplitude has the amplitude of 2 and the reduction ratio of peak to side-lobe S_R in dB can be calculated as [27].

$$S_R = 20log_{10}\left(\frac{amplitude\ of\ side\ noise\ which\ is\ maximum}{amplitude\ of\ main\ lobe}\right) = 20log_{10}\left(\frac{2}{19}\right) = -19.56\ (\text{approx})\ \text{dB} \tag{11}$$

From Figure 3 the maximum lobe for cross-correlation is 4, and Figure 4 shows the ambiguity function of R_{2CW} code. It is observed that there are no clear windows below 0.2 (normalized amplitude which is considered as standard threshold value), the clear windows are present above 0.3 normalized value.

4. Generation of 95-Bit and 190-Bit Sequence

The 190-bit and 95-bit code is generated from the matrix of Figure 1. Initially, the first row (R_{1CW}) is eliminated to get the matrix of 19×20, however one can eliminate the first column (R_{2CW}) to get 20×19 and test the codes. The codes of length 190-bit and 95-bit are generated as represented in Figures 6 and 7 respectively. The arrow in Figure 6 represents the matrix division vertically into two parts termed the first half (F_H) and the second half (S_H) respectively to obtain the desired length of codes.

```
R_1CW = 1 0 0 1 0 1 1 0 1 0 0 1 1 1 0 1 1 0 0

  1 0 1 1 0 1 0 0 1 0        1 1 0 0 0 1 0 0 1 1
  0 1 0 0 1 0 1 1 0 1        0 0 1 1 1 0 1 1 0 0
  0 1 0 0 1 0 1 1 0 1        0 0 1 1 1 0 1 1 0 0
  1 0 1 1 0 1 0 0 1 0        1 1 0 0 0 1 0 0 1 1
  0 1 0 0 1 0 1 1 0 1        0 0 1 1 1 0 1 1 0 0
  1 0 1 1 0 1 0 0 1 0        1 1 0 0 0 1 0 0 1 1
  1 0 1 1 0 1 0 0 1 0        1 1 0 0 0 1 0 0 1 1
  0 1 0 0 1 0 1 1 0 1        0 0 1 1 1 0 1 1 0 0
  1 0 1 1 0 1 0 0 1 0        1 1 0 0 0 1 0 0 1 1
  0 1 0 0 1 0 1 1 0 1        0 0 1 1 1 0 1 1 0 0
  0 1 0 0 1 0 1 1 0 1        0 0 1 1 1 0 1 1 0 0
  1 0 1 1 0 1 0 0 1 0        1 1 0 0 0 1 0 0 1 1
  1 0 1 1 0 1 0 0 1 0        1 1 0 0 0 1 0 0 1 1
  1 0 1 1 0 1 0 0 1 0        1 1 0 0 0 1 0 0 1 1
  0 1 0 0 1 0 1 1 0 1        0 0 1 1 1 0 1 1 0 0
  1 0      F_H      1 0      1 1      S_H      1 1
```

Figure 6. Generation of 190-bit code.

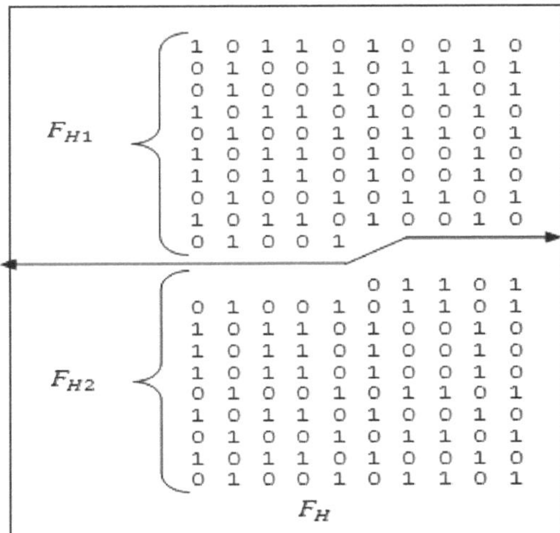

Figure 7. Generation of 95-bit code.

The sequence of 190-bit is obtained by concatenation of F_H (refer to Figure 5). The mathematical operations performed on this 190-bit sequence respectively are (i) complementing even terms, (ii) complementing the middle bit if three continuous 0's or 1's occurs and lastly (iii) complementing the bits present at the power of 2 positions. After performing all these operations the final code (F1) obtained can be represented as

$$F1 = 1\,1\,1\,1\,1\,0\,0\,0\,0\,1\,0\,1\,0\,1\,0\,0\,1\,0\,1\,0\,1\,0\,0\,1\,0\,1\,1\,0\,1\,0\,1\,1\,1\,0\,1\,0\,0\,1\,1\,0\,1\,0\,1\,0\,1$$
$$0\,0\,1\,0\,1\,0\,1\,0\,1\,0\,1\,1\,0\,1\,0\,1\,0\,0\,0\,0\,1\,1\,0\,1\,0\,1\,0\,1\,0\,1\,0\,0\,1\,0\,0\,1\,0\,1\,0\,1\,0\,0\,1\,0\,1\,0\,1\,0$$
$$1\,0\,1\,1\,0\,1\,0\,1\,0\,0\,1\,0\,1\,1\,0\,1\,0\,1\,0\,1\,0\,1\,0\,1\,0\,1\,0\,0\,1\,0\,1\,1\,1\,1\,0\,1\,0\,0\,1\,0\,1\,1\,0\,1\,0\,1$$
$$0\,1\,0\,1\,0\,0\,1\,0\,1\,0\,1\,0\,1\,0\,1\,0\,1\,0\,1\,0\,1\,0\,1\,1\,0\,1\,0\,1\,0\,1\,0\,1\,0\,0\,1\,0\,1\,0\,1\,0\,1\,0\,1\,1\,0\,1\,0.$$

The code obtained by concatenation of F_H performs better than the codes obtained by concatenation of S_H.

The sequence of 95-bit length is generated by horizontally dividing the first half (F_H) of the matrix equally as represented in Figure 7. Let the two equal parts be F_{H1} and F_{H2} respectively.

Figure 8 shows the auto-correlation property of code F1 and the maximum noise lobe is 30 the S_R value is -16.03. Figures 9 and 10 represent the cross-correlation and Doppler tolerance of sequence F1. The noise in Figure 10 is below 0.2 (normalized amplitude) and after 24 kHz frequency, the noise is almost zero.

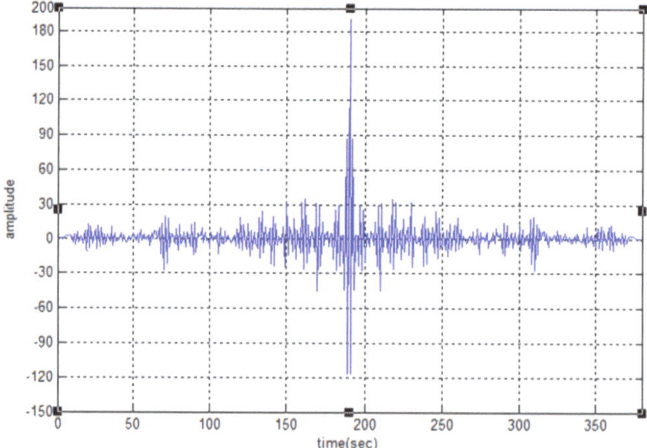

Figure 8. Auto-Correlation property of F1.

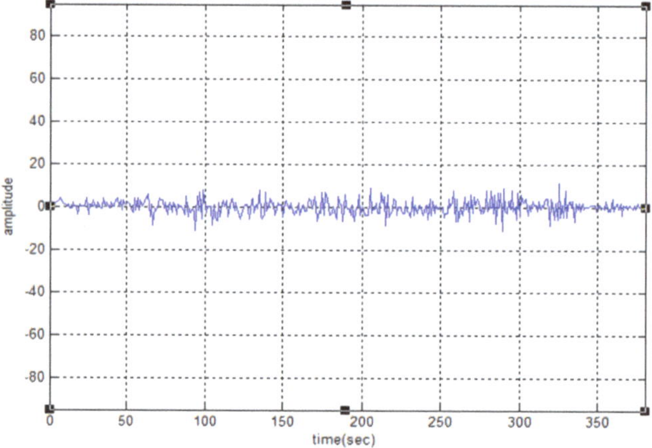

Figure 9. Cross-Correlation property of F1.

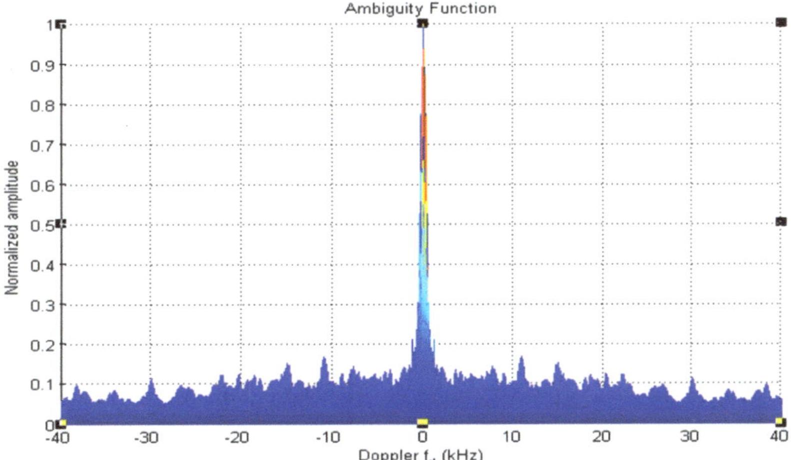

Figure 10. Ambiguity-Function of F1.

From Figure 7 the generation of 95-bits is shown (concatenation of F_{H1} and F_{H2}), as the concatenation of F_{H2} generates better properties than F_{H1}. The 95-bit sequence (F2) tested in this paper is obtained by complementing the even terms of the code generated by concatenation of F_{H2}.

(F2) = 0 0 1 1 1 0 1 1 1 0 0 0 0 1 1 0 0 0 0 0 0 1 0 1 1 0 0 1 1 1 1 0 1 0 0 1 1 0 1 0 0 0 1 1 1 0 0 0 1 1 1 1 0 1 1 1 1 0 1 1 0 0 0 0 1 0 1 1 1 1 0 0 0 0 1 0 1 1 1 1 1 1 0 0 0 1 1 1 0 0 0 0 0 0 1

Figures 11–13 depict the auto-correlation, cross-correlation, and ambiguity of sequence F2. The value of S_R for F2 is -15.48. In Figure 13 the noise amplitude is 0.2 at three frequencies and less than it for other frequencies with clear windows at 3–14 kHz, 16–21 kHz, and 24–38 kHz.

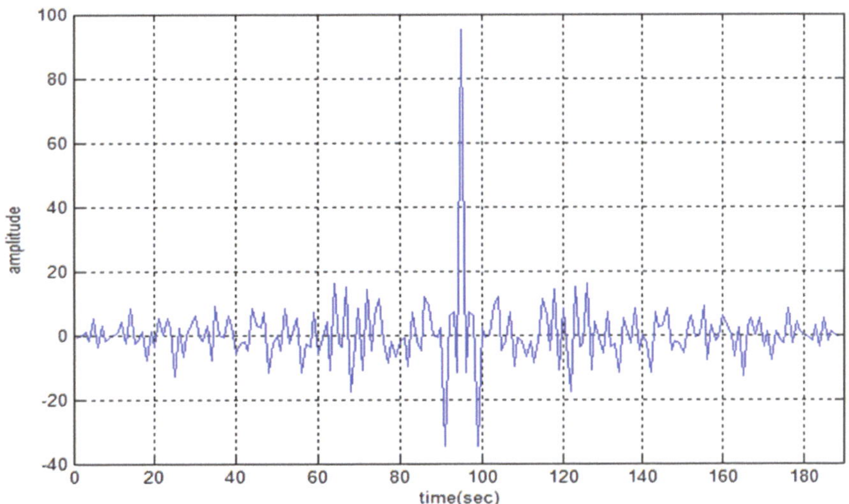

Figure 11. Auto-Correlation property of F2.

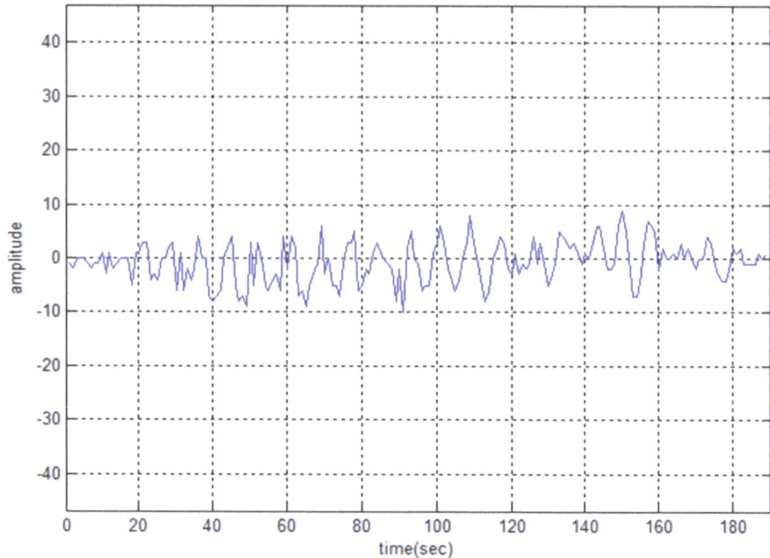

Figure 12. Cross-Correlation property of F2.

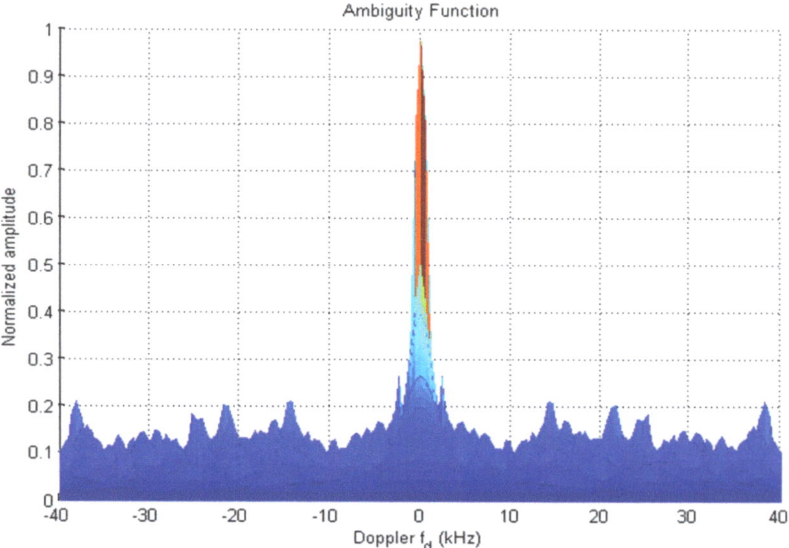

Figure 13. Ambiguity-Function of F2.

Figure 14 represents the noise peak (NP) versus the Doppler graph. It presents the comparative analysis of various existing approaches [26,27] with the proposed approach and it has been observed from the figure that the presented approach shows better performance with minimal side noise peaks when compared with the existing approaches as the presented approach obtained the digital codes with clear windows to detect the desired target in a multi-target environment.

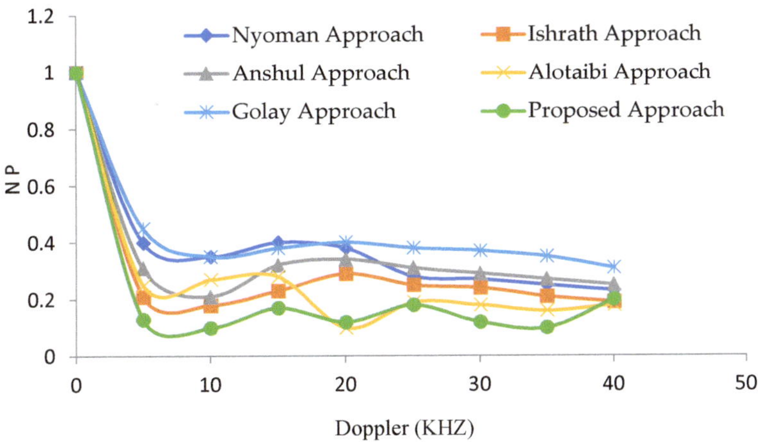

Figure 14. Noise peaks Vs Doppler.

Table 5 represents the length of the sequence and parameters achieved in terms of peak to side-lobe noise amplitude S_R in dB, cross-correlation noise amplitude, and clear windows at different Doppler's.

Table 5. Parameters of the obtained sequences.

Sequence Length	S_R in dB	Max. Noise Amplitude (Cross-Correlation)	Clear Windows of Ambiguity Figure
19	−19.56	4	above 0.3 normalized amplitude
95	−15.48	8	3–14 kHz, 16–21 kHz, and 24–38 kHz.
190	−16.03	9	0–40 Hz

5. Discussion

The designed sequences can be used to detect the static as well as moving targets in multi-target ambiance. From the simulated results it has been shown that the auto-correlation, cross-correlation, and Doppler tolerance properties of the designed codes are better in comparison with the existing approaches [28–30], as they are of a limited length and used for only static target detection. The Auto, cross-correlation and Doppler tolerance properties of the proposed approach adhere that the presented approach finds an extensive use to find the desired target in presence of Doppler. The proposed approach is simple as it generates the digital code sequences with different lengths. There is no limitation to the length of the sequences as one can design the codes of any length using any mathematical operations. The codes are cost-effective as they are obtained without employing any extra costly hardware components. As the presented digital codes are achieved from a single basic code i.e., binary hex equal-weighted code, by revising these codes using fundamental laws of mathematics and communication code theory, the stage of range gate in the phase of target detection is reduced which decreases the cost and hardware and minimizes the delay. Therefore can be employed to see the fast and multiple moving targets above MAC-3 i.e., latest generation combatant crafts by opting for a suitable code length with optimal properties to reduce all the noise amplitude peaks lower than 0.2 dB (normalized amplitude) which is the standard threshold value [23].

6. Conclusions

Radar systems utilize two or more antennas to detect the target in a multi-target environment. Hence, the interference of the sequence with itself and with other sequences should be optimal for static target detection and the sequences should be Doppler tolerant for the detection of moving targets. The clear windows obtained in Figures 9 and 12 are used to detect the moving targets at desire Doppler. Hence, the codes obtained in this paper can be used to detect the static and moving targets accurately with optimal interference and improvised range and resolution.

Funding: This work is supported by National Science & Technology Innovation Plan (NSTIP), Grant #14-ELE1448-10.

Institutional Review Board Statement: Not applicable.

Informed Consent Statement: Not applicable.

Data Availability Statement: Not applicable.

Conflicts of Interest: The author declares no conflict of interest.

References

1. Haimovich, A.M.; Blum, R.S.; Cimini, L.J. MIMO radar with widely separated antennas. *IEEE Signal Process. Mag.* **2007**, *25*, 116–129. [CrossRef]
2. Jian, L.; Stoica, P. MIMO Radar with Collocated Antennas. *IEEE Signal Process. Mag.* **2007**, *24*, 106–114.
3. Levanon, N.; Mozeson, E. *Radar Signals*; Wiley-IEEE Press: New York, NY, USA, 2004.
4. Antonio, G.S.; Fuhrmann, D.R.; Robey, F.C. MIMO radar ambiguity functions. *IEEE J. Sel. Top. Signal Process.* **2007**, *1*, 167–177. [CrossRef]
5. Chen, C.-Y.; Vaidyanathan, P.P. MIMO Radar Ambiguity Properties and Optimization Using Frequency-Hopping Waveforms. *IEEE Trans. Signal Process.* **2008**, *56*, 5926–5936. [CrossRef]
6. Deng, H. Polyphase Code Design for Orthogonal Netted Radar Systems. *IEEE Trans. Signal Process.* **2004**, *52*, 3126–3135. [CrossRef]
7. Deng, H. Synthesis of binary sequences with good autocorrelation and crosscorrelation properties by simulated annealing. *IEEE Trans. Aerosp. Electron. Syst.* **1996**, *32*, 98–107. [CrossRef]
8. Sharma, G.V.K.; Rajeswari, K.R. Four Phase Orthogonal Code Design for MIMO Radar Systems. In Proceedings of the IEEE National Conference on Communications, NCC-2012, Kharagpur, India, 3–5 February 2012.
9. Liu, B.; He, Z.; Zeng, J.; Liu, B. Polyphase Orthogonal Code Design for MIMO Radar Systems. In Proceedings of the 2006 CIE International Conference on Radar, Shanghai, China, 16–19 October 2006; pp. 1–4.
10. Lewis, B.L.; Kretschmer, F.F. A New Class of Polyphase Pulse Compression Codes and Techniques. *IEEE Trans. Aerosp. Electron. Syst.* **1981**, *AES-17*, 364–372. [CrossRef]
11. Lewis, B.L.; Kretschmer, F.F. Linear Frequency Modulation Derived Polyphase Pulse Compression Codes. *IEEE Trans. Aerosp. Electron. Syst.* **1982**, *AES-18*, 637–641. [CrossRef]
12. Lewis, B.L. Range-Time-Side lobe Reduction Technique for FM-Derived Polyphase PC Codes. *IEEE Trans. Aerosp. Electron. Syst.* **1993**, *AES-29*, 834–840.
13. Kretschmer, F.F.; Welch, L.R. Sidelobe Reduction Techniques for Polyphase Pulse Compression Codes. In Proceedings of the Record of the IEEE 2000 International Radar Conference, Alexandria, VA, USA, 7–12 May 2000; pp. 416–421.
14. Luszczyk, M.; Mucha, D. Kaiser-Bessel window weighting function for polyphase pulse compression code. In Proceedings of the 17th International Conference on Microwaves, Radar and Wireless Communications, Wroclaw, Poland, 19–21 May 2008; pp. 1–4.
15. Sahoo, A.K.; Panda, G. Doppler Tolerant Convolution Windows for Radar Pulse Compression. *Int. J. Electron. Commun. Eng.* **2011**, *4*, 145–152.
16. Sharma, G.V.K.; Rajeswari, K.R. MIMO Radar Ambiguity Optimization using Phase Coded Pulse Waveforms. *Int. J. Comput. Appl.* **2013**, *61*, 18–23. [CrossRef]
17. Reddy, R.K.; Anuradha, B. Improved SNR Of MST Radar Signals By Kaiser-Hamming And Cosh-Hamming Window Functions. *Int. J. Eng. Res. Appl.* **2015**, *5*, 32–38.
18. Kretschmer, F.F.; Lewis, B.L. Doppler Properties of Polyphase Coded Pulse Compression Waveforms. *IEEE Trans. Aerosp. Electron. Syst.* **1983**, *AES-19*, 521–531. [CrossRef]
19. Sindhura, S.K.; Reddy, S.N.; Kamaraj, P. Comparison of SNR Improvement for Lower Atmospheric Signals Using Wavelets. *Int. J. Adv. Res. Electr. Electron. Instrum. Eng.* **2015**, *4*, 8202–8209.
20. Sakhawat, S.; Usman, M.; Mahmood, B. Simulation of Bi-static Radar System Based on Reflected GPS L5 Signals. *Procedia Comput. Sci.* **2015**, *52*, 546–551. [CrossRef]
21. Rafiuddin, S.S.A.; Bhangdia, V.K. Empirical analysis on Doppler tolerant radar codes. *Int. J. Sci. Eng. Res.* **2013**, *4*, 1579–1582.
22. Singh, R.K.; Rani, D.E.; Ahmad, S.J. RSBHCWT: Re-Sampling Binary Hex Code Windowing Technique to Enhance Target Detection. *Indian J. Sci. Technol.* **2016**, *9*, 1–5.

3. Singh, R.K.; Rani, D.E.; Ahmad, S.J. HQECMT: Hex Quadratic Residue EX-OR Coded Matrix Technique to improve target detection in Doppler tolerant Radar. *Int. J. Sci. Res. (PONTE)* **2017**, *73*, 21–28.

4. Alotaibi, M. Improved target detection in Doppler tolerant radar using a modified hex coding technique. In *Lecture Notes in Electrical Engineering, Proceedings of the International Conference on Communications and Cyber Physical Engineering 2018, Hyderabad, India, 24–25 January 2018*; Springer: Singapore, 2018; pp. 63–72.

5. Aleem, M.; Singh, R.P.; Ahmad, S.J. Enhance multiple moving target detection in Doppler tolerant radar using IRAESC technique. In *Innovations in Electronics and Communication Engineering*; Springer: Singapore, 2019; Volume 33, pp. 417–426.

6. Alotaibi, M. Low noise moving target detection in high resolution radar using binary codes. *EURASIP J. Adv. Signal Process.* **2021**, *2021*, 8. [CrossRef]

7. Unissa, I.; Ahmad, S.J. *Long Binary Sequences with Good Auto and Cross-Correlation Properties*; LAP LAMBERT Academic Publishing: Saarbrücken, Germany, 2021; ISBN 978-6203856279.

8. Golay, M.J.E. Complementary series. *IRE Trans. Inform. Theory* **1961**, *7*, 82–87. [CrossRef]

9. Khatter, A.; Goyal, P. Design and analyze the various m-sequences codes in MATLAB. *Int. J. Emerg. Technol. Adv. Eng.* **2012**, *2*, 125–129.

10. Pramaita, I.N.; Diafari, I.G.A.G.K.; Negara, D.N.K.P.; Dharma, A. New Orthogonal Small Set Kasami Code Sequence. *Maj. Ilm. Teknol. Elektro* **2015**, *14*, 47–51. [CrossRef]

Article

An X-Band CMOS Digital Phased Array Radar from Hardware to Software

Yue-Ming Wu [1], Hao-Chung Chou [1], Cheng-Yung Ke [2], Chien-Cheng Wang [2], Chien-Te Li [2], Li-Han Chang [2], Borching Su [3], Ta-Shun Chu [1] and Yu-Jiu Wang [2,*]

[1] Department of Electrical Engineering, National Tsing Hua University, Hsinchu 30013, Taiwan; yueming@gapp.nthu.edu.tw (Y.-M.W.); s101061557@m101.nthu.edu.tw (H.-C.C.); tschu@ee.nthu.edu.tw (T.-S.C.)

[2] Tron Future Tech Inc., Hsinchu 300042, Taiwan; cyk@tronfuturetech.com (C.-Y.K.); ccw@tronfuturetech.com (C.-C.W.); ctl@tronfuturetech.com (C.-T.L.); lhc@tronfuturetech.com (L.-H.C.)

[3] Department of Electrical Engineering, National Taiwan University, Taipei 10617, Taiwan; borching@ntu.edu.tw

* Correspondence: yw@tronfuture.com

Citation: Wu, Y.-M.; Chou, H.-C.; Ke, C.-Y.; Wang, C.-C.; Li, C.-T.; Chang, L.-H.; Su, B.; Chu, T.-S.; Wang, Y.-J. An X-Band CMOS Digital Phased Array Radar from Hardware to Software. *Sensors* **2021**, *21*, 7382. https://doi.org/10.3390/s21217382

Academic Editors: Janusz Dudczyk and Piotr Samczyński

Received: 21 October 2021
Accepted: 3 November 2021
Published: 6 November 2021

Publisher's Note: MDPI stays neutral with regard to jurisdictional claims in published maps and institutional affiliations.

Abstract: Phased array technology features rapid and directional scanning and has become a promising approach for remote sensing and wireless communication. In addition, element-level digitization has increased the feasibility of complicated signal processing and simultaneous multi-beamforming processes. However, the high cost and bulky characteristics of beam-steering systems have prevented their extensive application. In this paper, an X-band element-level digital phased array radar utilizing fully integrated complementary metal-oxide-semiconductor (CMOS) transceivers is proposed for achieving a low-cost and compact-size digital beamforming system. An 8–10 GHz transceiver system-on-chip (SoC) fabricated in 65 nm CMOS technology offers baseband filtering, frequency translation, and global clock synchronization through the proposed periodic pulse injection technique. A 16-element subarray module with an SoC integration, antenna-in-package, and tile array configuration achieves digital beamforming, back-end computing, and dc–dc conversion with a size of $317 \times 149 \times 74.6$ mm^3. A radar demonstrator with scalable subarray modules simultaneously realizes range sensing and azimuth recognition for pulsed radar configurations. Captured by the suggested software-defined pulsed radar, a complete range–azimuth figure with a 1 km maximum observation range can be displayed within 150 ms under the current implementation.

Keywords: antenna-in-package (AiP); complementary metal-oxide-semiconductor (CMOS); digital beamforming (DBF); digital array radar; phased array; pulsed radar; radar signal detection; system-on-chip (SoC); transceiver; X-band

1. Introduction

Phased array technology has evolved over the past few decades. Categorized into analog, subarray digital, and element-level digital topologies [1], phased array systems accomplish spatial filtering and power combination through the synchronous excitation of each radiating element. In particular, element-level digitization provides opportunities for sophisticated signal processing and simultaneous reception of multiple beams, which are absent in both analog and subarray digital topologies [2–8]. Directional and fast-scanning characteristics render phased arrays an attractive approach for various applications, including field surveillance, wireless communication, and electronic warfare [9]. However, the price and occupied volume of beamforming systems limit their ubiquity [10]. From pioneering studies, potential methods of achieving low-cost compact-size element-level digital phased arrays can be classified into three categories: system-on-chip (SoC) integration, advanced packaging, and subarray module miniaturization.

First, advancements in semiconductor technology have enabled the integration of radio frequency (RF) front-end, baseband filtering, and data conversion systems on a single chip at competitive prices and reasonable performance levels [11,12]. Researchers developed a complementary metal-oxide-semiconductor (CMOS) Ku-band transceiver for frequency-modulated continuous-wave (FMCW) radar imaging [13,14], and in another study, they proposed an X-band CMOS four-channel phased array transceiver for synthetic aperture radar (SAR) imaging [15]. Through the use of RF phase-shifting and active switches, an eight-channel silicon-germanium (SiGe) receiver can configure the total number of simultaneous beams for a 2–16 GHz operating frequency [16]. From another perspective, an X-band CMOS FMCW radar transceiver including an on-chip quasi-circulator offers a single-antenna interface to reduce the system form factor [17,18].

Second, advanced packaging methods can enable the integration of active integrated circuits (ICs), passive filtering components, and embedded antennas into a limited space. Under the antenna-in-package scheme, a study presented an X-band active antenna module [19] with a commercial gallium arsenide (GaAs) transceiver, and other studies reported a 64-element phased array module for 5G wireless communication [20]. Another 3-D stacked packaging technique with a thermal-conductive aluminum substrate was developed for producing a high-power X-band T/R module and achieved a compact size of $20 \times 20 \times 3.7$ mm^3 [21].

Finally, because of the repetitive arrangement in phased array systems, subarray module miniaturization is a good approach for reducing the overall occupied volume. Although the dimensions of an antenna are fundamentally associated with operating frequency and available bandwidth, the use of embedded planar antennas in the tile array configuration can reduce the system form factor [22–25]. For example, studies have reported that incorporating a SiGe transceiver, a CMOS data converter, and a commercial digital processor can produce a compact subarray module tile for a digital beamforming system [26–28]. From another perspective, a study reported a Ku-band multiple-input multiple-output FMCW radar that enables a reduction in the total number of installed antennas through virtual array synthesis and demonstrates high-resolution 3-D imaging capability [29].

Studying essential considerations for realizing phased array hardware can facilitate the development of a low-cost compact-size digital beamforming system; additionally, the design methodology for signal processing back-end systems can be explored. Although element-level digitization enables the development of software-defined phased array radars, extremely high input/output (I/O) bandwidths are occupied by digitized T/R waveforms; consequently, powerful signal processors are required for real-time computing. Benefiting considerably from the shrinking CMOS technology node, field-programmable gate arrays (FPGAs) are equipped with excellent performance per watt, high-speed interfaces, and parallel computing capabilities at reasonable costs and time to market [30,31]. With features of low latency and reduced data throughput, a hierarchical digital beamforming topology [2,9,10,32] has been implemented in FPGAs and employed for real-time radar imaging.

With element-level digitization, digital beamformers can calibrate deterministic phase errors due to unbalanced routing length but still leave random phase fluctuations derived from the system-level clock distribution network. To further improve overall sidelobe rejection, we propose a novel periodic pulse injection circuit embedded in the CMOS transceiver to solve random phase fluctuations. Using this technique, we accomplished clock synchronization with non-overlapping frequency in our radar system using only two RF reference signals. This paper presents the design and implementation of an X-band element-level digital phased array radar utilizing fully integrated CMOS transceivers. We co-designed the radar system and the full-custom CMOS transceiver chips to optimize overall imaging performance. An 8–10 GHz transceiver SoC fabricated in 65 nm CMOS technology offers baseband filtering, frequency translation, and global clock synchronization through the proposed periodic pulse injection technique. Through CMOS SoC integration, antenna-in-package design, and tile array configuration, phased array systems can be miniaturized, and the reduced form factor can bypass volume limitations for various scenarios. The implemented 16-element subarray module with a 1×16 configuration accomplishes digital beamforming, back-end computing, and dc–dc conversion with a $317 \times 149 \times 74.6$ mm^3 size. The subarray modules provide scalability to create a large-scale software-defined beam-steering system through the suggested hierarchical back-end topology, which supports a maximum of 256 radiating elements in the current design.

This paper is an extension of a conference paper that has been accepted [33]. In this paper, beyond the circuit implementation of a fully integrated CMOS transceiver, an additional 16-element radar demonstrator with board-level-integrated front-end amplifiers is proposed, and the results of implementation and verification are presented. The rest of this paper is organized as follows. Section 2 describes the system architecture of the proposed digital phased array radar. Section 3 presents the detailed design methodology of the X-band CMOS transceiver. System integration of the suggested digital phased array radar is explained in Section 4. The experimental results of the CMOS transceiver SoC and radar demonstrator are reported in Section 5, including a performance comparison with other state-of-the-art works. Finally, the study conclusion is provided in Section 6.

2. System Architecture of the Proposed Radar

Figure 1 depicts the system architecture of the proposed element-level digital phased array radar. Consisting of two 16-element subarray modules and a hierarchical digital back-end, the radar demonstrator generates constructive interference at expected beam-steering directions and achieves range sensing under pulsed radar configuration. One of the subarray modules is operated in transmitting (TX) mode, and the other is operated in receiving (RX) mode. The TX subarray module comprises 16 linear-polarized patch antennas, 16 off-the-shelf front-end GaN power amplifiers, and 16 CMOS transceiver chips in a QFN package. The RX subarray module comprises 16 linear-polarized patch antennas, 16 commercial GaAs low-noise amplifiers, and 16 CMOS transceiver chips in a QFN package. The Qorvo TGA2598-SM GaN driver amplifier [34] and TGA2512-SM GaAs low-noise amplifier [35] are integrated into the RF signal path for the transmitting and receiving subarray module, respectively. Baseband filtering, frequency translation, and global clock synchronization are provided by 8–10 GHz transceiver SoCs through the proposed periodic pulse injection technique. In the transmitting (TX) subarray module, 16 CMOS transceiver chips are only operated in transmitting mode; in the receiving (RX) subarray module, 16 CMOS transceiver chips are only operated in receiving mode. On-chip 10-bit 100 MS/s analog-to-digital converters (ADCs) and 10-bit 100 MS/s digital-to-analog converters (DACs) are responsible for the quadrature-phased baseband signal digitization. The first-hierarchy FPGA is responsible for T/R module control, transmitted waveform excitation, received data acquisition, and digital signal processing (DSP). Digital beamformers implemented in first-hierarchy FPGAs collect relative magnitude and phase information from each array element and calculate essential complex-valued coefficients for the desired beamforming orientation. The second-hierarchy FPGA communicates with the computer through Ethernet and manages data transfer across the array hierarchy over SerDes interfaces. The Analog Devices' EVAL-ADF5355 PLL evaluation kit [36] is used to generate the necessary local oscillator (LO) reference signals at frequencies of $f_0/8$ and $2f_0$; f_0 represents the carrier frequency of each array element and was set to 8.5 GHz in this study for the proposed demonstrator. Moreover, FPGAs administer global timing regulation through a 100 MHz global clock and a global trigger. Four reference signals are distributed from the module input to each array element using active clock trees; system-level synchronization can be realized through these reference signals.

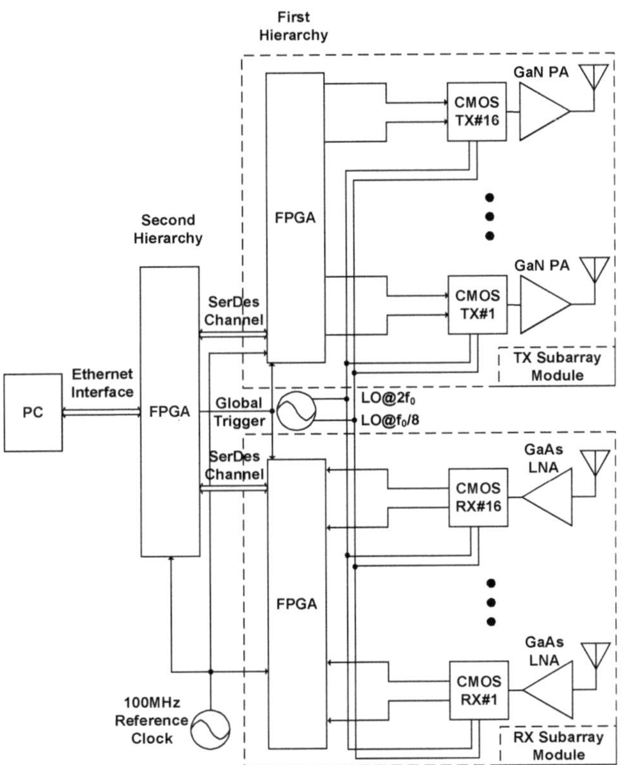

Figure 1. System architecture of the proposed digital phased array radar demonstrator.

3. Circuit Implementation of X-Band Transceiver

The proposed X-band CMOS transceiver serves as a hardware interface between the radiating antenna and back-end digital processor and provides analog signal processing and baseband-to-RF frequency translation. This section describes the circuit design and implementation of the transceiver SoC. We designed and simulated the CMOS circuits with the help of Cadence Virtuoso Analog Design Environment. Figure 2 illustrates a simplified block diagram of the CMOS transceiver circuitry, including the transmitter, receiver, local oscillator (LO) distribution, and quadrature clock generation. Packaging procedures for the reported CMOS chips are introduced in Section 4.1, and experimental results are summarized in Section 5.

3.1. Transmitter Design

As depicted in Figure 2, the implemented transmitter comprises baseband circuitry, a quadrature clock generator, a single-sideband (SSB) mixer, and a power amplifier. Off-chip signal processors deliver digitized data to on-chip 10-bit 100 MS/s current-steering DACs for in-phase and quadrature channels, the circuit schematic of which is presented in Figure 3. Incoming quadrature baseband signals undergo analog filtering and single-sideband modulation through the second-order low-pass filter and the SSB mixer. Subsequently, the power amplifier drives upconverted single-sideband signals to the external 50 Ω load. Quadrature clock generation circuitry executes frequency division and provides orthogonal LO signals at a frequency of f_0 to mixer switching pairs through $2f_0$ and $f_0/8$ reference signals; f_0 denotes the carrier frequency of each transceiver SoC and ranges from 8 to 10 GHz. The design process of the clock generator is reported in Section 3.3.

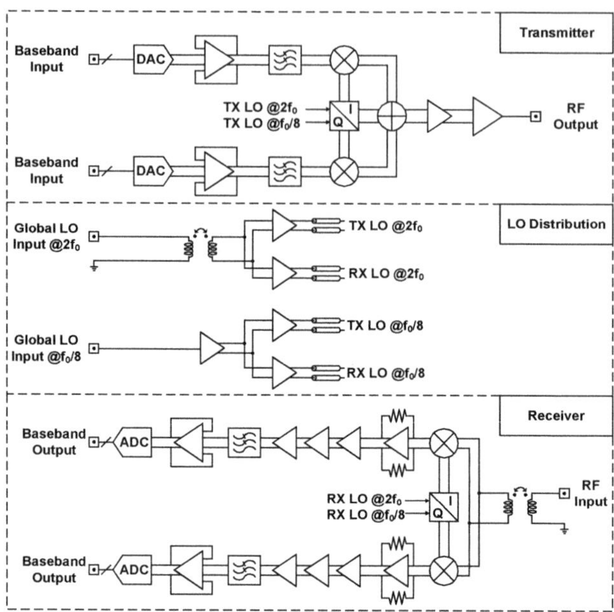

Figure 2. Simplified block diagram of the X-band CMOS transceiver.

The complete circuit schematic of an active single-sideband mixer is introduced in Figure 4. Differential baseband signals derived from the preceding low-pass filter are commuted in the form of current by the operational transconductance amplifier; a resistive source degeneration technique is employed to improve amplifier linearity, and shunt capacitors at the output nodes suppress LO feedthrough caused by mixer switching. Two current-driven double-balanced mixer pairs upconvert incoming baseband signals and achieve single-sideband modulation through the combination of quadrature RF currents; a resonant tank is applied to absorb the capacitive load of subsequent circuits. The subsequent power amplifier is driven by the current-mode logic (CML) buffer for SSB mixer differential outputs.

The integrated class-AB power amplifier boosts the single-sideband signal and delivers necessary RF power to the off-chip 50 Ω load, the circuit schematic of which is presented in Figure 5. A cascaded two-stage topology [37] that consists of a driver stage and power stage is adopted in the power amplifier design. Both the driver stage and power stage are implemented in the cascode common-source configuration to strengthen the available voltage swings associated with transistor breakdown voltage. The input matching network and the intermediate matching network apply resonant circuits to suppress out-of-band aggressors and absorb the transistor parasitic capacitors with compact areas. The power stage amplifier tolerates much higher voltage swing through the use of thick-gate-oxide common-gate devices and a 2.5 V drain bias voltage; subsequently, the transformer-based output network executes power combination and impedance conversion. The implemented power amplifier was determined to achieve a simulated drain efficiency of 18% and an output power of 18.3 dBm at 10 GHz.

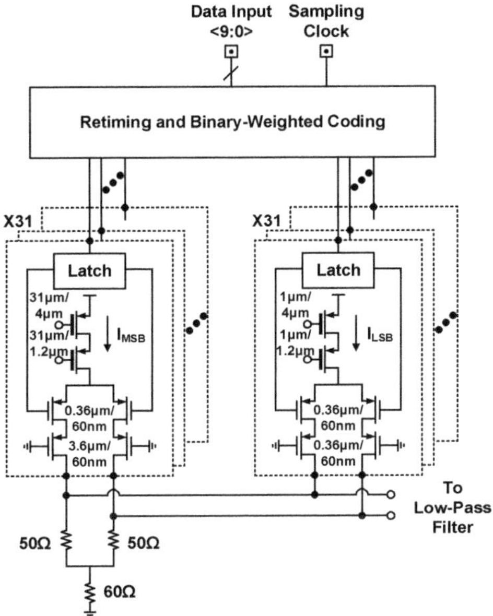

Figure 3. Circuit schematic of the digital-to-analog converter.

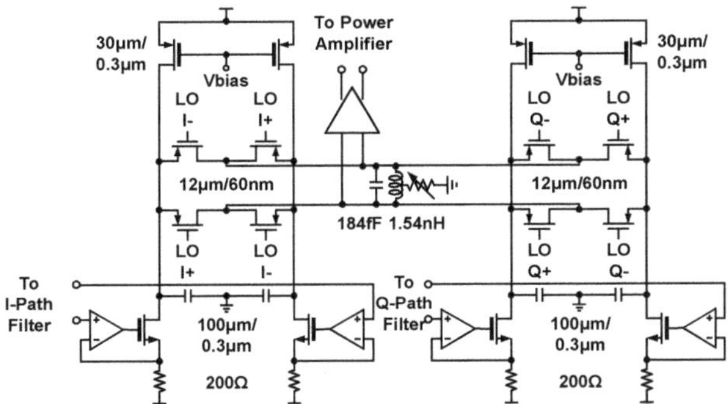

Figure 4. Circuit schematic of the single-sideband mixer.

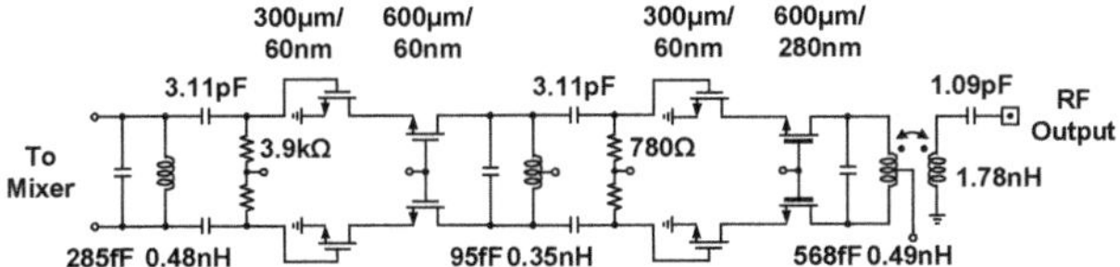

Figure 5. Circuit schematic of the class-AB power amplifier.

3.2. Receiver Design

As shown in Figure 2, the implemented direct-conversion quadrature receiver comprises a mixer-first RF front-end, a quadrature clock generator, and baseband circuitry. The incoming RF input signal undergoes frequency translation and analog signal processing; subsequently, the downconverted quadrature baseband signals are distributed to the off-chip signal processor through the digital interface. Both the transmitter and receiver are equipped with separate quadrature clock generation circuits to perform frequency division. The design is described in detail in Section 3.3.

The circuit schematic of the mixer-first RF front-end is illustrated in Figure 6. A transformer-based input matching network is adopted for single-ended-to-differential (S2D) conversion and dc current isolation. Current-driven passive mixers downconvert received RF signals and can tolerate simultaneous transmitter leakage. The transimpedance amplifiers (TIAs) provide current-to-voltage conversion for downconverted waveforms and drive subsequent baseband circuitry activity. However, in the absence of a low-noise amplifier, the input impedance of the TIAs is upconverted by a bidirectional passive mixer and can be directly observed at the receiver RF input port. Analyses of the impedance transparency feature in a mixer-first architecture are presented in [38,39]. Through the reported analysis procedures, the design of impedance matching and baseband load entails making tradeoffs among conversion gain, noise figure, and mixer LO driver power dissipation to achieve the desired 8–10 GHz operating frequency.

Figure 7 depicts the baseband circuitry [18]. Downconverted received signals undergo amplification and filtering through the programmable gain amplifier (PGA) and low-pass filter. With dc offset cancellation and feedback resistor control, a three-stage PGA accomplishes a simulated 60 dB gain tuning range and 1 dB gain tuning step, followed by a second-order multiple feedback low-pass filter with digital-controlled reconfigurable feedback resistors and capacitors. The overall filtering bandwidth is either 20 MHz or 40 MHz. Off-chip signal processors use sampled data from on-chip 10-bit 100 MS/s nonbinary redundant successive approximation register analog-to-digital converters (SAR ADCs) for in-phase and quadrature channels, the circuit schematic of which is presented in Figure 8.

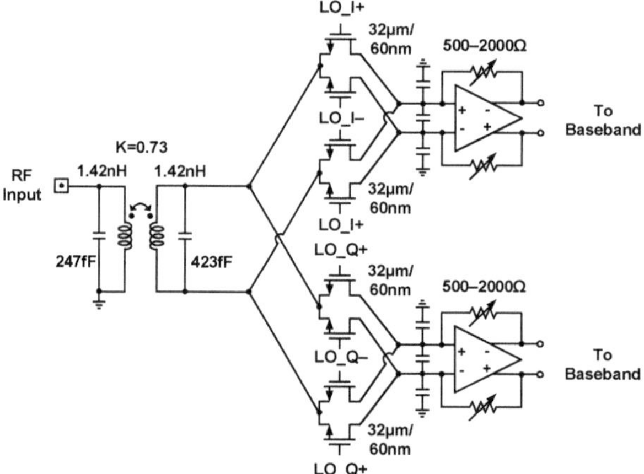

Figure 6. Circuit schematic of the mixer-first receiver front-end.

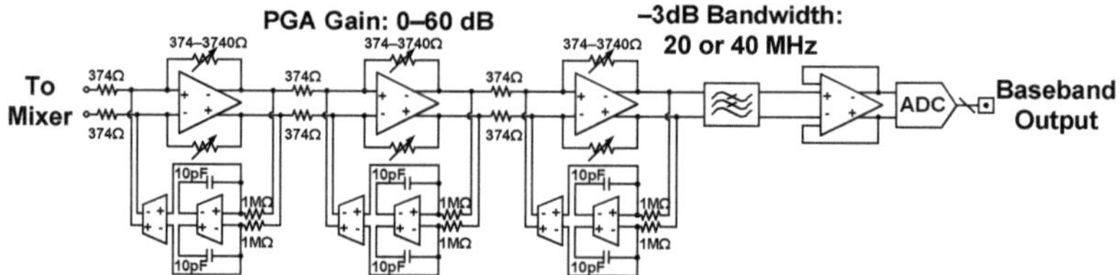

Figure 7. Circuit schematic of the baseband circuitry.

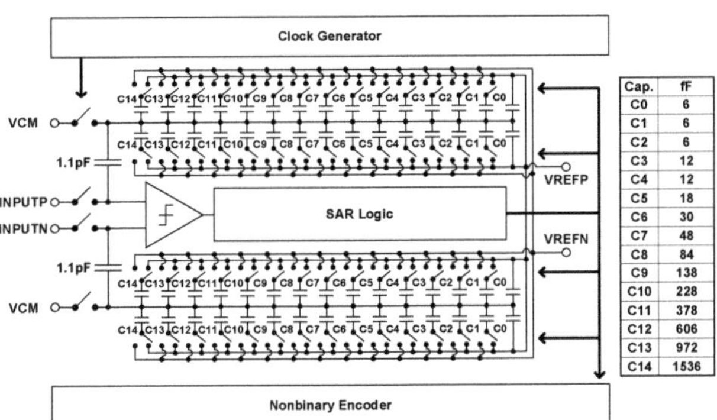

Figure 8. Circuit schematic of the analog-to-digital converter.

3.3. LO Distribution and Quadrature Clock Generation

In this study, two synchronized external LO signals at frequencies of $2f_0$ and $f_0/8$ were delivered to both the transmitter and receiver equipped with quadrature clock generation circuitry; f_0 denotes the carrier frequency of each transceiver SoC and ranges from 8 to 10 GHz. Clock generation circuits execute frequency division and provide orthogonal LO signals at a frequency of f_0 to the in-phase and quadrature (I/Q) mixers. Element-level digitization can calibrate deterministic phase errors while leaving random uncertainties, which are partially derived from temperature-induced phase fluctuations in the active LO distribution network and 180° phase uncertainty in the frequency division process. These problems can be solved by applying bandgap reference voltage biasing and the proposed periodic pulse injection technique. Detailed design processes are described as follows.

Figure 9a,b depict the circuit implementation of the LO distribution network composed of a $2f_0$ buffer chain and $f_0/8$ buffer chain. An on-chip bandgap reference circuit provides temperature-insensitive biasing voltages to CML buffers. Moreover, an inductive gain-peaking technique [40] is also introduced to compensate for power loss derived from signal routings. Conversely, with a relatively low frequency, the $f_0/8$ buffer chain applies a CML-based active S2D circuit and 50 Ω load CML buffers directly. Figure 10a,b present the simulated relationship between the relative phase and environment temperature for 16–20 GHz LO signals. In the 25–125 °C operating temperature range, the CML buffer chain exhibits 13.7° and 4.4° maximum phase deviations under activating and deactivating bandgap reference biasing configurations, respectively. Applying bandgap reference biasing to CML buffers is confirmed to reduce temperature-induced phase fluctuation in the LO path.

The $2f_0$ fed signal, along with the frequency division process, prevents substantial LO leakage at transceiver RF ports. However, the divide-by-2 process introduces 180° phase uncertainty to quadrature clock signals and causes difficulty in array element synchronization; that is, the output signals of the two synchronized CMOS SoCs may be either in phase or out of phase once the CMOS transceivers restart. Consequently, beamforming fails to operate because of arbitrary constructive or destructive interference. To solve the 180° phase ambiguity, our study proposes a quadrature clock generation circuit with a periodic pulse injection technique. A simplified block diagram is presented in Figure 11, where signals are presented in a single-ended form for clarity. A CML quadrature divider accomplishes frequency division and orthogonal clock generation from the buffered $2f_0$ signal delivered by the LO distribution network. Programmable CML delay lines drive either the transmitter or receiver mixers and enable I/Q phase imbalance calibration by altering the relative delay between I/Q clocks. Similarly, the CML circuits employ proportional-to-absolute-temperature biasing voltages to avoid temperature-induced phase fluctuation in the LO path.

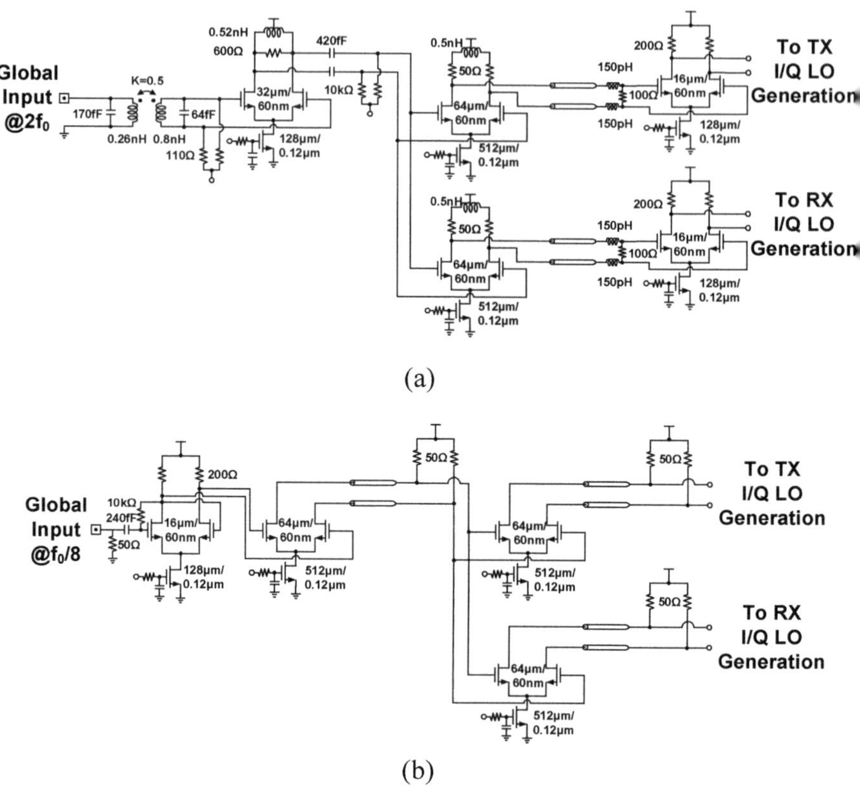

Figure 9. Circuit schematic of the clock distribution network composed of (**a**) $2f_0$ LO buffer chain and (**b**) $f_0/8$ LO buffer chain.

An additional $f_0/8$ path is introduced to eliminate phase uncertainty. Initially, the buffered $f_0/8$ signal undergoes a retiming process by the $2f_0$ signal for clock edge alignment. Subsequently, a periodic pulse signal with an adjustable duty cycle and delay is produced by the pulse generator circuit from the $f_0/8$ retiming clocks and is fed into the CML quadrature divider, as depicted in Figure 12. Finally, the pulse signal turns on a pMOS device and injects current into the divider output node periodically. The injected current increases the output voltage, consequently affecting the absolute phase of the quadrature signals, depending on the selected injection point. Simulated 10 GHz LO waveforms along with a 1.25 GHz pulse signal are depicted in Figure 13. Comparing the activating injecting case with the deactivating injecting case reveals a 180° phase difference after the pulse signal triggers the divider circuitry. Thus, the proposed quadrature divider can regulate absolute phases of output clocks through a chosen injection point (i.e., the relative delay between the $f_0/8$ retiming clock and generated pulse), and a well-defined relative phase relationship between any two transceiver chips is achievable. Furthermore, to prevent metastability in the retiming process, the proposed quadrature clock generation circuitry adopts an additional path to monitor the clock edge relationship between the $2f_0$ and $f_0/8$ signals.

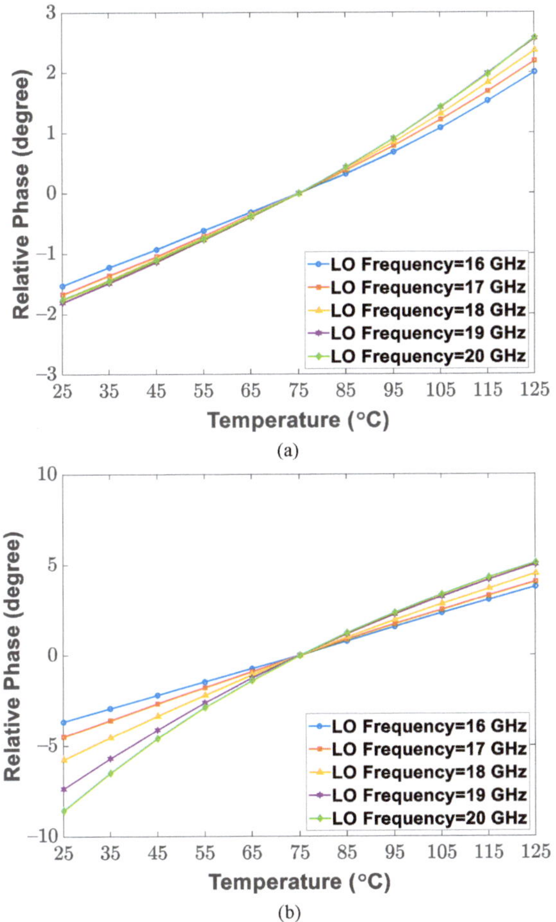

Figure 10. Simulated relationship between the relative phase of LO signal and environment temperature under (**a**) activating and (**b**) deactivating bandgap reference voltage biasing.

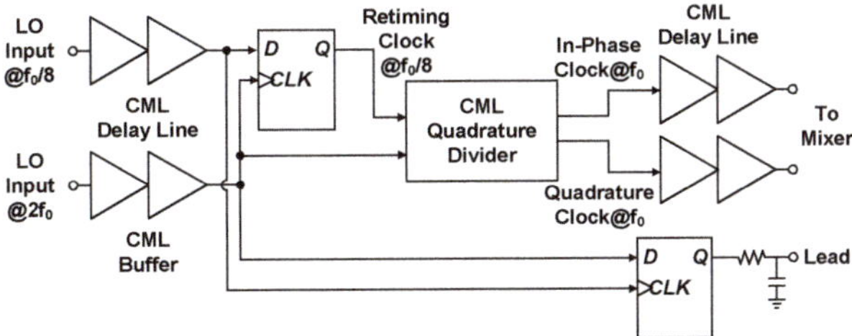

Figure 11. Simplified block diagram of the proposed quadrature clock generation circuit.

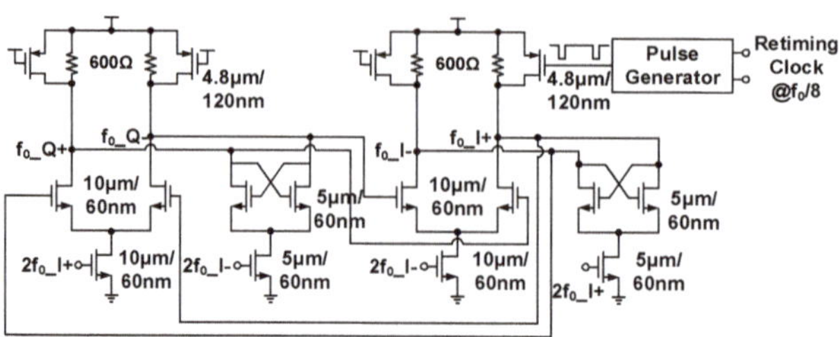

Figure 12. Circuit schematic of the CML quadrature divider with periodic pulse injection.

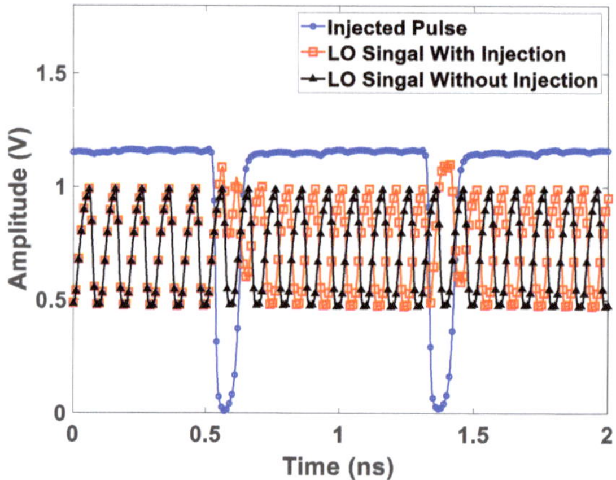

Figure 13. Simulated 10 GHz LO waveforms under deactivating and activating 1.25 GHz pulse injection.

4. System Integration of the Proposed Radar

This section describes the system integration of the proposed element-level digital phased array radar, which incorporates radiating element packaging and subarray module assembly. The corresponding experimental results are summarized in Section 5.

4.1. Array Element Packaging

An array element is the basic functional unit in a beamforming system and is responsible for transmitting or receiving expected signals through an embedded antenna. Each array element comprises a T/R module and the corresponding part of the interposer board on which the T/R module is mounted.

Our previous work [33] reported the antenna-in-package (AiP) method for T/R module packaging. The integration of the compact AiP module and high-performance front-end amplifiers can further improve the radiation power and noise figure at the cost of increased heat density. For example, the expected power density of the Qorvo TGA2598-SM GaN driver amplifier [34] is 0.48 W/mm^2. Because the top layer is occupied by a patch antenna and the bottom layer is epoxy encapsulated, external heat sinks are unavailable on the surface of the AiP module [33]; consequently, the heat generated by the active components spreads to the poor-thermal-conductivity ceramic substrate and the interposer board through solder balls. Additional off-chip front-end amplifiers increase the difficulty of heat dissipation in an ultrathin tile array. An alternative method for board-level integration of the CMOS T/R quad flat no-lead (QFN) module, off-the-shelf front-end amplifier, and linear-polarized patch antenna was adopted in this study and is explained as follows.

To ensure expedited shipping and preliminary verification processes on a limited budget, only an FR4 substrate was chosen for the QFN module and the front-end interposer board. As shown in Figure 14, the T/R QFN module includes a wire-bonded CMOS transceiver SoC and surrounding filtering capacitors; power rails, digital interfaces, baseband analog signals, and RF LO signals come into contact with the interposer board through bonding pads. On-chip 10-bit 100 MS/s ADCs and 10-bit 100 MS/s DACs are responsible for the baseband signal digitization. Under epoxy encapsulation, a packaged QFN module is $9.35 \times 7.75 \times 1.39$ mm^3 in size. Figure 15a,b depict the side view and top view of a 1×16 front-end interposer board including 16 array elements. Each array element is composed of the T/R QFN module, commercial front-end amplifier, and re-designed linear-polarized patch antenna. The edge-to-edge dimension of embedded patch antennas is set to 17.6 mm to avoid a grating lobe at 8.5 GHz operating frequency. The ground layer in the middle of the FR4 substrate serves as an antenna ground and provides electromagnetic interference shielding for RF signals. Due to limited resources, only one-layer Rogers RO4350B is embedded into the front-end interposer board to improve insertion loss between the commercial front-end amplifier and the patch antenna.

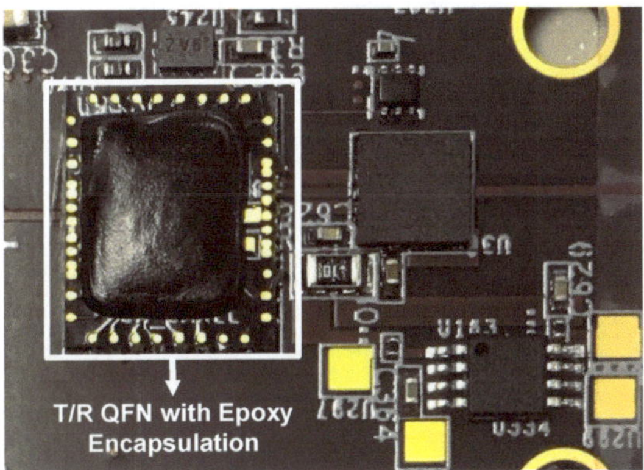

Figure 14. Top view of a CMOS T/R QFN module.

Detailed antenna dimensions are presented in Figure 16. Moreover, Figure 17 depicts the simulated input reflection coefficient and radiation efficiency of the patch antenna incorporating a vertical via and horizontal microstrip feed line. We designed and simulated the antenna with the help of Keysight Advanced Design System (ADS). Occupying a 7.8–9.9 GHz bandwidth, the implemented patch antenna achieves 78.4% and 52.2% radiation efficiency at the 8.5 GHz and 9.9 GHz operating frequencies, respectively. Figure 18a,b illustrate the far-field cut of the simulated radiation pattern for the E-plane and H-plane, respectively. Equipped with a maximum antenna gain of 4.741 dBi and 3.737 dBi for the E-plane and H-plane, the patch antenna accomplishes a half-power beamwidth (HPBW) of 78° for the E-plane and an HPBW of 120° for the H-plane. Furthermore, we simulated the radiation pattern of the implemented patch antenna array and present the simulation results of the 16-element antenna array at the 8.5 GHz center frequency. Figure 18c,d depict the simulated E-plane and H-plane radiation patterns for the 16-element antenna array under rectangular window extraction and 0° beamforming angle. For the E-plane cut, the 16-element antenna array achieved a maximum gain of 16.873 dBi and HPBW of 78°; for the H-plane cut, that is to say, our desired beamforming direction, the 16-element antenna array achieved a maximum gain of 15.879 dBi, HPBW of 6°, a peak sidelobe ratio of −14.52 dB, and a 12.142 dB gain improvement due to beamforming.

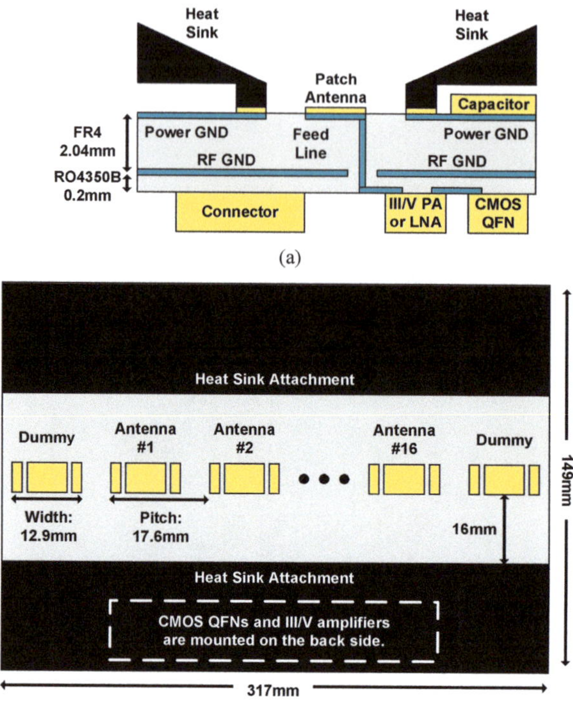

Figure 15. (**a**) Side view and (**b**) top view of a 1×16 interposer board.

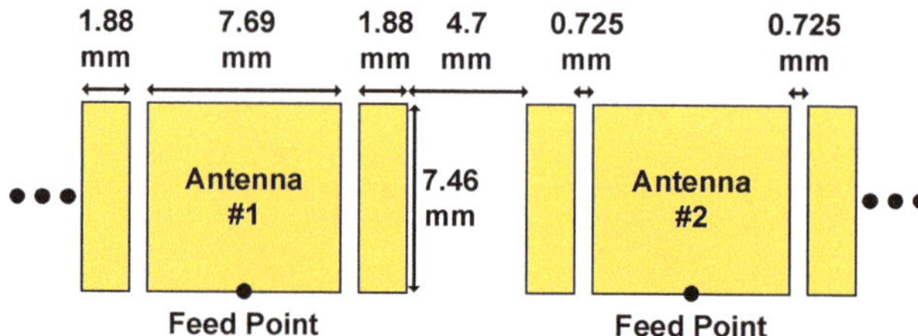

Figure 16. Dimensions for the patch antenna.

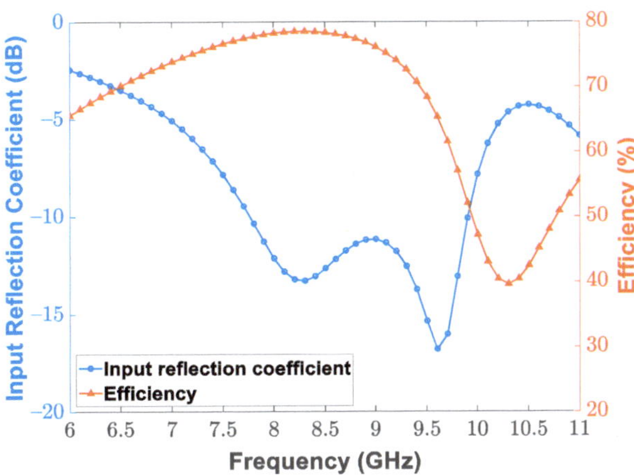

Figure 17. Simulated input reflection coefficient and radiation efficiency of the patch antenna.

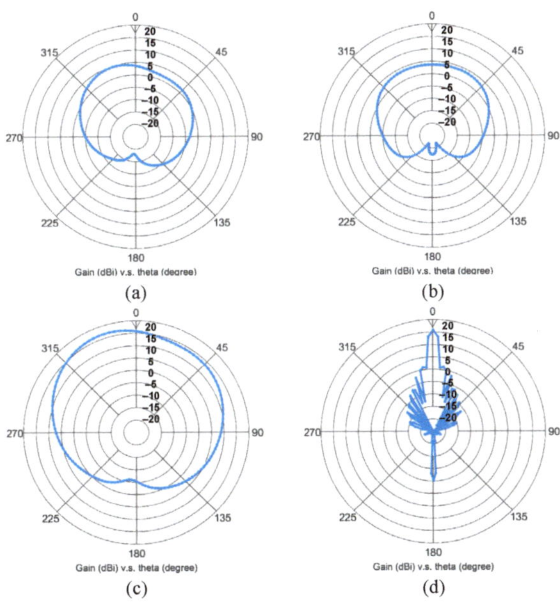

Figure 18. Simulated (**a**) E-plane and (**b**) H-plane radiation pattern for the single patch antenna; simulated (**c**) E-plane and (**d**) H-plane radiation pattern for the 16-element patch antenna array.

4.2. Assembly of 1 × 16 Subarray Module

Figure 19a,b depict the implemented 1 × 16 subarray module. The 1 × 16 subarray module comprises a front-end interposer board, a signal processor board, and two power module boards. In contrast to the 4 × 4 configuration [33], the 1 × 16 system replaces the T/R AiP module with the combination of the T/R QFN module, off-the-shelf front-end amplifier, and redesigned patch antenna for each radiating element. Sixteen equally spaced radiating elements execute baseband-to-RF frequency translation and produce constructive interference at the expected beam-steering directions. The signal processor, Xilinx FPGA SoC, communicates with radiating elements through physical routings in the interposer board and executes cross-hierarchy data transfer over the SerDes interface. Moreover, as mentioned in Section 3.3, each CMOS transceiver requires two external LO signals at frequencies of $f_0/8$ and $2f_0$ to eliminate phase uncertainty and generate quadrature clocks; the FPGAs also administer global timing regulation through a 100 MHz global clock and a global trigger. Four reference signals are distributed from the module input to each array element using active clock trees; system-level synchronization can be realized through these reference signals. Additionally, two customized power module boards generate essential supply voltages from a single 28 V dc input rail with high-efficiency dc–dc conversion. In our current implementation, the 1×16 subarray module achieves a size of $317 \times 149 \times 74.6$ mm^3, including a heat sink mounted on the interposer board for convection cooling. An improved thermal management system incorporating heat pipes and heat sinks may help further reduce the module volume.

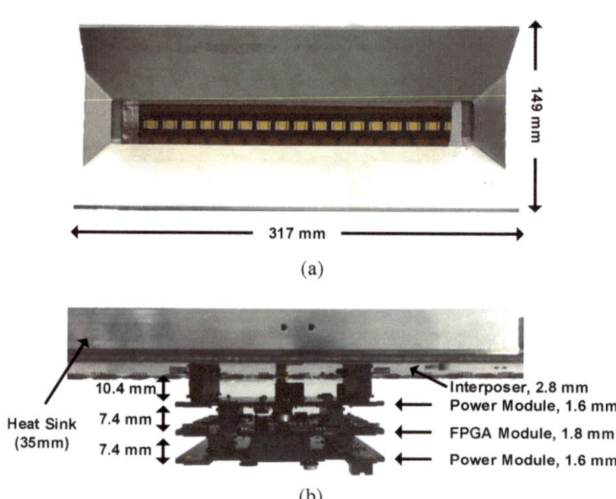

Figure 19. (**a**) Top view and (**b**) side view of a 1×16 subarray module.

5. Experimental Results

This section presents experimental results, including the circuit performance of the X-band CMOS transceiver and functional verification of the 1×16 subarray modules, for the proposed element-level digital phased array radar. After fabricating the transceiver SoCs, wafer-level measurements were performed to evaluate circuit specifications, and then qualified chips were packaged according to the assembly procedures detailed in Section 4. Subsequently, a finished radar demonstrator composed of two 1×16 subarray modules was subjected to power-on calibration, antenna pattern measurement, and range sensing experimentation.

5.1. CMOS Transceiver

This paper presents a fully integrated X-band transceiver SoC fabricated using 65 nm CMOS technology. Figure 20 presents a micrograph of the implemented chip. Occupying a chip area of 2.4×2 mm^2, the CMOS transceiver consumes a total power of 1.45 W in transmitting mode and 1.41 W in receiving mode on 1.2, 1.8, and 2.5 V dc power rails. The power consumption of key transmitting and receiving blocks is illustrated in Figure 21a,b. The substantial power dissipation is mainly derived from power-hungry CML circuits and on-chip low-dropout regulators (LDOs), which are employed to reduce temperature-induced phase fluctuations and eliminate system-level switching power supply noise, respectively.

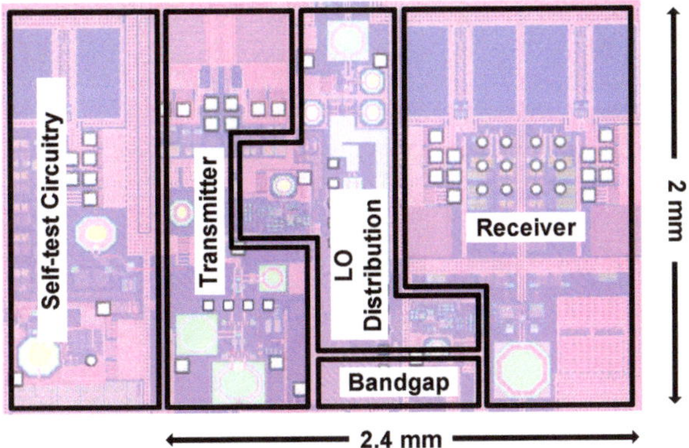

Figure 20. Micrograph of the implemented CMOS transceiver chip.

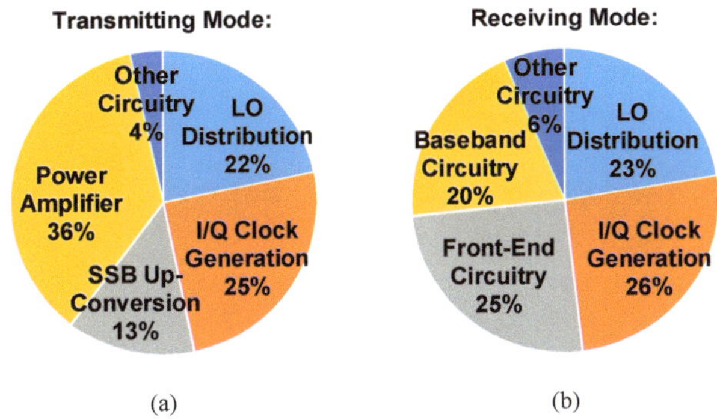

Figure 21. Power consumption of the CMOS transceiver operated in (**a**) transmitting mode and (**b**) receiving mode.

Circuit measurement specifications of the CMOS transceiver SoC under the wafer probing scheme are summarized as follows. By utilizing the mixer-first RF front-end, the receiver chain suppresses out-of-band interference through narrow-band noise performance and input matching; band switching operation is realized by controlling the reference frequency of the external LO signal. Figure 22a depicts the input reflection coefficient of the implemented receiver with a 50 Ω source impedance at LO frequencies ranging between 16 and 20 GHz; a Keysight E8257D analog signal generator provides −10 dBm RF power at the given LO frequencies. The minimum input reflection coefficient is achievable with a frequency that is near the given carrier frequency (i.e., half of the LO frequency), and the magnitude of the reflection coefficient increases as the measured frequency moves away from the selected operation band. The double-sideband (DSB) noise figure and conversion gain are illustrated in Figure 22b,c. The implemented receiver chain achieves a 14.6 dB DSB noise figure and maximum DSB conversion gain of 57.2 dB within the 40 MHz baseband bandwidth at a 20 GHz LO frequency. For linearity measurements, single-tone tests were performed at an 18 MHz baseband frequency and different LO frequencies. Figure 23a,b display measurement results of the receiver 1 dB compression point (P_{1dB}) under the lowest conversion gain of 3.8 dB and highest conversion gain of 57.2 dB. The input P_{1dB} reaches −14 and −66 dBm at an 18 GHz LO frequency for the lowest gain and highest gain cases, respectively. Moreover, two-tone tests were conducted at baseband frequencies of 18 and 20 MHz. The input third-order intercept points (IIP_3) of the implemented receiver are presented in Figure 24a,b, including those for both the lowest conversion gain and highest conversion gain cases. IIP_3 reaches -5.9 dBm for the lowest conversion gain case and -57.2 dBm for the highest conversion gain case at the 18 GHz LO frequency.

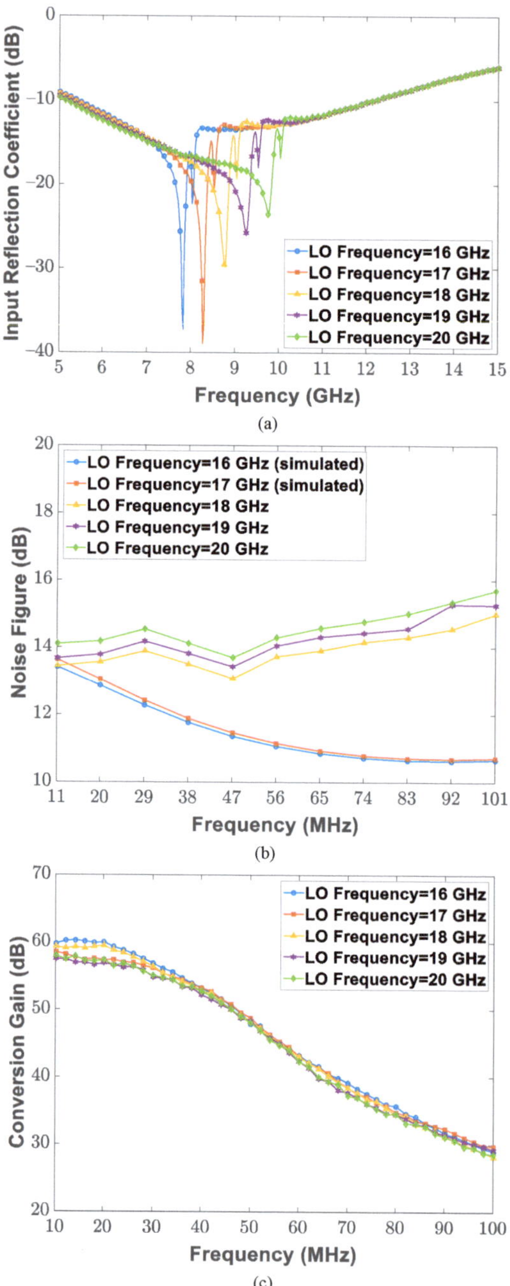

Figure 22. Circuit performance of the implemented receiver, including (**a**) input reflection coefficient, (**b**) noise figure, and (**c**) conversion gain.

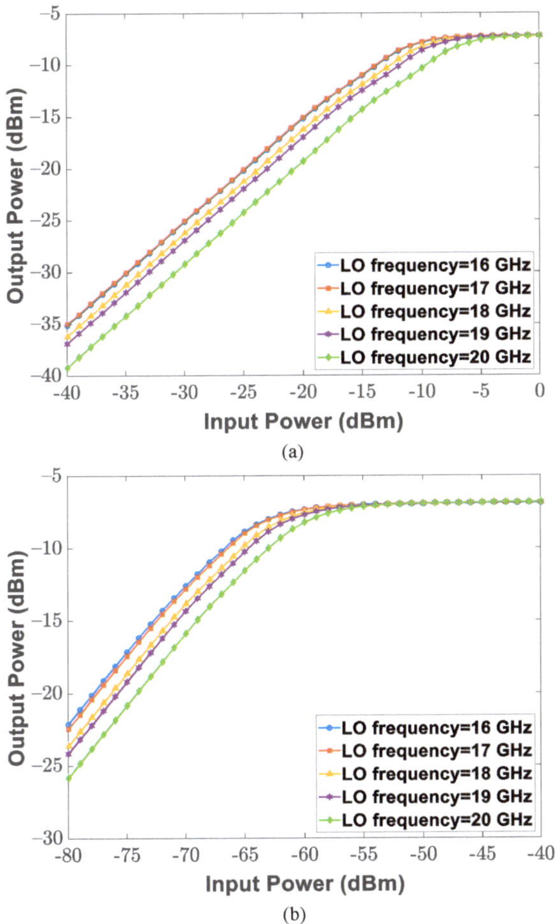

Figure 23. The 1 dB compression point (P_{1dB}) of the implemented receiver for (**a**) lowest conversion gain and (**b**) highest conversion gain.

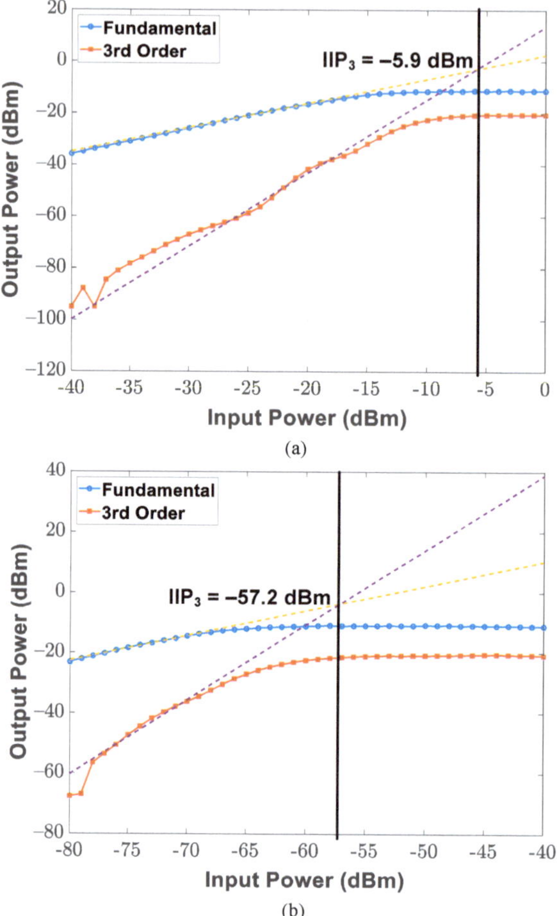

Figure 24. Third-order intercept point (IP_3) of the implemented receiver for (**a**) lowest conversion gain and (**b**) highest conversion gain.

The performance of the transmitter incorporating a spurious level, carrier leakage, saturated output power, and pulse-mode operation capability was examined. Figure 25a shows the output spectrum of the transmitter at 19 GHz LO and 5 MHz baseband input frequencies; lower-sideband (LSB) upconversion was demonstrated in the experiment. The implemented transmitter achieves a spurious level of −30.35 dBc and a carrier leakage of −33.95 dBc. The measured and simulated saturated output power among 8–10 GHz are presented in Figure 25b. In an 8–10 GHz operating frequency range, the CMOS power amplifier exhibits a peak saturated output power of 17.96 dBm and a peak drain efficiency of 11.9% at 8.5 GHz.

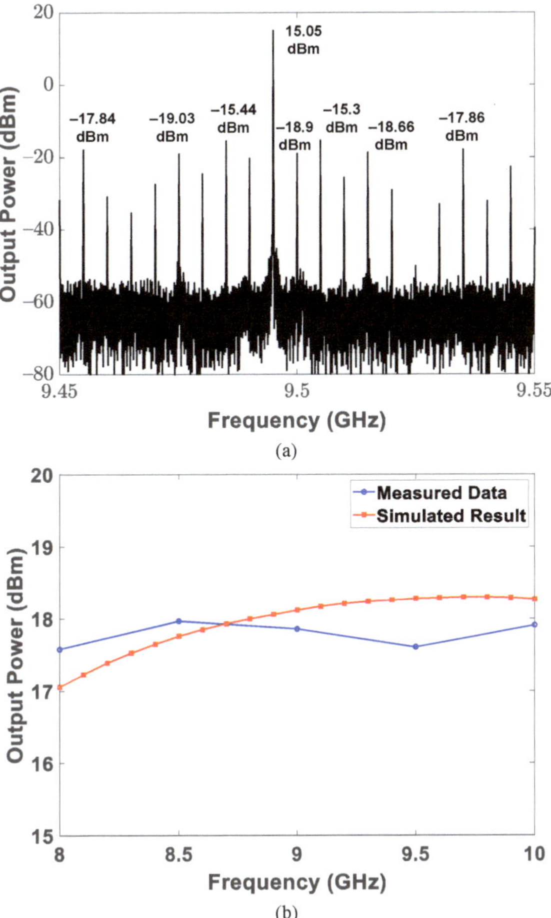

Figure 25. Circuit performance of the implemented transmitter, including (**a**) output spectrum at 19 GHz LO and 5 MHz baseband input frequencies, (**b**) saturated output power.

5.2. Radar Demonstrator with 1 × 16 Subarray Modules

This paper presents a 16-element radar demonstrator incorporating a hierarchical digital back-end and two 1 × 16 subarray modules with board-level integrated RF front-end amplifiers. As depicted in Figure 1, one of the 1 × 16 subarray modules is operated in transmitting mode, and the other is operated in receiving mode. Additionally, the module operated in transmitting mode is vertically arranged for elevation recognition, whereas that operating in receiving mode is horizontally configured for azimuth recognition through simultaneous multiangle scanning [41]. Digital beamformers implemented in the first-hierarchy FPGA are responsible for phase shifting and magnitude amplification, depending on the beam-steering orientation. After the successful execution of power-on calibrations [33], the performance of the digital beamformer was evaluated through the verification of antenna patterns. The measured E-plane antenna patterns for the demonstrator operating in receiving mode are presented in Figure 26, with the mainlobe being steered from −45° to 45° at 15° steps. By applying the kaiser window with a shape factor of 3, this study determined that the digital beamformer achieved an average sidelobe level of −25.64 dB and a half-power beamwidth of 8.06° at a 0° beamforming angle; additionally, at the same angle, an average sidelobe level of −13.11 dB and a half-power beamwidth of 6.27° were achieved without tapering. Benefiting from the digital beamforming scheme, the proposed 16-element demonstrator supports pulsed radar configuration; the corresponding experimental results are reported as follows.

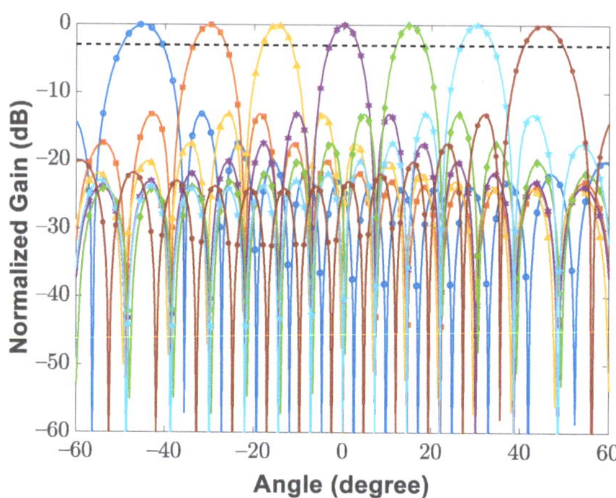

Figure 26. Measured antenna patterns for 16-element demonstrator operated in receiving mode.

A pulsed radar evaluates the target distance according to the time of flight between the radar-transmitted signal and target-reflected signal. The maximum unambiguous range of the pulsed radar can be expressed as follows. f_{PRF} and c represent the pulse repetition frequency and the speed of light, respectively.

$$\text{Maximum unambiguous range} = \frac{c}{2 \cdot f_{PRF}} \tag{1}$$

An up-chirp signal with a duration of 320 ns and sweep bandwidth of 20 MHz is employed for the transmitted pulse, and a 7.68 μs pulse repetition period and a 4.17% duty cycle are adopted for the radar demonstrator. Theoretically, a 1152 m maximum unambiguous range and 7.5 m range resolution can be achieved with the current system specifications. Because the transmitter and receiver are not simultaneously activated, the CMOS receiver avoids transmitter-induced baseband saturation and can increase the PGA gain for long-distance target sensing. Initially, each array element delivers digitized echo signals to the digital back-end. The DSP functional block implemented in the first-hierarchy FPGA executes waveform averaging for pulsed radar operation. Subsequently, processed data are delivered to MATLAB through an Ethernet interface. Matched filtering and digital beamforming are realized in MATLAB for time-of-flight calculation and spatial filtering with acceptable latency. Finally, the captured range–azimuth information for the given scene is displayed on the screen.

The 16-element radar demonstrator was placed on the seventh floor; the field of view seen by the demonstrator is depicted in Figure 27a, in which the target buildings and corresponding distance are labeled. An experiment was performed for functional verification of the pulsed radar, and the downconverted received waveform for a single pulse repetition period is displayed in Figure 27b. Once the radar demonstrator activates the receiver modules, the dc offset cancellation circuitry implemented in the CMOS PGA chain is triggered to update the bias point through analog feedback loops. During the settling process of the feedback loop, the receiver provides relatively little conversion gain; the blind range of the pulse radar demonstrator is broadened due to this effect.

Isometric and side views of the captured range–azimuth plot for labeled target buildings are presented in Figure 28a,b; the colored z-axis in the figure indicates the relative signal magnitude on a linear scale, and its value is normalized to the maximum receiving signal strength.

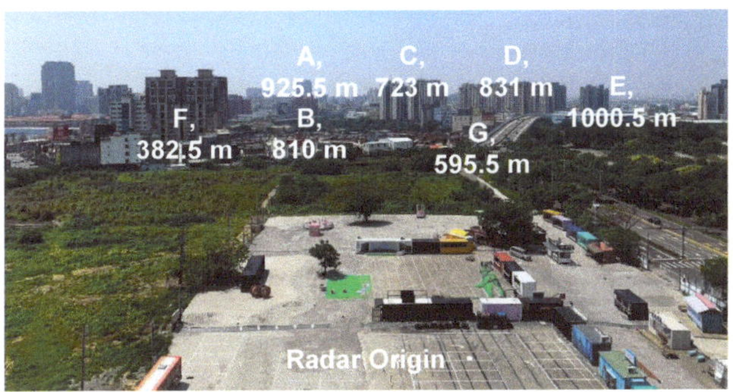

(a)

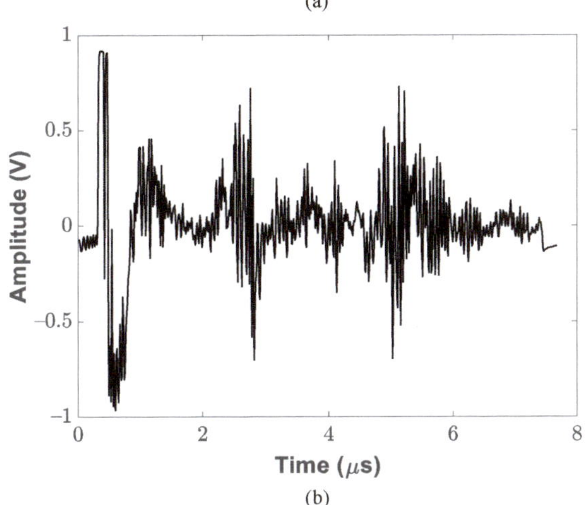

(b)

Figure 27. (**a**) Snapshot of target buildings. (**b**) Downconverted received waveform with pulse repetition period of 7.68 μs.

The target locations evaluated by the proposed radar demonstrator were confirmed to match the labeled buildings illustrated on the map. A 1000.5 m maximum observation range was accomplished for stationary buildings. To generate a complete pulsed radar image, the average latency between the user request and radar image display is 150 ms for the current hardware implementation, consisting of 2 ms for 256 consecutive measurements, 13 ms for data transfer through SerDes and Ethernet interfaces, and 135 ms for data processing in MATLAB. Further acceleration can be achieved by realizing the matched filtering and radar imaging algorithms in FPGAs. A comparison of the performance of the developed radar transceiver with other state-of-the-art works is summarized and reported in Table 1. Although the power consumption is not superior to that reported in prior studies, the performance of the proposed transceiver in terms of the spurious level and available conversion gain is comparable to that achieved in the state-of-the-art circuits. In addition, the radar demonstrator with scalable subarray modules simultaneously realizes range sensing and azimuth recognition for real-time pulsed radar imaging. Captured by a software-defined pulsed radar, a complete range–azimuth figure with a 1 km maximum observation range can be obtained within 150 ms under the current implementation. Users can construct a large-scale phased array radar with the presented topology of the array element and digital back-end system.

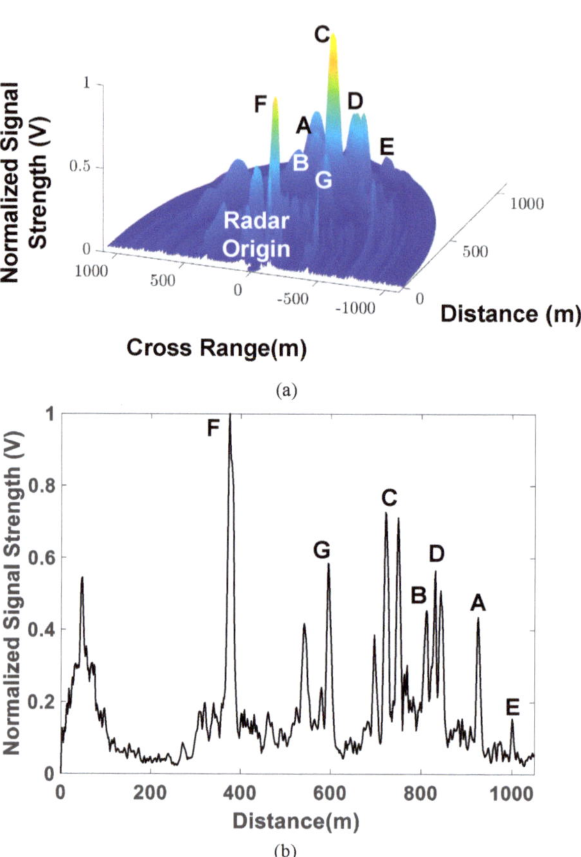

(b)

Figure 28. (**a**) Isometric view and (**b**) side view of the captured range–azimuth plot by using 16-element radar demonstrator.

Table 1. Performance comparison with state-of-the-art fully customized radar transceivers.

	TMTT2018 [18]	ISSCC2018 [15]	TMTT2017 [14]	TMTT2016 [16]	TMTT2016 [42]	This Work
Frequency Band	9.8–10.2 GHz	9.5–10.5 GHz	14.26–15.74 GHz	2–16 GHz	9–11 GHz	8–10 GHz
Technology	65 nm CMOS	65 nm CMOS	65 nm CMOS	0.13 μm SiGe BiCMOS	0.13 μm SiGe BiCMOS	65 nm CMOS
Die Size	$1.9 \times 2\ mm^2$	$2 \times 3.9\ mm^2$	$1.4 \times 2.9\ mm^2$	$2.5 \times 5\ mm^2$	$3 \times 5.2\ mm^2$	$2 \times 2.4\ mm^2$
SoC Integration	1TX + 1RX	4TX + 4RX	1TX + 1RX	8RX	1TX + 1RX	1TX + 1RX
TX Output Power	10.5 dBm	14.7 dBm	13.3 dBm	-	29.2 dBm	17.96 dBm
TX Spurious Level	-	-	-	-	-	−30.35 dBc
RX Conversion Gain	5–72 dB	Front-end: 15.3–28.6 dB Baseband: 0–60 dB	Front-end: 23.5 dB Baseband: 3–58 dB	6–11 dB	25 dB	3.8–57.2 dB
RX Noise Figure	16.5–18 dB	5.7–6.5 dB	5.6–6.3 dB	11.5–12.3 dB	3 dB	13.9–14.6 dB
RX Front-End Input P_{1dB}	2 dBm at 5 dB Gain −27 dBm at 32 dB Gain	−37 dBm	−33 dBm	−14 dBm	−18 dBm	−14 dBm
RX Front-End IIP_3	7 dBm at 5 dB Gain	-	-	-	-	−5.9 dBm

Table 1. *Cont.*

	TMTT2018 [18]	ISSCC2018 [15]	TMTT2017 [14]	TMTT2016 [16]	TMTT2016 [42]	This Work
RX Baseband Bandwidth	2 MHz	60–280 MHz	0.68–9.8 MHz	-	-	20 or 40 MHz
Power Consumption	147 mW	179 mW per TX 74 mW per RX	259.4 mW	250 mW per RX	4.128 W per TX 352 mW per RX	1.45 W per TX 1.41 W per RX
Number of Elements in the Demonstrator	-	4TX + 4RX	-	8RX	1TX + 1RX	16TX + 16RX
Beamforming Scheme	-	TX: RF Phase-Shifting, RX: Digital Beamforming	-	RF Phase-Shifting, Digital Beamforming	RF Phase-Shifting	Digital Beamforming
Modulation Type	Triangular Chirp	Pulsed Chirp	Sawtooth Chirp	-	-	Pulsed Chirp
Modulation Bandwidth	400 MHz (4%)	1 GHz (10%)	1.48 GHz (9.9%)	-	-	Programmable, 20 MHz for Pulsed Radar
Multibeamforming Capability	-	Yes (4 beams)	-	Yes (1, 2, or 4 beams)	-	Yes (16 beams)
Beam Steering Range	-	$\pm60°$	-	-	-	E-plane: $\pm45°$
Peak Sidelobe Ratio (PSLR)	-	−12.9 dB w/o tapering	−12.7 dB w/o tapering	-	-	−13.1 dB w/o tapering −25.6 dB w/i tapering
Radar Imaging Latency	-	off-line	-	-	-	150 ms per pulsed radar image

6. Conclusions

This paper describes the design and implementation of an X-band element-level digital phased array radar with fully integrated CMOS transceivers. Fabricated using 65 nm CMOS technology, an 8–10 GHz transceiver SoC provides analog signal processing, RF frequency translation, and global clock synchronization when the proposed periodic pulse injection technique is applied. Moreover, the scalable subarray module realizes system-level synchronization through four global reference signals and enhances the overall form factor through the use of vertically stacked printed circuit boards in a tile array configuration. Element-level digitization not only offers arbitrary weighting but also eliminates the deterministic magnitude and phase error of each radiating element. In summary, CMOS SoC integration, advanced packaging, and subarray module miniaturization result in a low-cost compact-size phased array system, and the proposed 1×16 subarray module accomplishes digital beamforming, back-end signal processing, and dc–dc conversion within dimensions of $317 \times 149 \times 74.6$ mm^3. Board-level integration of off-the-shelf front-end amplifiers further improves the performance of the 1×16 subarray module. Software-defined phased array radar demonstrators composed of the described subarray modules simultaneously fulfill range sensing and azimuth recognition for pulsed radar operation. A complete range–azimuth figure with a 1 km maximum distance can be captured within 150 ms by the pulsed radar demonstrator under the reported hardware implementation. In this study, waveform averaging was employed to reduce the data rate and improve the signal-to-noise ratio (SNR) for a stationary target and uncorrelated noise. The digital back-end averages out consecutively received waveforms and only delivers processed data to users. However, this technique fails to address the sensing of moving targets as a result of the time-variant relative phase in successive reflected signals. Additional hardware experiments and software algorithms should be developed for such scenarios in the future.

Author Contributions: Investigation, Y.-M.W., H.-C.C., C.-Y.K., C.-C.W., C.-T.L. and L.-H.C.; Project administration, B.S., T.-S.C. and Y.-J.W.; Writing original draft, Y.-M.W. All authors have read and agreed to the published version of the manuscript.

Funding: This research received no external funding.

Institutional Review Board Statement: Not applicable.

Informed Consent Statement: Not applicable.

Acknowledgments: The authors thank the Taiwan Semiconductor Research Institute for providing measurement assistance.

Conflicts of Interest: The authors declare no conflict of interest.

References

1. Herd, J.S.; Conway, M.D. The Evolution to Modern Phased Array Architectures. *Proc. IEEE* **2016**, *104*, 519–529. [CrossRef]
2. Talisa, S.H.; O'Haver, K.W.; Comberiate, T.M.; Sharp, M.D.; Somerlock, O.F. Benefits of Digital Phased Array Radars. *Proc. IEEE* **2016**, *104*, 530–543. [CrossRef]
3. Gebert, N.; Krieger, G.; Moreira, A. Digital Beamforming on Receive: Techniques and Optimization Strategies for High-Resolution Wide-Swath SAR Imaging. *IEEE Trans. Aerosp. Electron. Syst.* **2009**, *45*, 564–592. [CrossRef]
4. Huber, S.; Younis, M.; Patyuchenko, A.; Krieger, G.; Moreira, A. Spaceborne Reflector SAR Systems with Digital Beamforming. *IEEE Trans. Aerosp. Electron. Syst.* **2012**, *48*, 3473–3493. [CrossRef]
5. Wijayaratna, S.; Madanayake, A.; Wijenayake, C.; Bruton, L.T. Digital VLSI Architectures for Beam-Enhanced RF Aperture Arrays. *IEEE Trans. Aerosp. Electron. Syst.* **2015**, *51*, 1996–2011. [CrossRef]
6. Babur, G.; Manokhin, G.O.; Geltser, A.A.; Shibelgut, A.A. Low-Cost Digital Beamforming on Receive in Phased Array Radar. *IEEE Trans. Aerosp. Electron. Syst.* **2017**, *53*, 1355–1364. [CrossRef]
7. Pulipati, S.K.; Ariyarathna, V.; Madanayake, A.; Wijesekara, R.T.; Edussooriya, C.U.S.; Bruton, L.T. A 16-Element 2.4-GHz Multibeam Array Receiver Using 2-D Spatially Bandpass Digital Filters. *IEEE Trans. Aerosp. Electron. Syst.* **2019**, *55*, 3029–3038. [CrossRef]
8. Zhao, Y.; Chen, L.; Zhang, F.; Li, Y.; Wu, Y. A Novel MIMO-SAR System Based on Simultaneous Digital Beam Forming of Both Transceiver and Receiver. *Sensors* **2020**, *20*, 6604. [CrossRef]
9. Fulton, C.; Yeary, M.; Thompson, D.; Lake, J.; Mitchell, A. Digital Phased Arrays: Challenges and Opportunities. *Proc. IEEE* **2016**, *104*, 487–503. [CrossRef]
10. Puglielli, A.; Townley, A.; LaCaille, G.; Milovanović, V.; Lu, P.; Trotskovsky, K.; Niknejad, A.M. Design of Energy- and Cost-Efficient Massive MIMO Arrays. *Proc. IEEE* **2016**, *104*, 586–605. [CrossRef]
11. Nguyen, V.V.; Nam, H.; Choe, Y.J.; Lee, B.H.; Park, J.D. An X-band Bi-Directional Transmit/Receive Module for a Phased Array System in 65-nm CMOS. *Sensors* **2018**, *18*, 2569. [CrossRef]
12. Ha, J.K.; Noh, C.K.; Lee, J.S.; Kang, H.J.; Kim, Y.M.; Kim, T.H.; Jung, H.N.; Lee, S.H.; Cho, C.S.; Kim, Y.J. RF Transceiver for the Multi-Mode Radar Applications. *Sensors* **2021**, *21*, 1563. [CrossRef]
13. Wang, Y.; Tang, K.; Zhang, Y.; Lou, L.; Chen, B.; Liu, S.; Zheng, Y. A Ku-band 260 mW FMCW synthetic aperture radar TRX with 1.48 GHz BW in 65 nm CMOS for micro-UAVs. In Proceedings of the 2016 IEEE International Solid-State Circuits Conference (ISSCC), San Francisco, CA, USA, 31 January–4 February 2016; pp. 240–242.
14. Wang, Y.; Lou, L.; Chen, B.; Zhang, Y.; Tang, K.; Qiu, L.; Zheng, Y. A 260-mW Ku-band FMCW transceiver for synthetic aperture radar sensor with 1.48-GHz bandwidth in 65-nm CMOS technology. *IEEE Trans. Microw. Theory Tech.* **2017**, *65*, 4385–4399. [CrossRef]
15. Lou, L.; Tang, K.; Chen, B.; Guo, T.; Wang, Y.; Wang, W.; Zheng, Y. A 253 mW/channel 4TX/4RX pulsed chirping phased-array radar TRX in 65 nm CMOS for X-band synthetic-aperture radar imaging. In Proceedings of the 2018 IEEE International Solid-State Circuits Conference-(ISSCC), San Francisco, CA, USA, 11–15 February 2018; pp. 160–162.
16. Sayginer, M.; Rebeiz, G.M. An eight-element 2–16-GHz programmable phased array receiver with one, two, or four simultaneous beams in SiGe BiCMOS. *IEEE Trans. Microw. Theory Tech.* **2016**, *64*, 4585–4597. [CrossRef]
17. Kao, Y.H.; Chou, H.C.; Peng, C.C.; Wang, Y.J.; Su, B.; Chu, T.S. A single-port duplex RF front-end for X-band single-antenna FMCW radar in 65 nm CMOS. In Proceedings of the 2017 IEEE International Solid-State Circuits Conference (ISSCC), San Francisco, CA, USA, 5–9 February 2017; pp. 318–320.
18. Chou, H.C.; Kao, Y.H.; Peng, C.C.; Wang, Y.J.; Chu, T.S. An X-Band Frequency-Modulated Continuous-Wave Radar Sensor System With a Single-Antenna Interface for Ranging Applications. *IEEE Trans. Microw. Theory Tech.* **2018**, *66*, 883–891. [CrossRef]
19. Khandelwal, N.; Jackson, R.W. Active Antenna Module for Low-Cost Electronically Scanned Phased Arrays. *IEEE Trans. Microw. Theory Tech.* **2008**, *56*, 2286–2292. [CrossRef]
20. Sadhu, B.; Tousi, Y.; Hallin, J.; Sahl, S.; Reynolds, S.K.; Renström, Ö.; Valdes-Garcia, A. A 28-GHz 32-element TRX phased-array IC with concurrent dual-polarized operation and orthogonal phase and gain control for 5G communications. *IEEE J. Solid-State Circuits* **2017**, *52*, 3373–3391. [CrossRef]
21. Yeo, S.K.; Chun, J.H.; Kwon, Y.S. A 3-D X-band T/R Module Package With an Anodized Aluminum Multilayer Substrate for Phased Array Radar Applications. *IEEE Trans. Microw. Theory Tech.* **2010**, *33*, 883–891. [CrossRef]
22. Chappell, W.; Fulton, C. Digital Array Radar panel development. In Proceedings of the 2010 IEEE International Symposium on Phased Array Systems and Technology, Waltham, MA, USA, 12–15 October 2010; pp. 50–60.
23. Ortiz, J.A.; Díaz, J.; Aboserwal, N.; Salazar, J.L.; Jeon, L.; Sim, S.; Chun, J. Ultra-compact universal polarization X-band unit cell for high-performance active phased array radar. In Proceedings of the 2016 IEEE International Symposium on Phased Array Systems and Technology (PAST), Waltham, MA, USA, 18–21 October 2016; pp. 1–5.

24. Urzaiz, F.I.; de Quevedo, Á.D.; Ayuso, A.M.; Machado, Á.G.; Menoyo, J.G.; López, A.A. Design, implementation and first experimental results of an X-band ubiquitous radar system. In Proceedings of the 2018 IEEE Radar Conference (RadarConf18), Oklahoma City, OK, USA, 23–27 April 2018; pp. 1150–1155.
25. Iijima, K.; Okada, Y.; Kumamoto, T.; Kawaguchi, T.; Ikeuchi, H.; Sawahara, Y.; Shinonaga, M. Compact Superconducting Sub-array Module for X-band Phased Array Antenna. In Proceedings of the 2018 IEEE Radar Conference (RadarConf18), Oklahoma City, OK, USA, 23–27 April 2018; pp. 77–82.
26. Hoffmann, T.; Fulton, C.; Yeary, M.; Saunders, A.; Thompson, D.; Murmann, B.; Guo, A. IMPACT—A common building block to enable next generation radar arrays. In Proceedings of the 2016 IEEE Radar Conference (RadarConf), Philadelphia, PA, USA, 2–6 May 2016; pp. 1–4.
27. Hoffmann, T.; Fulton, C.; Yeary, M.; Saunders, A.; Thompson, D.; Murmann, B.; Guo, A. Measured performance of the IMPACT common module—A building block for next generation phase arrays. In Proceedings of the 2016 IEEE International Symposium on Phased Array Systems and Technology (PAST), Waltham, MA, USA, 18–21 October 2016; pp. 1–7.
28. Hoffmann, T.; Livadaru, M.; Jensen, D. IMPACT common module and S-band planar array beamforming measurements. In Proceedings of the 2018 IEEE Radar Conference (RadarConf18), Oklahoma City, OK, USA, 23–27 April 2018; pp. 588–592.
29. Ganis, A.; Navarro, E.M.; Schoenlinner, B.; Prechtel, U.; Meusling, A.; Heller, C.; Ziegler, V. A portable 3-D imaging FMCW MIMO radar demonstrator with a 24 × 24 antenna array for medium-range applications. *IEEE Trans. Geosci. Remote Sens.* **2018**, *56*, 298–312. [CrossRef]
30. Miller, L.A. The Role of FPGAs in the Push to Modern and Ubiquitous Arrays. *Proc. IEEE* **2016**, *104*, 576–585. [CrossRef]
31. Tsao, K.C.; Lee, L.; Chu, T.S.; Huang, Y.H. A Two-Stage Reconstruction Processor for Human Detection in Compressive Sensing CMOS Radar. *Sensors* **2018**, *18*, 1106. [CrossRef]
32. Tian, H.; Guo, S.; Zhao, P.; Gong, M.; Shen, C. Design and Implementation of a Real-Time Multi-Beam Sonar System Based on FPGA and DSP. *Sensors* **2021**, *21*, 1425. [CrossRef]
33. Wu, Y.M.; Ke, C.Y.; Wang, C.C.; Tang, Y.H.; Chen, Y.W.; Li, C.T.; Wang, Y.J. An X-band Scalable 4 × 4 Digital Phased Array Module using RF SoC and Antenna-in-Package. In Proceedings of the 2019 IEEE Radar Conference (RadarConf), Boston, MA, USA, 22–26 April 2019; pp. 1–6.
34. Qorvo. TGA2598-SM. 2019. Available online: https://www.qorvo.com/products/p/TGA2598-SM (accessed on 2 November 2021).
35. Qorvo. TGA2512-SM. 2019. Available online: https://www.qorvo.com/products/p/TGA2512-1-SM (accessed on 2 November 2021).
36. Analog Devices. ADF5355 Evaluation Board. 2019. Available online: https://www.analog.com/en/design-center/evaluation-hardware-and-software/evaluation-boards-kits/eval-adf5355.html (accessed on 2 November 2021).
37. Wang, H.; Sideris, C.; Hajimiri, A. A CMOS Broadband Power Amplifier With a Transformer-Based High-Order Output Matching Network. *IEEE J. Solid-State Circuits* **2010**, *45*, 2709–2722. [CrossRef]
38. Andrews, C.; Molnar, A.C. A Passive Mixer-First Receiver With Digitally Controlled and Widely Tunable RF Interface. *IEEE J. Solid-State Circuits* **2010**, *45*, 2696–2708. [CrossRef]
39. Homayoun, A.; Razavi, B. A low-power CMOS receiver for 5 GHz WLAN. *IEEE J. Solid-State Circuits* **2015**, *50*, 630–643. [CrossRef]
40. Mohan, S.S.; del Mar Hershenson, M.; Boyd, S.P.; Lee, T.H. Bandwidth extension in CMOS with optimized on-chip inductors. *IEEE J. Solid-State Circuits* **2000**, *35*, 346–355. [CrossRef]
41. Harter, M.; Hildebrandt, J.; Ziroff, A.; Zwick, T. Self-Calibration of a 3-D-Digital Beamforming Radar System for Automotive Applications With Installation Behind Automotive Covers. *IEEE Trans. Microw. Theory Tech.* **2016**, *64*, 2994–3000. [CrossRef]
42. Liu, C.; Li, Q.; Li, Y.; Deng, X.D.; Li, X.; Liu, H.; Xiong, Y.Z. A fully integrated X-band phased-array transceiver in 0.13-μm SiGe BiCMOS technology. *IEEE Trans. Microw. Theory Tech.* **2016**, *64*, 575–584. [CrossRef]

MDPI AG
Grosspeteranlage 5
4052 Basel
Switzerland
Tel.: +41 61 683 77 34

Sensors Editorial Office
E-mail: sensors@mdpi.com
www.mdpi.com/journal/sensors